CRC Series in Chromatography

Editors-in-Chief

Gunter Zweig, Ph.D. and Joseph Sherma, Ph.D.

General Data and Principles
Gunter Zweig, Ph.D. and
Joseph Sherma, Ph.D.

Lipids
Helmut K. Mangold, Dr. rer. nat.

Hydrocarbons
Walter L. Zielinski, Jr., Ph.D.

Carbohydrates
Shirley C. Churms, Ph.D.

Inorganics
M. Qureshi, Ph.D.

Drugs
Ram Gupta, Ph.D.

Phenols and Organic Acids
Toshihiko Hanai, Ph.D.

Terpenoids
Carmine J. Coscia, Ph.D.

Amino Acids and Amines
S. Blackburn, Ph.D.

Steroids
Joseph C. Touchstone, Ph.D.

Polymers
Charles G. Smith,
Norman E. Skelly, Ph.D.,
Carl D. Chow, and Richard A. Solomon

Pesticides and Related Organic Chemicals
Joseph Sherma, Ph.D. and
Joanne Follweiler, Ph.D.

Plant Pigments
Hans-Peter Köst, Ph.D.

Nucleic Acids and Related Compounds
Ante M. Krstulovic, Ph.D.

CRC Handbook of Chromatography

Amino Acids and Amines

Volume II

Author
Stanley Blackburn, Ph.D., C.Chem., F.R.S.C.
Leeds, England

CRC Series in Chromatography
Editors-in-Chief

Gunter Zweig, Ph.D.
President
Zweig Associates
Arlington, Virginia
(Deceased)

Joseph Sherma, Ph.D.
Professor of Chemistry
Lafayette College
Easton, Pennsylvania

CRC Press, Inc.
Boca Raton, Florida

Library of Congress Cataloging-in-Publication Data
(Revised for volume 2)

Blackburn, S. (Stanley)
Amino acids and amines.

(CRC handbook of chromatography) (CRC series in chromatography)
Includes bibliographies and indexes.
1. Chromatographic analysis--Tables. 2. Chromatographic analysis--Handbooks, manuals, etc. I. Title.
II. Series.
QD79.C4B55 1983 543'.089 82-9561
ISBN-08493-3064-5 (v. 1)

Direct all inquires to CRC Press, Inc., 2000 Corporate Blvd., N.W., Boca Raton, Florida, 33431.

International Standard Book Number 0-8493-3066-1

Library of Congress Card Number 82-9561
Printed in the United States

CRC SERIES IN CHROMATOGRAPHY

SERIES PREFACE

This is the third volume in this series written by Dr. Stanley Blackburn. In 1983, the first volume on amino acids and amines was published, and in 1986, Volume I on peptides appeared. The current volume updates the coverage of methods and data for the gas and liquid chromatography of amino acids and amines. It includes a large amount of material on the resolution of optical isomers, which has been one of the most important areas of research for many compound types in recent years.

It is hoped that Dr. Blackburn will produce Volume II on peptides when this is justified by the amount of available new data.

I would appreciate hearing from readers who can offer corrections or comments on the present volume as well as suggestions for topics and authors of future volumes in the CRC Handbook of Chromatography series.

Joseph Sherma, Ph.D.
Easton, PA

PREFACE

The phenomenal rapid increase in the growth of chemical and biochemical research and information is now a matter of common knowledge. This increase in research has been accompanied by a comparable increase in the field of amino acid analysis. The original *Handbook of Chromatography* provided tables relating to the chromatography of a very wide range of chemical compounds. In the early 1970s such a compilation in one volume was no longer feasible and individual volumes dealing with specific classes of compounds including amino acids and amines (Volume I) were necessary. The continuing escalation of research in the amino acid and amine field is shown by the publication of *Amino Acids and Amines, Volume II* after a relatively short interval. During this period the ion exchange method pioneered by Moore and Stein, and its later development, is still the method of choice for the majority of research workers wishing to determine the amino acid composition of proteins and peptides. Alternative techniques involving high performance liquid chromatography of amino acid derivatives, however, are being intensively studied and may well be increasingly used in the future. The material in the present volume represents a continuation of that presented in *Amino Acids and Amines, Volume I*, and principally covers the literature since 1981. In addition to the extensive tables of chromatographic data, methods of sample preparation and derivatization and methods of detection are described. A further section of the volume reviews techniques used for the chromatographic separation of free amino acids and their enantiomers, gas chromatographic separations, and the separation of amino acid derivatives. Specific methods for the determination of proline and hydroxyproline are described. A section listing chromatographic materials draws the reader's attention to commercial sources of supply, especially for materials relevant to separations described in the present volume.

S. Blackburn

THE EDITORS-IN-CHIEF

Gunter Zweig, Ph.D., received his undergraduate training at the University of Maryland, College Park, where he was awarded the Ph.D. in biochemistry in 1952. Two years following his graduation, Dr. Zweig was affiliated with the late R. J. Block, pioneer in paper chromatography of amino acids. Zweig, Block, and Le Strange wrote one of the first books on paper chromatography, which was published in 1952 by Academic Press and went into three editions, the last one authored by Gunter Zweig and Dr. Joe Sherma, the co-Editor-in-Chief of this series. *Paper Chromatography* (1952) was also translated into Russian.

From 1953 to 1957, Dr. Zweig was research biochemist at the C. F. Kettering Foundation, Antioch College, Yellow Springs, Ohio, where he pursued research on the path of carbon and sulfur in plants, using the then newly developed techniques of autoradiography and paper chromatography. From 1957 to 1965, Dr. Zweig served as lecturer and chemist, University of California, Davis and worked on analytical methods for pesticide residues, mainly by chromatographic techniques. In 1965, Dr. Zweig became Director of Life Sciences, Syracuse University Research Corporation, New York (research on environmental pollution), and in 1973 he became Chief, Environmental Fate Branch, Environmental Protection Agency (EPA) in Washington, D.C. From 1980 to 1984 Dr. Zweig was Visiting Research Chemist in the School of Public Health, University of California, Berkeley, where he was doing research on farmworker safety as related to pesticide exposure.

During his government career, Dr. Zweig continued his scientific writing and editing. Among his works are (many in collaboration with Dr. Sherma) the now 11-volume series on *Analytical Methods for Pesticides and Plant Growth Regulators* (published by Academic Press); the pesticide book series for CRC Press; co-editor of *Journal of Toxicology and Environmental Health;* co-author of basic review on paper and thin-layer chromatography for *Analytical Chemistry* from 1968 to 1980; co-author of applied chromatography review on pesticide analysis for *Analytical Chemistry,* beginning in 1981.

Among the scientific honors awarded to Dr. Zweig during his distinguished career were the Wiley Award in 1977, the Rothschild Fellowship to the Weizmann Institute in 1963/64; and the Bronze Medal by the EPA in 1980.

Dr. Zweig authored or co-authored over 80 scientific papers on diverse subjects in chromatography and biochemistry, besides being the holder of three U.S. patents. In 1985, Dr. Zweig became president of Zweig Associates, Consultants in Arlington, Va.

Following his death on January 27, 1987, the Agrochemicals Section of the American Chemical Society posthumously elected him a Fellow and established the Gunther Zweig Award for Young Chemists in his honor.

Joseph Sherma, Ph.D., received a B.S. in Chemistry from Upsala College, East Orange, N.J., in 1955 and a Ph.D. in Analytical Chemistry from Rutgers University in 1958, carrying on his thesis research in ion exchange chromatography under the direction of the late William Rieman III. Dr. Sherma joined the faculty of Lafayette College in September, 1958, and is presently Charles A. Dana Professor and Head of the Chemistry Department.

Dr. Sherma, independently and with others, has written over 300 research papers, chapters, books, and reviews involving chromatography and other analytical methodology. He is editor for residues and trace elements of the *Journal of the Association of Official Analytical Chemists* and a member of the advisory board of the *Journal of Planar Chromatography.* He is a consultant on analytical methodology for many companies and government agencies.

Dr. Sherma has received two awards for superior teaching at Lafayette College and the 1979 Distinguished Alumnus Award from Upsala College for outstanding achievements as an educator, researcher, author, and editor. He is a member of the ACS, Sigma Xi, Phi Lambda Upsilon, SAS, AIC, and AOAC. Dr. Sherma's current interests are in quantitative TLC, mainly applied to clinical analysis, pesticide residues, and food additives.

THE AUTHOR

Dr. Stanley Blackburn gained his Honours B.Sc. degree in chemistry and his Ph.D. in organic chemistry at the University of Leeds, England. He is a former research scientist at the Wool Industries Research Association, Leeds, where his research interests included the development of chromatographic techniques, the structure and amino acid sequence of the proteins of wool keratin, and the end group determination of peptides and proteins.

Dr. Blackburn is a Chartered Chemist, a Fellow of the Royal Society of Chemistry, and a member of the Biochemical Society and the American Chemical Society. His current work is centered on scientific writing and documentation. He has written more than 30 scientific papers dealing with protein analysis and structure and is the author of several texts, including *Amino Acid Determination: Methods and Techniques*, *Protein Sequence Determination: Methods and Techniques*, and *Enzyme Strucutre and Function*, all published by Marcel Dekker, Inc. He is also author of two volumes of the *CRC Handbook of Chromatography*, *Amino Acids Amines, Volume I* and *Peptides, Volume I*.

TABLE OF CONTENTS

Section I
Tables

Section I

TABLES

Wherever possible tables are arranged according to classes of chemical compounds. This was not always possible when different types of compounds were chromatographed under the same experimental conditions. The reader is referred to the compound index for specific compounds which may appear in different tables.

Table GC 1
THE INFLUENCE OF THE ESTER GROUP ON THE SEPARATION OF *N*-TRIFLUOROACETYL AMINO ACID ESTERS

Column packing	P1	P1	P1	P1
Temperature	T1	T1	T1	T1
Gas Flow rate (mℓ/min)	He, 30	He, 30	He, 30	He, 30
Column				
Length (ft)	6	6	6	6
Diameter (mm, I.D.)	2	2	2	2
Material	G	G	G	G
Detector	FID	FID	FID	FID

	t_r(min)			
Amino acid	**Isopropyl ester**	***n*-Propyl ester**	**Isobutyl ester**	***n*-Butyl ester**
Alanine	0.90	1.11	1.38	1.60
Arginine	5.82	6.18	6.48	6.72
Glutamic acid	4.29	4.87	5.50	5.97
Leucine	1.82	2.22	2.60	2.87
Methionine	3.48	3.92	4.28	4.54
Norleucine	2.13	2.54	2.93	3.20
Phenylalanine	4.08	4.67	5.00	5.25
Proline	2.52	2.98	3.40	3.66
Serine	1.29	1.56	1.87	2.12
Tryptophan-1	8.45	9.20	9.81	10.44
Tryptophan-2	7.09	7.55	7.90	8.31

Note: The incomplete diacylation of tryptophan resulted in two derivatives.

Column packing: P1 = 2% OV-17/1% OV-210 on Supelcoport®, 100-120 mesh
Temperature: T1 = the temperature was raised from 125—135°C at 10°C/min, thereafter at 20°C/min to 220°C where it was held constant for 10 min

REFERENCE

1. **Gamerith, G.,** *J. Chromatogr.*, 268, 403, 1983.

Table GC 2
THE INFLUENCE OF THE ESTER GROUP ON THE SEPARATION OF *N*-HEPTAFLUOROBUTYRYL AMINO ACID ESTERS

Column packing	P1	P1	P1	P1
Temperature	T1	T1	T1	T1
Gas Flow rate (mℓ/min)	He, 30	He, 30	He, 30	He, 30
Column				
Length (ft)	6	6	6	6
Diameter (mm, I.D.)	2	2	2	2
Material	G	G	G	G
Detector	FID	FID	FID	FID

Table GC 2 (continued)
THE INFLUENCE OF THE ESTER GROUP ON THE SEPARATION OF *N*-HEPTAFLUOROBUTYRYL AMINO ACID ESTERS

	t_r(min)			
Amino acid	**Isopropyl ester**	***n*-Propyl ester**	**Isobutyl ester**	***n*-Butyl ester**
Alanine	1.19	1.52	1.94	2.30
Arginine		9.31	9.50	9.74
Glutamic acid	6.50	7.51	8.27	8.86
Leucine	2.64	3.32	3.98	4.51
Methionine	5.56	6.25	6.70	7.08
Norleucine	3.20	3.94	4.61	5.11
Phenylalanine	6.74	7.25	7.63	7.98
Proline	3.79	4.64	5.24	5.68
Serine	2.35	2.91	3.44	3.97
Tryptophan-1	11.16	11.77	12.22	12.73
Tryptophan-2	10.05	10.44	10.70	11.06

Note: The incomplete diacylation of tryptophan resulted in two derivatives.

Column packing: P1 = 2% OV-1/1% OV-210 on Supelcoport®, 100-120 mesh
Temperature: T1 = 120°C for 0.5 min, then raised to 135°C at 5°C/min, then to 220°C at 15°C/min, the final hold time being 13 min

REFERENCE

1. **Gamerith, G.,** *J. Chromatogr.*, 268, 403, 1983.

Table GC 3
THE INFLUENCE OF THE ESTER GROUP ON THE SEPARATION OF *N*-HEPTAFLUOROBUTYRYL AMINO ACID ESTERS

Column packing	P1	P1	P1	P1
Temperature	T1	T1	T1	T1
Gas Flow rate (mℓ/min)	He, 30	He, 30	He, 30	He, 30
Column				
Length (ft)	6	6	6	6
Diameter (mm, I.D.)	2	2	2	2
Material	G	G	G	G
Detector	FID	FID	FID	FID

	t_r(min)			
Amino acid	**Isopropyl ester**	***n*-Propyl ester**	**Isobutyl ester**	***n*-Butyl ester**
Alanine	1.44	1.94	2.49	2.88
Glutamic acid	7.20	7.95	8.56	8.87
Leucine	3.05	3.62	4.43	4.84
Methionine	6.21	6.70	7.06	7.36
Norleucine	3.23	4.02	4.76	5.07
Phenylalanine	6.82	7.32	7.50	7.81
Proline	4.18	4.89	5.45	5.82
Serine	0.91	0.64	1.44	1.66
Tryptophan-2	13.86	15.00	15.90	16.59

Table GC 3 (continued)
THE INFLUENCE OF THE ESTER GROUP ON THE SEPARATION OF *N*-HEPTAFLUOROBUTYRYL AMINO ACID ESTERS

Note: The incomplete diacylation of tryptophan resulted in two derivatives.

Column packing: P1 = 0.31% Carbowax® 20M/0.28% Silar® 5 CP/0.06% Lexan® on Chromosorb® W AW, 120-140 mesh
Temperature: T1 = 120°C for 0.5 min, raised at 5°C/min to 135°C and then at 15°C/min to 220°C and held constant for 10 min

REFERENCE

1. **Gamerith, G.,** *J. Chromatogr.*, 268, 403, 1983.

Table GC 4
THE INFLUENCE OF THE ACYL GROUP ON THE SEPARATION OF *N*-ACYL AMINO ACID *n*-PROPYL ESTERS

Column packing	P1	P1
Temperature	T1	T1
Gas Flow rate (mℓ/min)	He, 30	He, 30
Column		
Length (ft)	6	6
Diameter (mm, I.D.)	2	2
Material	G	G
Detector	FID	FID

	t_r(min)	
Amino acid	**TFA**	**HFB**
Alanine	1.22	1.52
Glutamic acid	7.30	7.51
Leucine	2.84	3.32
Methionine	5.86	6.25
Norleucine	3.38	3.94
Phenylalanine	6.96	7.25
Proline	4.19	4.64
Serine	1.86	2.91

Abbreviations: TFA = trifluoroacetyl; HFB = heptafluorobutyryl

Column packing: P1 = 2% OV-17/1% OV-210 on Supelcoport®, 100-120 mesh
Temperature: T1 = 120°C for 0.5 min, then raised to 135°C at 5°C/min, then to 220°C at 15°C/min, the final hold time being 13 min

REFERENCE

1. **Gamerith, G.,** *J. Chromatogr.*, 268, 403, 1983.

Table GC 5
THE INFLUENCE OF THE ACYL GROUP ON THE SEPARATION OF *N*-ACYL AMINO ACID ISOPROPYL ESTERS

Column packing	P1	P1
Temperature	T1	T1
Gas Flow rate (mℓ/min)	He, 30	He, 30
Column		
Length (ft)	6	6
Diameter (mm, I.D.)	2	2
Material	G	G
Detector	FID	FID

	t_r(min)	
Amino acid	**TFA**	**HFB**
Alanine	1.58	2.05
Glutamic acid	7.89	8.12
Leucine	3.75	4.46
Methionine	6.95	7.37
Norleucine	4.49	5.19
Phenylalanine	8.10	8.41
Proline	5.25	5.77
Serine	2.52	4.08

Abbreviations: TFA = trifluoroacetyl; HFB = heptafluorobutyryl

Column packing: P1 = 0.65% EGA on Chromosorb® W AW, 80-100 mesh
Temperature: T1 = held at 115°C for 1 min, then raised to 118°C at 1.5°C/min and to 210°C at 15°C/min, then held at this temperature for 15 min

REFERENCE

1. **Gamerith, G.,** *J. Chromatogr.*, 268, 403, 1983.

Table GC 6
THE INFLUENCE OF THE ACYL GROUP ON THE SEPARATION OF *N*-ACYL AMINO ACID *n*-BUTYL ESTERS

Column packing	P1	P1	P2	P2
Temperature	T1	T1	T2	T2
Gas Flow rate (mℓ/min)	He, 30	He, 30	He, 30	He, 30
Column				
Length (ft)	6	6	6	6
Diameter (mm, I.D.)	2	2	2	2
Material	G	G	G	G
Detector	FID	FID	FID	FID

Table GC 6 (continued)
THE INFLUENCE OF THE ACYL GROUP ON THE SEPARATION OF *N*-ACYL AMINO ACID *n*-BUTYL ESTERS

Amino acid	t_r(min)			
	TFA	HFB	TFA	HFB
Alanine	1.12	1.58	3.82	2.55
Glutamic acid	7.42	7.69	9.42	7.97
Leucine	2.62	3.39	4.44	3.16
Methionine	5.40	5.94	7.93	6.72
Norleucine	3.19	4.00	4.81	3.56
Phenylalanine	6.39	6.83	8.53	7.30
Proline	3.99	4.59	6.41	5.24
Serine	1.66	2.95	8.84	7.63

Abbreviations: TFA = trifluoroacetyl; HFB = heptafluorobutyryl

Column packing: P1 = 3% SP 2100 on Supelcoport®, 100-120 mesh
P2 = 5% Carbowax® 20M on Supelcoport®, 100-120 mesh
Temperature: T1 = 120°C increased to 130°C at 5°C/min and to 220°C at 15°C/min, then held at this temperature for 15 min
T2 = 130°C for 0.5 min, then raised to 135°C at 5°C/min and to 220°C at 15°C/min and held at this temperature for 13 min

REFERENCE

1. **Gamerith, G.,** *J. Chromatogr.*, 268, 403, 1983.

Table GC 7
ENANTIOMERS OF *N*-TRIFLUOROACETYL ISOPROPYL ESTERS OF α- AND γ-AMINO ACIDS ON *N*-LAUROYL-L-VALINE *TERT*-OCTYLAMIDE

Column packing		P1		P1
Gas Flow rate (mℓ/min)		He, 3		He, 0.5—1.5
Column				
Length (m)		50		9
Diameter (mm, I.D.)		0.5		0.35
Material		SS[a]		G[b]
Detector		FID		FID

Amino acid	Isomer	t_r(min)	T(°C)	t_r(min)	T(°C)
		Protein α-Amino Acids			
Ala	D	5.50	130	6.96	115
	L	6.20		7.98	
Thr	D	8.70	130	11.34	115
	L	9.50		12.60	
Val	D	8.68	130	11.43	115
	L	9.38		12.48	
Gly		10.80	130	14.22	115
allo-Ile	D	12.50	130	16.59	115
	L	13.80		18.48	
Ile	D	13.80	130	18.48	115
	L	15.04		20.46	
Ser	D	18.28	130	23.94	115
	L	19.92		26.31	

Table GC 7 (continued)
ENANTIOMERS OF *N*-TRIFLUOROACETYL ISOPROPYL ESTERS OF α- AND γ-AMINO ACIDS ON *N*-LAUROYL-L-VALINE *TERT*-OCTYLAMIDE

Amino acid	Isomer	t_r(min)	T(°C)	t_r(min)	T(°C)
		Protein α-Amino Acids			
Leu	D	18.04	130	25.68	115
	L	21.40		31.62	
Pro	D	19.30	130	26.13	115
	L	20.00		26.73	
Asp	D	55.70	130	84.15	115
	L	57.36		86.85	
Met	D	21.18	170	31.26	150
	L	23.06		34.38	
Glu	D	28.08	170	44.28	150
	L	30.10		47.73	
Phe	D	29.80	170	44.70	150
	L	31.80		48.37	
Orn	D	104.80	180		
	L	114.80			
Lys	D	150.00	180		
	L	161.50			
		Nonprotein α-Amino Acids			
α-Aminobutanoic acid	D	7.90	130	9.90	115
	L	8.74		11.28	
α-Aminopentanoic acid	D	12.70	130	17.01	115
	L	14.40		19.86	
α-Aminohexanoic acid	D	21.28	130	28.56	115
	L	24.30		33.69	
α-Aminoheptanoic acid	D	27.10	140	51.36	115
	L	30.40		60.36	
α-Aminooctanoic acid	D	45.90	140	93.24	115
	L	51.32		109.71	
tert-Leucine	D	22.00	100		
	L	22.84			
Phenylglycine	D	33.90	150	23.40	150
	L	35.50		24.30	
		γ-Amino Acids			
γ-Aminopentanoic acid	L	30.64	130	40.86	115
	D	31.90		42.45	
γ-Amino-δ-methylhexanoic acid	L	62.86	130	28.02	142
	D	66.20		29.82	
γ-Amino-ε-methylheptanoic acid	L	108.50	130	46.80	142
	D	115.88		49.59	

[a] Capillary column coated by the plug method with a 5% solution of the stationary phase in chloroform.

[b] Whisker-walled capillary washed with chloroform and dichloromethane and dried with nitrogen gas. The capillary was then coated using a 10% solution of the stationary phase in dichloromethane and was preconditioned for 1 day at 170°C.

Column packing: P1 = *N*-lauroyl-L-valine *tert*-octylamide

Table GC 7 (continued)
ENANTIOMERS OF *N*-TRIFLUOROACETYL ISOPROPYL ESTERS OF α- AND γ-AMINO ACIDS ON *N*-LAUROYL-L-VALINE *TERT*-OCTYLAMIDE

REFERENCE

1. **Charles, R. and Watabe, K.,** *J. Chromatogr.*, 298, 253, 1984.

Table GC 8
ENANTIOMERS OF *N*-TRIFLUOROACETYL ISOPROPYL ESTERS OF α- AND γ-AMINO ACIDS ON *N*-DOCOSANOYL-L-LEUCINE *TERT*-OCTYLAMIDE

Column packing	P1	P1
Gas Flow rate (mℓ/min)	He, 3	He, 0.5—1.5
Column		
Length (m)	50	9
Diameter (mm, I.D.)	0.5	0.35
Material	SS[a]	G[b]
Detector	FID	FID

Amino acid	Isomer	t_r(min)	T(°C)	t_r(min)	T(°C)
		Protein α-Amino Acids			
Ala	D	4.30	120	3.18	130
	L	5.10		3.63	
Thr	D	4.84	120	4.20	130
	L	5.40		4.52	
Val	D	5.40	120	5.19	130
	L	6.04		5.70	
Gly		7.30	120	5.97	130
allo-Ile	D	8.20	120	7.68	130
	L	9.10		8.46	
Ile	D	9.10	120	8.46	130
	L	10.10		9.18	
Ser	D	9.50	120	8.78	130
	L	10.30		9.31	
Leu	D	11.30	120	10.32	130
	L	13.70		12.12	
Pro	D	14.00	120	13.29	130
	L	14.00		13.29	
Asp	D	36.84	120	46.14	130
	L	38.24		48.18	
Met	D	11.60	170	22.32	160
	L	12.60		24.18	
Glu	D	16.06	170	30.00	160
	L	17.16		32.22	
Phe	D	18.24	170	32.88	160
	L	19.30		35.19	
Tyr	D	24.20	170	57.00	160
	L	26.00		62.46	
Orn	D	22.44	200	32.49	195
	L	23.80		33.87	
Lys	D	31.26	200	47.55	195
	L	32.68		48.81	

Table GC 8 (continued)
ENANTIOMERS OF *N*-TRIFLUOROACETYL ISOPROPYL ESTERS OF α- AND γ-AMINO ACIDS ON *N*-DOCOSANOYL-L-LEUCINE *TERT*-OCTYLAMIDE

Amino acid	Isomer	t_r(min)	T(°C)	t_r(min)	T(°C)
Protein α-Amino Acids					
Nonprotein α-Amino Acids					
α-Aminobutanoic acid	D	5.02	120	4.68	130
	L	5.74		5.25	
α-Aminopentanoic acid	D	8.10	120	7.62	130
	L	9.60		8.76	
α-Aminohexanoic acid	D	13.10	120	12.45	130
	L	15.70		14.28	
α-Aminoheptanoic acid	D	15.00	130	21.69	130
	L	17.30		24.87	
α-Aminooctanoic acid	D	25.90	130	37.80	130
	L	29.86		43.44	
tert-Leucine	D	9.30	100		
	L	10.06			
Phenylglycine	D	18.26	150	17.70	160
	L	19.16		18.36	
γ-Amino Acids					
γ-Aminopentanoic acid	L	12.60	130	18.00	130
	D	12.90		18.54	
γ-Amino-δ-methylhexanoic acid	L	24.30	130	36.45	130
	D	26.10		39.45	
γ-Amino-ε-methylheptanoic acid	L	41.40	130	61.83	130
	D	44.50		66.24	

[a] Capillary column coated by the plug method with a 5% solution of the stationary phase in chloroform.

[b] Whisker-walled capillary washed with chloroform and dichloromethane and dried with nitrogen gas. The inner surface of the capillary was deactivated with benzyltriphosphonium chloride at 350°C for 3 hr and coated using a 10% solution of the stationary phase in dichloromethane. It was then preconditioned for 1 day at 200°C.

Column packing: P1 = *N*-docosanoyl-L-leucine *tert*-octylamide

REFERENCE

1. **Charles, R. and Watabe, K.,** *J. Chromatogr.*, , 298, 253, 1984.

Table GC 9
ENANTIOMERS OF *N*-TRIFLUOROACETYL ISOPROPYL ESTERS OF α-AMINO ACIDS ON *N*-LAUROYL-L-VALINE *TERT*-OCTYLAMIDE AT 115°C

Column packing	P1	P1
Temperature (°C)	115	115
Gas Flow rate (mℓ/min)	He, 3	He, 0.5—1.5

Table GC 9 (continued)
ENANTIOMERS OF *N*-TRIFLUOROACETYL ISOPROPYL ESTERS OF α-AMINO ACIDS ON *N*-LAUROYL-L-VALINE *TERT*-OCTYLAMIDE AT 115°C

Column		
Length (m)	50	9
Diameter (mm, I.D.)	0.5	0.35
Material	SS[a]	G[b]
Detector	FID	FID

Amino acid	t_r(min)	
D-Alanine	8.70	6.96
L-Alanine	10.10	7.98
D-Allo-isoleucine	21.20	16.59
L-Allo-isoleucine	23.60	18.48
D-Isoleucine	23.60	18.48
L-Isoleucine	26.26	20.46
D-Leucine	32.40	25.68
L-Leucine	40.20	31.62
D-Proline	31.00	26.13
L-Proline	32.40	26.73
D-Valine	14.40	11.43
L-Valine	15.80	12.48

[a] Capillary column coated by the plug method with a 5% solution of the stationary phase in chloroform.

[b] Whisker-walled capillary washed with chloroform and dichloromethane and dried with nitrogen gas. The capillary was then coated using a 10% solution of the stationary phase in dichloromethane and was preconditioned for 1 day at 170°C.

Column packing: P1 = *N*-lauroyl-L-valine *tert*-octylamide

REFERENCE

1. **Charles, R. and Watabe, K.,** *J. Chromatogr.*, 298, 253, 1984.

Table GC 10
(R)- AND (*S*)-*N*-TRIFLUOROACETYL AMINO ACID ISOPROPYL ESTERS

Column packing	P1
Temperature (°C)	120
Gas Flow rate (mℓ/min)	N_2, 20
Column	
Length (m)	4
Diameter (mm, I.D.)	2

Table GC 10 (continued)
(R)- AND (*S*)-*N*-TRIFLUOROACETYL AMINO ACID ISOPROPYL ESTERS

Amino acid	t_r(min) R	t_r(min) S
Alanine	22	26
Leucine	78	99
Phenylalanine[a]	290	345
Valine	34	40

[a] Temperature = 130°C, flow rate = 30 mℓ/min.

Column packing: P1 = 10% optically active SP-300 on Supelcoport® 100-120 mesh

REFERENCE

1. **Deschenaux, R. and Bernauer, K.,** *Helv. Chim. Acta,* 67, 373, 1984.

Table GC 11
TRIFLUOROACETYL AMINO ACID *N*-BUTYL ESTERS

Column packing	P1
Temperature	T1
Gas Flow rate (mℓ/min)	H_2, 2
Column	
Length (m)	30
Diameter (mm)	0.33
Form	Capillary
Detector	FID

Amino acid	I
N-α-Acetyllysine	2019
Alanine	1159
2-Aminoadipic acid	1868
α-Aminoisobutyric acid	1235
β-Aminoisobutyric acid	1282
Aspartic acid + asparagine	1676
Creatinine	1440
Glutamic acid + glutamine	1807
Glycine	1173
Glycylproline	1931
Histidine (monoacyl)	1983
Hydroxyproline	1489
Isoleucine	1358
Leucine	1349
Lysine	1811
Methionine	1568
1-Methylhistidine + 3-methylhistidine	1888

Table GC 11 (continued)
TRIFLUOROACETYL AMINO ACID *N*-BUTYL ESTERS

Amino acid	I
N-α-Methyllysine	1851
N-Methylproline	1328
Nicotinic acid	1384
Norleucine[a]	1389
Ornithine	1690
Phenylalanine	1678
Phenylglycine	1584
Prolylhydroxyproline	2058
Pyroglutamic acid	1505
Serine	1257
Threonine	1246
N-TFA-Tyrosine	1976
N(*O*)-TFA-tyrosine	1769
Valine	1280

[a] Internal standard.

Column packing: P1 = OV-101 coated capillary column
Temperature: T1 = temperature programmed from 80—280°C at 3°C/min

REFERENCE

1. **Schneider, K., Neupert, M., Spiteller, G., Henning, H. V., Matthaei, D., and Scheler, F.,** *J. Chromatogr.*, 345, 19, 1985.

Table GC 12
N(*O*)-HEPTAFLUOROBUTYRYL AMINO ACID *n*-BUTYL ESTERS

Column packing	P1
Temperature	T1
Gas Flow rate (mℓ/min)	H_2, 2
Column	
Length (m)	50
Diameter (mm)	0.22
Form	Capillary
Material	Silica
Detector	FID

Amino acid	**t_r(min)**
Aca	43.03
Alanine	14.01
β-Aminoisobutyric acid	19.50
Arginine	43.21
Aspartic acid + asparagine	32.99
Creatinine + phenylalanine	34.00
Cycloleucine	23.56
Cysteine	25.95
Glutamic acid + glutamine	38.10
Glycine	16.76

Table GC 12 (continued)
N(O)-HEPTAFLUOROBUTYRYL AMINO ACID *n*-BUTYL ESTERS

Amino acid	t_r(min)
Histidine	53.20
Isoleucine	20.11
Leucine	19.85
Lysine	43.52
Methionine	31.13
1-Methylhistidine	44.49
3-Methylhistidine	46.66
Ornithine	41.51
Proline	24.99
Serine	19.27
Threonine	17.06
Tyrosine	39.03
Valine	16.57

Abbreviations: Aca = *trans*-4-(aminomethylcyclohexane) carboxylic acid used as an internal standard

Column packing: P1 = OV-1701 chemically bonded fused-silica capillary column

Temperature: T1 = temperature programmed from 130—260°C at 3°C/min

REFERENCE

1. **Schneider, K., Neupert, M., Spiteller, G., Henning, M. V., Matthaei, D., and Scheler, F.,** *J. Chromatogr.*, 345, 19,1985.

Table GC 13
N(O,S)-PENTAFLUOROBENZOYL AMINO ACID ISOBUTYL ESTERS

Column packing	P1
Temperature	T1
Gas Flow rate (mℓ/min)	He, 2
Column	
Length (m)	15
Form	Narrow bore Capillary
Detector	ECD

Amino acid	t_r(min)
Alanine	5.53
β-Alanine	7.00
γ-Aminobutyric acid	9.17
Aspartic acid	12.64
Citrulline	18.50
Cysteine	16.58
Glutamic acid	14.14
Glycine	6.00

Table GC 13 (continued)
N(O,S)-PENTAFLUOROBENZOYL AMINO ACID ISOBUTYL ESTERS

Amino acid	t_r (min)
Histidine	18.90
Hydroxyproline	11.78
Isoleucine	8.45
Leucine	8.31
Lysine	20.25
Methionine	12.23
α-Methyl-*p*-tyrosine (internal standard)	18.12
Ornithine	19.26
Phenylalanine	13.57
Serine	14.51
Threonine	8.93
Tryptophan	19.81
Tyrosine	21.08
Valine	7.22

Column packing: P1 = SE-54 (1% vinyl, 5% phenylmethyl polysiloxane)

Temperature: T1 = initial temperature of 140°C for 0.5 min, increasing at a rate of 5°C/min to 180°C, held for 0.1 min and increasing at 8°C/min to 310°C where it is maintained for 3 min

REFERENCE

1. **Yeung, J. M., Baker, G. B., and Coutts, R. T.,** *J. Chromatogr.*, 378, 293, 1986.

Table GC 14
N-TRIFLUOROACETYL ISOPROPYL ESTERS OF α-, β- AND γ-AMINO ACIDS ON DIAMIDE STATIONARY PHASES

Column packing	P1	P2
Temperature (°C)	140	130
Gas	He	He
Column		
Length (ft)	150	150
Diameter (in.)	0.02	0.02
Form	Capillary	Capillary
Material	SS	SS
Detector	FID	FID

Amino acid	r[a]	$r_{D/L}$[b]	$r_{D/L}$
1[c]	D 0.17	0.924	0.797
	L 0.18		
2	L 0.41	1.027	0.958[d]
	D 0.42		
3	L 1.02	1.053	1.094[e]
	D 1.07		
4	D 0.29	1.00	0.816
	L 0.29		

Table GC 14 (continued) *N*-TRIFLUOROACETYL ISOPROPYL ESTERS OF α-, β- AND γ-AMINO ACIDS ON DIAMIDE STATIONARY PHASES

Amino acid	r^a	$r^b_{D/L}$	$r_{D/L}$
5	L 0.84 D 0.86	1.023	0.928[d]
6	L 1.89 D 2.04	1.080	1.049[d]
7	D 0.54 L 0.62	0.870	0.737
8	L 1.16 D 1.20	1.030	0.945[d]
9	L 3.37 D 3.62	1.077	1.047[d]

[a] Corrected relative time with respect to decyl acetate, which had a corrected retention time of 12 min at 140°C and 21 min at 130°C.
[b] $r_{D/L}$ = resolution factor = ratio of the corrected retention time of the D- over that of the L-enantiomer.
[c] The formulas of the amino acids are shown below.
[d] Column 250 ft × 0.02 in.
[e] Temperature = 120°C.

1 

2 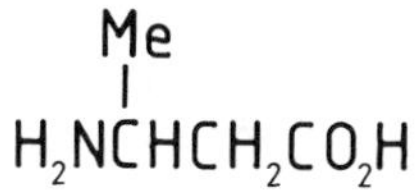

3 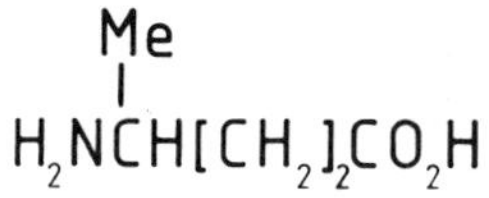

4 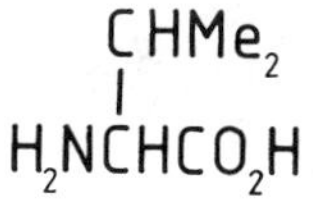

5 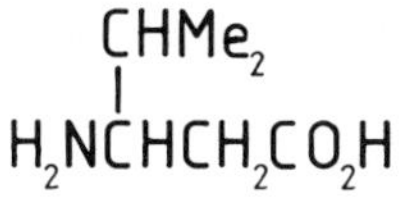

6 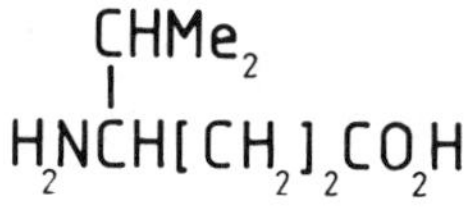

7 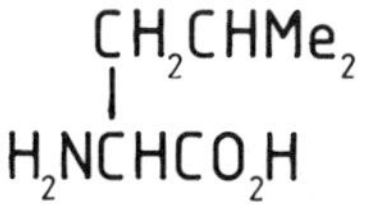

8 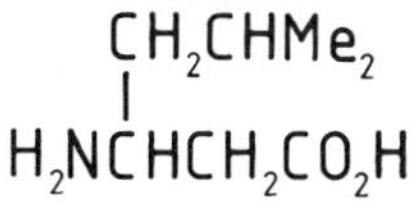

9 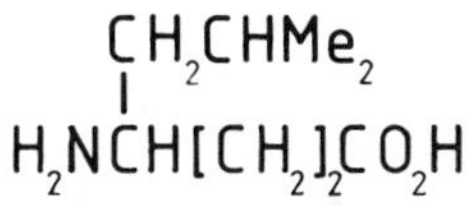

Packing: P1 = *N*-dodecanoyl-L-valine 6-undecylamide
P2 = *N*-dodecanoyl-L-valine *t*-butylamide

REFERENCE

1. **Feibush, B., Balan, A., Altman, B., and Gil-Av, E.,** *J. Chem. Soc. Perkin Trans.*, 2, 1230, 1979.

Table GC 15
ENANTIOMERS OF *N*-TRIFLUOROACETYL AMINO ACID ISOPROPYL ESTERS ON DIAMIDE STATIONARY PHASES

Column packing		P1		P2		P3		P4	
Amino acid	**Isomer**	**t_r(min)**	**T(°C)[a]**	**t_r(min)**	**T(°C)[a]**	**t_r(min)**	**t(°C)[a]**	**t_r(min)**	**T(°C)[a]**
Alanine	D	5.94	120	12.80	130	14.60	120	4.82	130
	L	6.48		15.76		12.60		5.84	
	D	5.00	130	8.46	150	9.80	130		
	L	5.40		9.84		8.60			
α-Aminobutyric acid	D	8.46	120	11.00	130	12.80	130	10.00	120
	L	9.44		13.90		11.00		7.92	
	D	8.60	130					9.14	130
	L	9.20						7.36	
β-Aminobutyric acid	D			21.88	130	22.80	130	13.52	120
	L	No separation		22.50		21.80		14.60	
α-Aminoheptanoic acid	D	42.88	120	53.60	130	58.80	130	43.92	120
	L	46.74		68.20		49.60		56.48	
	D	29.70	130					34.68	130
	L	31.80						43.92	
α-Aminohexanoic acid	D	24.64	120	30.90	130	34.40	130	25.08	120
	L	27.14		39.60		29.00		32.32	
	D	17.40	130					20.70	130
	L	18.70						26.02	
β-Amino-δ-methyl-hexanoic acid	D			63.90	130	60.40	130	41.40	120
	L			67.4		56.80		46.40	
γ-Amino-δ-methylhexanoic acid	D	52.90	130	97.90	130				
	L	51.90		92.68		No separation		No separation	
γ-Amino-ϵ-methylheptanoic acid	D	98.20	130	166.64	130				
	L	94.70		159.20		No separation		No separation	
β-Amino-γ-methylpentamoic acid	D			44.84	130	43.00	130	29.00	120
	L	No separation		46.96		40.00		33.40	
α-Aminooctanoic acid	D	75.26	120	93.20	130	101.20	130	75.76	120
	L	82.02		119.00		85.20		96.52	
	D	51.50	130					59.58	130
	L	55.10						75.76	

α-Aminopentanoic acid	D	14.56	120	18.20	130	21.80	130	13.84	120
	L	16.28		23.60		18.60		18.64	
	D	10.64	130					12.04	130
	L	11.50						15.34	
γ-Aminovaleric acid	D	28.20	130	56.60	130				
	L	28.20		56.60		No separation		No separation	
Aspartic acid	D	29.64	140	52.28	160	38.00	150	26.86	160
	L	30.30		54.46		36.80		27.88	
	D	15.10	160	41.08	170	26.60	160		
	L	15.30		42.50		26.00			
Glutamic acid	D	71.34	140	114.14	160	96.00	150	58.14	160
	L	75.48		126.64		87.60		64.98	
	D	34.80	160	85.70	170	63.60	160		
	L	36.20		95.50		59.80			
	D			61.14	180				
	L			66.64					
	D			40.28	200				
	L			43.20					
Glycine		13.40	120	24.36	130	19.20	130	9.94	130
		10.70	130						
Isoleucine	D	13.10	120	30.24	130	33.00	120	11.72	130
	L	14.16		36.40		28.40		14.42	
	D	9.86	130	19.14	150	21.60	130		
	L	10.66				18.88			
Allo-isoleucine	D	11.86	120	27.50	130	30.20	120	10.70	130
	L	13.10		33.44		25.80		13.00	
	D	9.24	17.58	150		19.68	130		
	L	10.10		20.08		17.20			
Leucine	D	19.60	120	38.80	130	47.92	120	17.20	120
	L	21.76		52.06		38.68		23.24	
	D	15.70	130	24.30	150	30.20	130	15.64	130
	L	17.20		29.54		25.20		20.74	
	D			10.70	170	12.60	150	6.50	150
	L			12.30		11.00		7.90	
	D							5.00	160
	L							5.88	
Methionine	D	51.12	140	81.76	160	69.60	150	42.96	160

Table GC 15 (continued)
ENANTIOMERS OF *N*-TRIFLUOROACETYL AMINO ACID ISOPROPYL ESTERS ON DIAMIDE STATIONARY PHASES

Amino acid	Isomer	t_r(min)	T(°C)[a]	t_r(min)	T(°C)[a]	t_r(min)	t(°C)[a]	t_r(min)	T(°C)[a]
	L	54.32		93.63		62.80		49.06	
	D	24.90	160	63.14	170	47.20	160		
	L	26.00		70.56		43.20			
	D			30.84	200	95.20	150	64.34	160
	L			32.86		90.40		72.06	
Phenylalanine	D	76.64	140	118.06	160	95.20	150	64.34	160
	L	81.94		133.52		90.40		72.06	
	D	39.00	160	90.06	170	66.40	160		
	L	40.70		99.44		60.80			
Phenylglycine	D	37.08	140	64.96	160	51.40	150	33.16	160
	L	37.74		69.90		47.20		35.96	
	D	19.70	160	50.08	170	35.40	160		
	L	19.70		53.08		33.00			
Proline	D	20.46	120	46.04	130	40.40	130	22.42	130
	L	21.20		47.92		40.40		22.42	
	D	17.70	130	29.02	150				
	L	17.70		30.06					
Serine	D	15.80	130	38.64	130	24.20	130	14.24	130
	L	16.80		45.40		22.08		16.04	
tert-Leucine	D	8.30	120	12.70	130	11.60	130	7.20	130
	L	8.80		13.90		11.00		8.02	
	D	5.90	130						
	L	6.10		20.12	130	12.40	130	7.16	130
Threonine	D	10.00	120	23.64		11.60		8.26	
	L	11.20		11.64	150				
	D	6.90	130	13.04					
	L	7.56		20.50	130	20.60	120	7.42	130
Valine	D	8.06	120	24.34		17.80		8.96	
	L	8.76		12.50	150	13.40	130		
	D	6.60	130	14.22		12.00			
	L	7.00							

[a] Column temperature.

Conditions	Packings:	P1, P2, P3, and P4 are chiral diamide phases of formula $C_{11}H_{23}$ CONHCH(R″)CONH-*tert*-C_4H_9 derived from L-alanine (P1,R″=CH_3), L-leucine (P2,R″=$CH_2CH[CH_3]_2$), D-phenylglycine (P3,R″=C_6H_5), and L-phenylalanine (P4,R″=$CH_2C_6H_5$)
	Columns:	Stainless steel capillary columns 100 ft × 0.02 in. I.D. were coated with P1 and P2, and columns 150 ft × 0.02 in. I.D. were coated with P3 and P4
	Gas:	He, flow rate 2.8—3.0 mℓ/min
	Detector:	FID

REFERENCE

1. **Chang, S.-C., Charles, R., and Gil-Av, E.,** *J. Chromatogr.*, 235, 87, 1982.

Table GC 16
N-TRIFLUOROACETYL AMINO ACID ISOPROPYL ESTERS ON DIAMIDE STATIONARY PHASES

Column packing	P1	P2	P3
Temperature (°C)	130	130	130
Gas Flow rate (mℓ/min)	He, 2.8—3.0	He, 2.8—3.0	He, 2.8—3.0
Column			
Length (ft)	150	100	150
Diameter (in., I.D.)	0.02	0.02	0.02
Material	SS	SS	SS
Detector	FID	FID	FID

Amino acid	$r_{D/L}$	$4_{L/D}$	$r_{L/D}$
Alanine	1.224	1.231	1.212
Allo-isoleucine	1.188	1.216	1.215
α-Aminopentanoic acid	1.244	1.297	1.274
Isoleucine	1.175	1.204	1.230
Leucine	1.288	1.342	1.326
tert-Leucine	1.094	1.095	1.114
Valine	1.172	1.187	1.207

Note: $r_{L/D}$ (for L-stationary phases) and $r_{D/L}$ (for D-stationary phases) = resolution factor = ratio of the corrected retention time of the second over that of the first emerging enantiomer.

Column packings: P1, P2, and P3 are *N*-lauroyl-*tert*-butylamide stationary phases derived from D-valine, L-leucine, and L-phenylalanine, respectively.

REFERENCE

1. **Chang, S.-C., Charles, R., and Gil-Av, Ed.,** *J. Chromatogr.*, 235, 87, 1982.

Table GC 17
THE ELUTION ORDERS OF *N*-PENTAFLUOROPROPIONYL AMINO ACID *n*-PROPYL ESTERS ON DIFFERENT COLUMNS

	Elution order		
Derivative of	**OV-101**	**Chiral**	**OV-101 and chiral**
Alanine	1	1	1
allo Isoleucine	7	4	6
Arginine	18	19	18
Aspartic acid	12	10	11
Cystine	10	11	10
Glutamic acid	14	14	14
Glycine	2	5	2
Histidine	19	18	
Isoleucine	8	6	5
Leucine	6	8	8
Lysine	15	17	17
Methionine	11	12	12
Ornithine	17	16	16
Phenylalanine	13	13	13

Table GC 17 (continued)
THE ELUTION ORDERS OF *N*-PENTAFLUOROPROPIONYL AMINO ACID *n*-PROPYL ESTERS ON DIFFERENT COLUMNS

	Elution order		
Derivative of	**OV-101**	**Chiral**	**OV-101 and chiral**
Proline	9	7	9
Serine	5	9	7
Threonine	4	3	3
Tryptophan	20	20	
Tyrosine	16	15	15
Valine	3	2	4

Conditions: Columns: Glass capillary coated with OV-101
Glass capillary coated with Chirasil®-val stationary phase
Two-column system; precolumn 80°C, isothermal for 5 min, then raised at 10°C/min to 130°C, then at 15°C/min to 200°C: main column 75°C raised at 5°C/min to 180°C
Carrier gas: Hydrogen
Detector: Flame ionization

REFERENCE

1. **Chinghai, W., Frank, H., Guanghua, W., Liangmo, Z., Bayer, E., and Peichang, L.,** *J. Chromatogr.*, 262, 352, 1983.

Table GC 18
THE EFFECT OF THE RATE OF TEMPERATURE PROGRAMMING ON THE ELUTION ORDER OF *N*-PENTAFLUOROPROPIONYL AMINO ACID *n*-PROPYL ESTERS

	Temperature programming rate	
Derivative of	**A**	**B**
Alanine	1	1
Alloisoleucine	4	5
Glycine	5	4
Isoleucine	6	6
Leucine	8	7
Proline	7	8
Serine	9	9
Threonine	2	2
Valine	3	3

Conditions: In the two-column system, column 1 is a glass capillary coated with OV-101, column 2 is a glass capillary coated with Chirasil®-val stationary phase; in temperature program A, the temperature of column 1 is increased from 80—200°C at 10°C/min, that of column 2 from 75—100°C at 2.5°C/min, then at 7°C/min to 200°C; in program B, the

Table GC 18 (continued)
THE EFFECT OF THE RATE OF TEMPERATURE PROGRAMMING ON THE ELUTION ORDER OF *N*-PENTAFLUOROPROPIONYL AMINO ACID *n*-PROPYL ESTERS

temperature of column 1 is increased from 40—200°C at 8°C/min, that of column 2 from 75—120°C at 2.5°C/min, is isothermal for 0.1 min and then is increased at 3°C/min to 200°C

Detector: Flame ionization

REFERENCE

1. **Chinghai, W., Frank, H., Guanghua, W., Liangmo, Z., Bayer, E., and Peichang, L.,** *J. Chromatogr.*, 262, 352, 1983.

Table GC 19
N-TERT-BUTYL-UREIDO-DL-AMINO ACID ISOPROPYL ESTERS

Amino acid	α	Column temp (°C)
Alanine	1.09	180
Aminobutyric acid	1.08	180
Aspartic acid	1.08	180
Alloisoleucine	1.09	180
Allothreonine (OTMS)[a]	1.05	180
Glutamic acid	1.07	200
Isoleucine	1.08	180
Isovaline	1.02	170
Leucine	1.09	180
Methionine	1.07	200
Norleucine	1.09	180
Norvaline	1.09	180
Phenylalanine	1.07	200
Phenylglycine	1.02	200
Pipecolic acid	1.03	160
Proline	1.03	180
Pyroglutamic acid	1.02	160
Serine (OTMS)[a]	1.11	180
Threonine (OTMS)[a]	1.09	180
Valine	1.08	180

Note: The D-enantiomers are eluted first.

[a] Side chain hydroxy group was silylated with MSTFA (*N*-methyl-*N*-trimethylsilyltrifluoroacetamide).

Conditions: A 25-m glass capillary column coated with XE-60-L-valine-(S)-α-phenylethylamide
Carrier gas = hydrogen

REFERENCE

1. **König, W. A., Benecke, I., Lucht, N., Schmidt, E., Schulze, J., and Sievers, S.,** *J. Chromatogr.*, 279, 555, 1983.

Table GC 20
N-METHYL-*N*-*TERT*-BUTYL-UREIDO-D,L-AMINO ACID *TERT*-BUTYLAMIDES (A) AND *N*-ISOPROPYLUREIDO-*N*-METHYL-D,L-AMINO ACID ISOPROPYLAMIDES (B)

Amino acid	α	Column temp (°C)
	A	
N-Methylalanine	1.023	160
N-Methylalloisoleucine	1.033	160
N-Methylaminobutyric acid	1.031	160
N-Methylhomophenylalanine	1.008	160
N-Methylisoleucine	1.028	160
N-Methylleucine	1.037	160
N-Methylphenylalanine[a]	1.009	160
N-Methylvaline	1.037	160
	B	
N-Methylalanine	1.014	170
N-Methylalloisoleucine	1.037	170
N-Methylaminobutyric acid	1.032	170
N-Methylhomophenylalanine	1.007	180
N-Methylisoleucine	1.047	170
N-Methylleucine	1.049	170
N-Methylphenylalanine[a]	1.000	180
N-Methylvaline	1.038	170

Note: L-Enantiomers are eluted first.

[a] D-Enantiomer is eluted first.

Conditions: A 25-m Pyrex® glass capillary column coated with XE-60-L-valine-(R)-α-phenylethylamide
Carrier gas = hydrogen

REFERENCE

1. **König, W. A., Benecke, I., Lucht, N., Schmidt, E., Schulze, J., and Sievers, S.,** *J. Chromatogr.*, 279, 555, 1983.

Table GC 21
t-BUTYLDIMETHYLSILYL DERIVATIVES OF AMINO ACIDS

Amino acid	Mono/di/tri derivative	Column temp (°C)	t_r(min)
Alanine	Di	185[a]	5.1
(3,3,3-$^2H_3$) Alanine	Di	185[a]	5.1
Histidine	Tri	250[a]	11.9
(2,3,3-$^2H_3$) Histidine	Tri	250[a]	11.9
Leucine	Di	180	4.7
(5,5,5-$^2H_3$) Leucine	Di	180	4.7
(1-^{13}C) Leucine	Di	180	4.7
(4,5,5,5,6,6,6-$^2H_7$) Leucine	Di	180	4.6
1-Methylhistidine	Di	250[a]	7.0
3-Methylhistidine	Di	250[a]	5.4

Table GC 21 (continued)
t-BUTYLDIMETHYLSILYL DERIVATIVES OF AMINO ACIDS

[a] Column length of 2 m.

Conditions: Packing: 3% OV-11
Column: 4 m × 6 mm
Gas: He; flow rate = 30 mℓ/min
Detection: Mass spectrometry

REFERENCE

1. **Schwenk, W. F., Berg, P. J., Beaufrere, B., Miles, J. M., and Haymond, M. W.,** *Anal. Biochem.*, 141, 101, 1984.

Table GC 22
AMINO ACIDS AS THEIR *TERT*-BUTYLDIMETHYLSILYL DERIVATIVES

Column packing	P1	P2	P3	P4
Temperature	T1	T1	T2	T2
Gas Flow rate (mℓ/min)	N_2, 18	N_2, 18	He, 5	He, 5
Column				
Length	6 ft	6 ft	25 m	12 m
Diameter (I.D.)	1.8 mm	1.8 mm	0.32 mm	0.32 mm
Form			Fused silica capillary	WCOT capillary
Material	G	G		
Detector	FID	FID	FID	FID

Amino acid		**t_r(min)**		
Alanine	7.69	9.15	8.48	6.94
α-Aminobutyric acid	8.50	9.92	9.37	7.57
γ-Aminobutyric acid	11.48	14.55	11.47	9.25
Arginine I	24.28	28.30	18.45	16.32
Arginine II				20.52
Aspartic acid	18.92	22.14	15.47	13.40
Carboxymethylcysteine	24.48	29.08	18.98	16.65
Cysteine	19.69	23.21	16.06	13.90
Cystine I			23.89	21.69
Cystine II			25.72	
Glutamic acid	20.90	24.37	16.58	14.50
Glycine	7.33	9.92	8.99	6.68
Histidine	25.43	31.15	19.30	16.95
Isoleucine	11.08	12.98	10.77	9.02
Leucine	10.50	12.20	10.32	8.67
Lysine I	22.78	25.58	17.90	15.51
Lysine II		26.21	17.41	
Methionine	15.29	19.33	14.05	11.42
Norleucine	11.48	13.46	10.98	9.25
Phenylalanine	17.42	21.75	15.34	12.61
Proline	11.48	14.55	11.64	9.25
Serine	16.12	18.19	13.48	11.87
Threonine	16.67	18.56	13.72	12.18
Tryptophan I	26.41	33.23	20.11	17.53
Tryptophan II		36.78	22.10	
Valine	9.63	11.39	9.90	8.16

Table GC 22 (continued)
AMINO ACIDS AS THEIR *TERT*-BUTYLDIMETHYLSILYL DERIVATIVES

Packing: P1 = 3.0% SE-30 on Supelcoport®, 100-120 mesh
P2 = 3.0% SP-2250 on Supelcoport®, 100-120 mesh
P3 = fused silica capillary column with 0.25-μm bonded methylphenyl (50%) silicone
P4 = WCOT capillary column coated with 0.5 μm SE-30

Temperature: T1 = after an initial hold of 1 min at 110°C the column is temperature programmed at 5°C/min to 260°C
T2 = after an initial hold of 2 min at 100°C the column is temperature programmed at 10°C/min to 280°C

REFERENCE

1. **Mawhinney, T. P., Robinett, R. S. R., Atalay, A., and Madson, M. A.,** *J. Chromatogr.*, 358, 231, 1986.

Table GC 23
ENANTIOMERS OF *N*-TRIFLUOROACETYL DIISOPROPYL ESTER DERIVATIVES OF SUBSTITUTED GLUTAMIC ACIDS

Column packing	P1
Gas, inlet pressure	He, 1.5 bar
Column	
Length (m)	50
Diameter (mm, I.D.)	0.25
Material	Fused silica
Detector	FID

Racemate	**Temp (°C)**	**First isomer eluted**	**t_r (min)[a]**	**α**
Glutamic acid	165	R	11.7	1.082
2-Methylglutamic acid	160	—	7.8	1.000
	165	—	6.5	1.000
4-Methyleneglutamic acid	165	R	9.9	1.086
threo-4-Hydroxyglutamic acid	165	2R,4R	14.3	1.058
erythro-4-Hydroxyglutamic acid	165	2R,4S	12.7	1.071
threo-3-Methylglutamic acid	160	2R,3S	10.9	1.092
erythro-3-Methylglutamic acid	160	2R,3R	10.9	1.073
threo-4-Methylglutamic acid	160	2R,4R	11.8	1.076
erythro-4-Methylglutamic acid	160	2R,4S	12.4	1.105
threo-3-Fluoroglutamic acid[b]	175	2S,3S	30.2	1.065
erythro-3-Fluoroglutamic acid[b]	175	2S,3R	40.0	1.050
threo-4-Fluoroglutamic acid	165	2R,4R	15.9	1.063
erythro-4-Fluoroglutamic acid	165	2R,4S	15.0	1.053

[a] Retention time (min) from the solvent peak; retention time of the solvent peak 5.0 min.
[b] As *N*-acetyl *O,O'*-diisopropyl ester.

Packing: P1 = polysiloxane XE-60-(*S*)-valine-(*S*)-phenylethylamide coated on Chrompack® column

REFERENCE

1. **Maurs, M., Ducrocq, C., Righini-Tapie, A., and Azerad, R.,** *J. Chromatogr.*, 325, 444, 1985.

Table GC 24
THE EFFECT OF ELECTRON CAPTURE DETECTION (ECD) ON AMINO ACID RMRs

Amino acid	RMR_{ECD}/RMR_{FID}
Glutamic acid	3.9
Histidine	2.5
Hydroxyproline	1.8
Methionine	1.4
Serine	2.0
Tryptophan	6.0
Tyrosine	1.4

Note: RMR is the molar response relative to norleucine. The RMRs are calculated for each amino acid by peak-height measurement. The sensitivity of electron capture detection and flame ionization detection is compared by using the quotient between RMRs.

REFERENCE

1. **Bengtsson, G., Odham, G., and Westerdahl, G.,** *Anal. Biochem.*, 111, 163, 1981.

Table GC 25
VOLATILE AMINES

Column packing	P1	P2
Temperature (°C)	80	50
Gas Flow rate (mℓ/min)	He, 17.6[a]	He, 17.6[a]
Column		
Length (m)	2.10	2.10
Diameter (mm, O.D., I.D.)	6.35, 2.0	6.35, 2.0
Material	Glass[b]	Glass[b]
Detection	FID	FID

Amine	t_r(min:sec)	
Methylamine		0.54
Dimethylamine		2.00
Ethylamine		2.48
Trimethylamine	0.54	3.24
Isopropylamine	1.21	
n-Propylamine	1.54	
Isobutylamine	3.12	
n-Butylamine	4.09	
Pyrrolidine	6.30	
2-Methylbutylamine	6.45	
3-Methylbutylamine	7.00	
n-Pentylamine	8.30	
Piperidine	12.00	

[a] Helium loaded with ammonia.
[b] Borosilicate glass.

Column packing: P1 = 28% Pennwalt® 223 + 4% KOH on Gas-Chrom® R, 80-100 mesh

Table GC 25 (continued)
VOLATILE AMINES

P2 = 4.8% PEG 20M + 0.3% KOH on Carbopack® B, 100-120 mesh

REFERENCE

1. **Pons, J.-L., Rimbault, A., Darbord, J. C., and Leluan, B.,** *J. Chromatogr.*, 337, 213, 1985.

Table GC 26
ALIPHATIC AMINES

Column packing	P1
Temperature	T1
Gas Flow rate (mℓ/min)	N_2, 45
Column	
Length (m)	2
Diameter (mm, I.D.)	2
Material	G
Detection	D1

Amine	**t_r(min)**
Methylamine	1.08 ± 0.005
Dimethylamine	1.93 ± 0.004
Ethylamine	2.39 ± 0.003
Trimethylamine	2.71 ± 0.001
Isopropylamine	4.46 ± 0.007
n-Propylamine	5.63 ± 0.007
tert-Butylamine	6.54 ± 0.008
Diethylamine	7.02 ± 0.010
sec-Butylamine	8.22 ± 0.007
Isobutylamine	8.73 ± 0.005
n-Butylamine	9.93 ± 0.011

Column packing: P1 = SEPABEAD GHP-1 (GHP-1) 60-80 mesh, a spherical styrene-divinylbenzene copolymer (Mitsubishi Chemical, Tokyo); 10 g of GHP-1 is extracted with 100 mℓ of methanol for 3 hr in a Soxhlet® apparatus; it is then mixed with a solution of 1.0 g of KOH in 30 mℓ of methanol, and the mixture is dried under reduced pressure below 50°C in a rotary evaporator; the treated material is packed into the chromatographic column, which is conditioned at 200°C for 2 days; during the conditioning, 10 μℓ of water is injected 15—20 times to ensure a stable column; before use, 10 μℓ of 1 *N* KOH solution is injected twice at 170°C

Temperature: T1 = the column temperature is isothermal at 140°C for 3 min, is increased to 170°C at 10°C/min, and is maintained at this temperature for 10 min

Detection: D1 = nitrogen-phosphorus flame ionization detector

Table GC 26 (continued)
ALIPHATIC AMINES

REFERENCE

1. **Kuwata, K., Akiyama, E., Yamazaki, Y., Yamasaki, H., and Kuge, Y.,** *Anal. Chem.*, 55, 2199, 1983.

Table GC 27
LIGAND-EXCHANGE GAS CHROMATOGRAPHY OF LOWER ALIPHATIC AMINES

	Adjusted retention time (min)[a] (NH_3 conc in mobile phase [μmol/mℓ])							
Amine	**1.04**	**2.79**	**4.68**	**6.51**	**8.60**	**10.80**	**13.21**	**15.39**
Ethylamine	11.4	3.7	2.3	1.6	1.2	1.0	0.8	0.7
n-Propylamine	15.6	4.9	2.9	2.2	1.7	1.5	1.2	1.0
Isopropylamine	4.1	1.3	0.9	0.7	0.6	0.5	0.4	0.4
n-Butylamine	33.5	10.8	6.4	4.7	3.7	3.2	2.7	2.4
Isobutylamine	13.3	4.5	2.7	2.0	1.6	1.4	1.1	0.9
sec-Butylamine	6.5	2.3	1.5	1.2	1.0	0.8	0.7	0.6
tert-Butylamine	2.1	0.9	0.6	0.5	0.4	0.4	0.3	0.3
n-Amylamine	74.6	23.7	14.3	10.5	8.4	7.1	6.1	5.5
Isoamylamine	52.7	17.0	10.3	7.7	6.1	5.1	4.4	4.2
Neopentylamine	9.6	3.1	2.0	1.5	1.2	1.1	0.9	0.8
tert-Amylamine	4.4	1.7	1.1	0.9	0.8	0.7	0.6	0.5
n-Hexylamine		55.5	32.7	24.1	18.8	16.7	14.2	12.4
n-Heptylamine		127.0	74.8	54.7	43.2	37.7	32.6	28.5
2-Ethylhexylamine		83.2	48.2	35.9	29.0	24.8	21.5	19.1
1,1,3,3-Tetramethylbutylamine	16.0	5.3	3.9	3.3	3.0	2.7	2.6	2.4
Dimethylamine	5.2	1.4	0.9	0.6	0.5	0.5	0.4	0.3
Diethylamine	1.7	0.7	0.6	0.5	0.5	0.4	0.4	0.3
Di-*n*-propylamine	2.8	1.5	1.2	1.1	1.0	1.0	0.9	0.9
Diisopropylamine	0.7	0.5	0.5	0.4	0.4	0.4	0.4	0.3
N-Ethyl-*n*-butylamine	4.6	2.2	1.7	1.4	1.3	1.2	1.1	1.0
N-Ethyl-*tert*-butylamine	0.9	0.7	0.6	0.6	0.5	0.5	0.5	0.4
Di-*n*-butylamine	11.7	5.9	4.7	4.3	4.0	3.8	3.6	3.5
Diisobutylamine	2.5	1.8	1.7	1.6	1.6	1.5	1.5	1.5
Di-*n*-amylamine	58.8	26.7	20.7	18.4	17.0	16.0	15.4	15.4
Diisoamylamine	30.0	14.3	11.2	10.0	9.3	8.8	8.5	8.3
Trimethylamine	0.4	0.2	0.2	0.1	0.1	0.1	0.1	0.1
Triethylamine	0.7	0.5	0.5	0.5	0.5	0.5	0.4	0.4
Tri-*n*-propylamine	2.3	2.2	2.2	2.2	2.2	2.2	2.1	2.1
Tri-*n*-butylamine	15.0	14.6	14.4	14.4	14.3	14.0	14.0	13.9

[a] Retention time of methane was used as t_o.

Conditions:
- Column: 3 m × 4 mm I.D.
- Packing: 5% Cu(II) stearate coated on Chromosorb® G AW/DMCS, 80-100 mesh

Chromosorb® G AW/DMCS is added to a benzene solution of metal stearate and the solvent is evaporated with a rotary vacuum evaporator; the dried packing material is resieved to 80-100 mesh and equilibrated with ammonia vapor for 6 hr to form metalammine complex by spreading in a dish, which is placed in a desiccator containing concentrated aqueous ammonia at the bottom

Table GC 27 (continued)
LIGAND-EXCHANGE GAS CHROMATOGRAPHY OF LOWER ALIPHATIC AMINES

Glass spiral columns are prepared and conditioned for at least 12 hr by passing nitrogen containing ammonia and water vapor through them at a temperature 5°C or more higher than the operating temperature
Carrier gas: N_2, H_2O concentration in gas 0.75—112 μmol/mℓ
Flow rate: 20.0 mℓ/min
Temperature: 80°C
Detector: Hydrogen flame ionization detector

REFERENCE

1. **Fujimura, K., Kitanaka, M., Takayanagi, H., and Ando, T.,** *Anal. Chem.*, 54, 918, 1982.

Table GC 28
LOWER ALIPHATIC AMINES

Column packing	P1	P2	P3
Temperature (°C)	80	80	80
Gas Flow rate (mℓ/min)	N_2, 20	N_2, 20	N_2, 20
Column			
Length (m)	3	3	3
Diameter (mm, I.D.)	4	4	4
Form	Spiral	Spiral	Spiral
Material	G	G	G
Detector	D1	D1	D1
NH_3 concentration[a]	2.50	2.49	2.50
H_2O concentration[b]	0.35	0.38	0.53

Amine	**Net retention volume V_N^c**		
Ethylamine	82.6	36.7	82.2
n-Propylamine	92.9	39.3	90.0
Isopropylamine	13.7	18.4	41.7
n-Butylamine	188.2	81.3	183.9
Isobutylamine	92.9	35.4	71.7
sec-Butylamine	35.6	24.9	65.2
tert-Butylamine	6.9	13.1	31.3
n-Amylamine	408.6	171.7	327.3
Isoamylamine	298.4	128.5	237.2
tert-Amylamine	19.5	28.8	50.9
n-Hexylamine	929.5	397.2	596.0
Diethylamine	5.7	19.7	52.2
Di-*n*-propylamine	16.1	38.0	82.3
Diisopropylamine	4.6	13.1	28.7
Di-*n*-butylamine	101.0	173.0	279.1
Diisobutylamine	20.7	66.9	80.9
Di-*n*-amylamine	441.8	901.7	989.3
Diisoamylamine	219.2	464.0	530.8
Trimethylamine	1.1	3.9	7.8
Triethylamine	4.6	17.0	27.4
Tri-*n*-propylamine	24.1	104.9	75.6
Tri-*n*-butylamine	163.0	977.9	366.4

[a] NH_3 concentration in gas phase (μmol/mℓ).
[b] H_2O concentration in gas phase (μmol/mℓ).
[c] V_N values are calculated from $V_N = jV_{R'}$ where j is the pressure drop correction factor and $V_{R'}$ is the adjusted retention volume.

Table GC 28 (continued)
LOWER ALIPHATIC AMINES

Packing: P1 = 5% Cu(II) stearate on Chromosorb® G AW/DMCS, 80-100 mesh

P2 = 5% sodium stearate on Chromosorb® G AW/DMCS, 80-100 mesh

P3 = 5% PEG 20 M on Chromosorb® G AW/DMCS, 80-100 mesh

Packing materials are prepared by adding Chromosorb® G AW/DMCS to a benzene solution of Cu(II) stearate or an absolute methanol solution of sodium stearate; the solvent is evaporated using a rotary vacuum evaporator and resieving the dried packing materials to 80-100 mesh; the material is then equilibrated with ammonia vapor for 6 hr to form the metal-amine complex by spreading it in a dish which is placed in a desiccator containing concentrated aqueous ammonia at the bottom; the chromatography columns are prepared and conditioned for at least 12 hr by passing nitrogen-containing ammonia and water vapor through them at a temperature 5°C or more higher than the operating temperature

Detector: D1 = hydrogen flame ionization detector

REFERENCE

1. **Fujimura, K., Kitanaka, M., Takayanagi, H., and Ando, T.,** *Anal. Chem.*, 54, 918, 1982.

Table GC 29
THE EFFECT OF WATER VAPOR IN THE GAS PHASE OF THE RETENTION OF ALIPHATIC AMINES

Column packing	P1	P1
Temperature (°C)	80	80
Gas Flow rate (mℓ/min)	N_2, 20.0	N_2, 20.0
Column		
Length (m)	3	3
Diameter (mm, I.D.)	4	4
Form	Spiral	Spiral
Material	G	G
Detector	D1	D1
NH_3 concentration[a]	5.03	5.03
H_2O concentration[b]	0	0.80
Amine	**Adjusted retention time (min)[c]**	
Ethylamine	3.8	2.0
n-Propylamine	3.9	2.8
n-Butylamine	8.8	6.0
Isobutylamine	3.6	2.5
sec-Butylamine	1.7	1.4
tert-Butylamine	0.9	0.6
n-Amylamine	19.2	13.5
n-Hexylamine	43.6	30.2
n-Heptylamine	101.3	67.6
Di-*n*-propylamine	1.2	1.2
Di-*n*-butylamine	4.6	4.6
Di-*n*-amylamine	20.3	19.5
Triethylamine	0.4	0.5
Tri-*n*-propylamine	2.3	2.2
Tri-*n*-butylamine	15.1	14.5

Table GC 29 (continued)
THE EFFECT OF WATER VAPOR IN THE GAS PHASE OF THE RETENTION OF ALIPHATIC AMINES

[a] NH_3 concentration in gas phase (μmol/mℓ).
[b] H_2O concentration in gas phase (μmol/mℓ).
[c] Retention time of methane was used as t_o.

Column packing: P1 = 5% Cu(II) stearate on Chromosorb® G AW/DMCS, 80-100 mesh
Chromosorb® G AW/DMCS is added to a benzene solution of Cu(II) stearate and the solvent evaporated in a rotary vacuum evaporator; after the dried packing material is resieved to 80-100 mesh it is equilibrated with ammonia vapor for 6 hr to form the metal-amine complex by spreading it on a dish, which is placed in a desiccator containing concentrated aqueous ammonia at the bottom; the chromatography column is prepared and conditioned for at least 12 hr by passing nitrogen containing ammonia and water vapor through the column at a temperature 5°C or more higher than the operating temperature

Detector: D1 = hydrogen flame ionization detector

REFERENCE

1. **Fujimara, K., Kitanaka, M., Takayanagi, H., and Ando, T.,** *Anal. Chem.*, 54, 918, 1982.

Table GC 30
AMMONIA, DIMETHYLAMINE, AND HYDRAZINE

Column packing	P1
Gas Flow rate (mℓ/min)	H_2, 100
Column	
Length (ft)	8
Diameter (in.)	0.25
Detection	D1

Compound	t_r(min)	Column temp (°C)
Ammonia	3.40	60
	1.52	110
Dimethylamine	5.80	60—125[a]
	1.90	110
Hydrazine	40.80	110
	19.29	125
Isobutanol	13.80	110
Monomethylhydrazine	22.00	110
	10.20	125
Unsymmetrical dimethylhydrazine	7.20	110
	3.00	125
Water	21.52	60—125[a]
	14.84	110
	7.20	125

Table GC 30 (continued)
AMMONIA, DIMETHYLAMINE, AND HYDRAZINE

[a] Temperature programmed at 15°C/min.

Column packing: P1 = 30% Carbowax®-400 on Anakrom® AB, 90-100 mesh
Detection: D = thermal conductivity

REFERENCE

1. **Salvapathy, G. S. and Subba Rao, Y. V.,** *J. Chromatogr. Sci.*, 21, 14, 1983.

Table GC 31
LIGAND EXCHANGE SEPARATION OF ANILINE BASES

Column packing	P1	P2	P3	P4
Temperature (°C)	75	75	75	75
Gas Flow rate (mℓ/min)	N_2, 10[a]	N_2, 10[b]	N_2, 10	N_2, 10[c]
Column				
Length (m)	3	3	3	3
Diameter (mm, I.D.)	4	4	4	4
Form	Spiral	Spiral	Spiral	Spiral
Material	G	G	G	G
Detector	FID	FID	FID	FID
Aniline base		**t_r(min)**		
Aniline	2.9	2.5	5.6	0.45
N-Methylaniline	4.0	3.5	6.3	0.40
N-Ethylaniline	6.2	5.4	9.4	0.54
N-*n*-Propylaniline	12.2	10.7	18.7	0.94
N-Isopropylaniline	6.7	5.8	10.3	0.58
N-*n*-Butylaniline	27.0	23.5	41.0	1.84
N-Isobutylaniline	16.7	15.1	25.8	1.23
N-*sec*-Butylaniline	13.0	11.8	20.1	1.00
N,N-Dimethylaniline	4.5	3.8	6.8	0.37
N,N-Diethylaniline	11.3	9.8	17.3	0.83
N,N-Di-*n*-propylaniline	37.8	32.7	57.6	2.40
N,N-Di-*n*-butylaniline				7.80
o-Fluoroaniline	1.9	1.6	3.0	0.25
m-Fluoroaniline	3.5	3.1	6.2	0.48
p-Fluoroaniline	3.8	3.2	8.0	0.63
o-Chloroaniline	7.2	6.5	11.7	0.61
m-Chloroaniline	16.5	14.8	29.8	1.56
p-Chloroaniline	19.0	16.9	39.0	1.96
o-Toluidine	5.2	4.4	8.6	0.61
m-Toluidine	6.6	5.4	11.8	0.80
p-Toluidine	6.9	5.8	14.6	0.91
2,3-Xylidine	14.9	12.2	25.8	1.46
2,4-Xylidine	12.3	10.0	21.9	1.23
2,5-Xylidine	11.9	9.7	19.3	1.11
2,6-Xylidine	9.8	8.0	14.8	0.83
3,4-Xylidine	19.7	15.4	40.3	2.25
3,5-Xylidine	14.9	11.9	26.7	1.47

Table GC 31 (continued)
LIGAND EXCHANGE SEPARATION OF ANILINE BASES

[a] NH_3 concentration in mobile phase, 1.20 μmol/mℓ, H_2O concentration 0.13 μmol/mℓ.
[b] NH_3 concentration in mobile phase, 4.69 μmol/mℓ, H_2O concentration 0.15 μmol/mℓ.
[c] NH_3 concentration in mobile phase, 1.14 μmol/mℓ, H_2O concentration, 0.11 μmol/mℓ.

Column packing: P1, P2, and P3 are Chromosorb® G AW DMCS, 80-100 mesh coated with 3% Mn(II) stearate; P4 is uncoated Chromosorb® G AW DMCS; coating is performed thusly: a solution of manganese stearate in refluxing benzene is cooled to room temperature and added to Chromosorb® G AW DMCS; the mixture is stirred gently for 10 min and the solvent evaporated slowly under reduced pressure using a rotary evaporator; after rinsing with *n*-hexane and drying in air at room temperature, the packing is equilibrated with ammonia vapor for 24 hr in a dish in a desiccator containing 14% ammonia at the bottom; after air drying the packing is sieved to 80-100 mesh and placed in the column using suction; the column is then conditioned at 80°C for 24 hr using nitrogen containing ammonia gas and water vapor at a flow rate of 15 mℓ/min

REFERENCE

1. **Fujimura, K., Kitanaka, M., and Ando, T.,** *J. Chromatogr.*, 241, 295, 1982.

Table GC 32
DICHLOROANILINE ISOMERS

Column packing	P1	P2	P3	P4	P5	P6
Temperature (°C)	140	140	140	140	140	140
Gas Flow rate (mℓ/min)	N_2, 25	N_2, 25	N_2, 25	N_2, 25	N_2, 25	N_2, 25
Column						
Length (m)	2.25	2.25	2.25	2.25	2.25	2.25
Diameter (mm)	3	3	3	3	3	3
Form	U	U	U	U	U	U
Material	SS	SS	SS	SS	SS	SS
Detector	FID	FID	FID	FID	FID	FID
Isomer			$r_{2,6\text{-dichloroaniline}}$			
2,6-Dichloroaniline	1.00	1.00	1.00	1.00	1.00	1.00
2,5-Dichloroaniline	2.77	2.88	3.17	3.25	2.81	2.82
2,4-Dichloroaniline	3.89	4.00	4.33	5.25	4.31	4.00
2,3-Dichloroaniliine	4.31	4.00	4.33	5.25	4.31	4.00
3,5-Dichloroaniline	7.89	7.75	9.00	10.00	8.00	8.82
3,4-Dichloroaniline	10.89	11.50	12.17	13.75	11.38	11.82

Column packings: P1, P2, P3, P4, P5, and P6 are acid washed Sil-O-Cel®, C_{22} firebrick, 60-80 mesh, coated with 20% (1) nicotinic acid, (2) ^{1}H benztriazole, (3) nicotinamide, (4) imidazole, (5) benzimidazole, and (6) phthalimide, respectively

Table GC 32 (continued)
DICHLOROANILINE ISOMERS

REFERENCE

1. **Ono, A.,** *J. Chromatogr. Sci.*, 20, 367, 1982.

Table GC 33
HEPTAFLUOROBUTYRYL DERIVATIVES OF AMINES

Column packing	P1	P2	P3	P4
Temperature (°C)	75	85	65	100
Gas	na	na	na	na
Column				
Length (m)	1.52	1.52	1.52	1.52
Diameter (cm, I.D.)	0.4	0.4	0.4	0.4
Material	G	G	G	G
Detector	FID	FID	FID	FID

Amine		$r_{isobutylamine}$		
Ethylamine	0.52			
Isopropylamine	0.82			0.47
n-Propylamine		0.73	0.81	
Isobutylamine	1.00	1.00	1.00	1.00
n-Butylamine	1.34	1.38	1.47	
2-Methylbutylamine	1.83	1.75	1.88	
Isoamylamine		1.75	2.06	1.67
n-Amylamine				1.98
Pyrrolidine	2.57	3.44	4.50	
n-Hexylamine		5.70		
Ethanolamine	2.55	2.05	2.87	

Column packing: P1 = 3% SE-30 coated on acid-washed dimethylchlorosilane-treated Chromosorb® W, 85-100 mesh
P2 = 5% Apiezon® N coated on acid-washed dimethylchlorosilane-treated Chromosorb® W, 85-100 mesh
P3 = 3% OV-17 coated on acid-washed dimethylchlorosilane-treated Chromosorb® W, 85-100 mesh
P4 = 10% PEGA coated on acid-washed dimethylchlorosilane-treated Chromosorb® W, 85-100 mesh

REFERENCE

1. **Tavakkol, A. and Drucker, D. B.,** *J. Chromatogr. Sci.*, 22, 12, 1984.

Table GC 34
TRIFLUOROACETYL DERIVATIVES OF AMINES

Column packing	P1	P2	P3	P4	P3
Temperature	T1	T2	T3	T4	T5
Gas	na	na	na	na	na
Column					
Length (m)	1.52	1.52	1.52	1.52	2.74
Diameter (cm, I.D.)	0.4	0.4	0.4	0.4	0.2
Material	G	G	G	G	G
Detector	FID	FID	FID	FID	FID
Amine			$r_{isobutylamine}$		
Ethylamine	0.58		0.54	0.75[a]	0.47
Isopropylamine					0.38
n-Propylamine	0.79	0.77	0.83	1.00[a]	0.84
sec-Butylamine					0.79
Isobutylamine	1.00	1.00	1.00		1.00
n-Butylamine					1.34
2-Methylbutylamine	1.79	2.14	1.93	1.56[a]	
Isoamylamine		2.07	2.16	1.82[a]	1.47
n-Amylamine	1.94		2.90		1.63
Pyrrolidine	2.34	3.54	4.29	1.57[a]	1.78
n-Hexylamine					1.80
Di-*n*-butylamine					2.08
1,3-Diaminopropane					2.39
β-Phenylethylamine					2.67
Putrescine					2.92
Cadaverine					3.34
Tyramine					3.45
Indole					3.69
Spermidine					4.00
Spermine					4.88

[a] Retention relative to *n*-propylamine.

Column packing: P1 = 3% SE-30 coated on acid-washed dimethylchlorosilane-treated Chromosorb® W, 85-100 mesh
P2 = 5% Apiezon® N coated on acid-washed dimethylchlorosilane-treated Chromosorb® W, 85-100 mesh
P3 = 3% OV-17 coated on acid-washed dimethylchlorosilane-treated Chromosorb® W, 85-100mesh
P4 = 10% PEGA coated on acid-washed dimethylchlorosilane-treated Chromosorb® W, 85-100 mesh

Temperature: T1 = 60°C
T2 = 80°C
T3 = 70°C
T4 = 120°C
T5 = after an initial isothermal period of 7 min, the column is heated from 50—280°C at 6.75°C/min

REFERENCE

1. **Tavakkol, A. and Drucker, D. B.,** *J. Chromatogr. Sci.*, 22, 12, 1984.

Table GC 35
TRIFLUOROACETYL DERIVATIVES OF BIOLOGICAL AMINES

Column packing	P1	P2
Temperature	T1	T2
Gas Flow rate (mℓ/min)	A-CH_4, 90:10; 30, or 40	He
Column		
Length (m)	1.8	10
Diameter (mm, I.D.)	4	
Form		Capillary
Material	G	G
Detector	D1	D1

Amine	t_r(min)	
2-Phenylethylamine		4.9
m-Tyramine		8.8
p-Tyramine	2.4	10.1
p-Octopamine	1.8	11.2
Normetanephrine	9.1	13.7
3-Methoxytyramine	5.5	14.1
Benzylamine[a]		3.6
3-Phenylpropylamine[a]		7.8
Tranylcypromine[a]		9.6
2-(4-Chlorophenyl)ethylamine[a]	3.6	10.5

[a] Internal standard.

Column packing: P1 = 3% OV-17 on Gas-Chrom® Q, 100-120 mesh
P2 = grade AA WCOT SP 2100 glass capillary column

Temperature: T1 = an initial temperature of 155°C was maintained for 4.7 min followed by an increase of 5°C/min to a final temperature of 195°C
T2 = an initial temperature of 80°C was maintained for 0.6 min followed by a temperature increase of 30°C/min to 120°C; at a run time of 11.6 min the column temperature was increased to 160°C at a similar rate

Detector: D1 = ^{63}Ni source linear electron-capture detector

REFERENCE

1. **LeGalt, D. F., Baker, G. B., and Coutts, R. T.,** *J. Chromatogr.*, 225, 301, 1981.

Table GC 36
ACETYL DERIVATIVES OF AMINES

Column packing	P1	P2	P3	P4
Temperature (°C)	90	130	105	150
Gas	na	na	na	na
Column				
Length (m)	1.52	1.52	1.52	1.52
Diameter (cm, I.D.)	0.4	0.4	0.4	0.4
Material	G	G	G	G
Detector	FID	FID	FID	FID

Amine		$r_{isobutylamine}$		
Methylamine				0.55
Isopropylamine	0.53	0.52	0.49	0.65
Isobutylamine	1.00	1.00	1.00	1.00
n-Butylamine	1.40	1.36	1.40	1.33
2-Methylbutylamine			1.88	
Isoamylamine	1.80	1.85	1.88	1.54
n-Amylamine	2.47	2.36	2.57	1.98
Pyrrolidine	1.80	2.77	3.21	
Ethanolamine		4.13	5.53	
1,3-Diaminopropane		6.37		

Column packing: P1 = 3% SE-30 coated on acid-washed dimethylchlorosilane-treated Chromosorb® W, 85-100 mesh
P2 = 5% Apiezon® N coated on acid-washed dimethylchlorosilane-treated Chromosorb® W, 85-100 mesh
P3 = 3% OV-17 coated on acid-washed dimethylchlorosilane-treated Chromosorb® W, 85-100 mesh
P4 = 10% PEGA coated on acid-washed dimethylchlorosilane-treated Chromosorb® W, 85-100 mesh

REFERENCE

1. **Tavakkol, A. and Drucker, D. B.,** *J. Chromatogr. Sci.*, 22, 12, 1984.

Table GC 37
SEPARATION OF ENANTIOMERS OF AMINES AS ISOPROPYL UREA DERIVATIVES

Amine	α	Column temp (°C)
2-Aminopentane	1.021	170
2-Amino-3-methylpentane	1.027	170
(diastereoisomers)	1.024	
2-Aminohexane	1.025	170
2-Amino-5-methylhexane	1.033	170
2-Aminoheptane	1.030	170
2-Amino-6-methylheptane	1.033	170
2-Aminooctane	1.034	170
2-Phenylethylamine	1.058	200

Note: The L-enantiomers have longer retention times than the D-enantiomers.

Conditions: Packing: XE-60-*S*-valine-*S*-α-phenylethylamide
Column: 40 m Pyrex® glass capillary
Carrier gas: H_2
Detection: FID

REFERENCE

1. **König, W. A., Benecke, I., and Sievers, S.,** *J. Chromatogr.,* 238, 427, 1982.

Table GC 38
ENANTIOMERS OF AMINES AND AMINO ALCOHOLS

Column packing	P1	P2
Gas, pressure	H_2, 0.7 bar	H_2, 0.7 bar
Column		
Length (m)	35	35
Diameter (mm, I.D.)	0.2	0.2

Racemate	α	Temp (°C)	α	Temp (°C)
D,L-Alanine	1.00	70	1.013	80
D,L-Valine	1.026	80	1.015	80
D,L-allo-Isoleucine	1.011	80	1.00	80
D,L-Isoleucine	1.022	80	1.009	80
D,L-Leucine	1.017	80	1.00	80
R,S-2-Aminopentane	1.009	80	1.00	80
R,S-2-Amino-3-methylpentane	1.007	80	1.00	80
	1.010	80	1.005	80
R,S-2-Aminohexane	1.015	80	1.010	80
R,S-2-Amino-5-methylhexane	1.019	80	1.011	80
R,S-2-Aminoheptane	1.017	80	1.012	80
R,S-2-Amino-6-methylheptane	1.017	80	1.012	80
R,S-2-Aminooctane	1.020	80	1.015	80

Table GC 38 (continued)
ENANTIOMERS OF AMINES AND AMINO ALCOHOLS

Racemate	α	Temp (°C)	α	Temp (°C)
R,S-Phenylethylamine	1.028	100	1.023	100
D,L-Alaninol	1.00	80	1.00	100
D,L-Aminobutanol			1.011	100
D,L-Valinol	1.008	80	1.018	100
D,L-Norvalinol	1.018	80	1.017	100
D,L-Leucinol	1.025	80	1.022	100
D,L-Norleucinol	1.023	80	1.019	100
D,L-Norephedrine			1.013	130
D,L-Mandelic acid	1.006	90	1.007	90

Note: With the amino alcohols the D-enantiomers have longer retention times than the corresponding L-forms on both the Z-Val and Z-Leu phases. For the amines the *S*-enantiomers have longer retention times in the case of 2-aminopentane and phenylethylamine. The other 2-aminoalkanes were available only as racemates.

Column packing: P1 = Z-L-Val-OV-225 coated on Pyrex® glass capillary
P2 = Z-L-Leu-OV-225 coated on Pyrex® glass capillary

REFERENCE

1. **König, W. A. and Benecke, I.,** *J. Chromatogr.*, 209, 91, 1981.

Table GC 39
ACETYLATED HEPTAFLUOROBUTYRYL DERIVATIVES OF HISTAMINE AND ANALOGS

Column packing	P1	P1
Temperature (°C)	170	170
Gas Flow rate (mℓ/min)	He, 40	He, 40
Column		
Length (m)	2	2
Diameter (mm)	2	2
Form	Coil	Coil
Material	G	G
Detector	D1	D1

Heptafluorobutyryl derivative	t_r(min) Unacylated	Acylated
Histamine	15.82	9.41
N^α-Methylhistamine	13.49	6.52
N^π-Methylhistamine	25.64	
N^τ-Methylhistamine	2.58	
2-Methylhistamine	8.72	5.20
4-Methylhistamine	13.48	7.53

Column packing: P1 = 3% OV-275 on Chromosorb® W
Detector: D1 = nitrogen-phosphorus detector

Table GC 39 (continued)
ACETYLATED HEPTAFLUOROBUTYRYL DERIVATIVES OF HISTAMINE AND ANALOGS

REFERENCE

1. **Wollin, A. and Navert, H.,** *Anal. Biochem.*, 145, 73, 1985.

Table GC 40
INFLUENCE OF CHEMICAL COMPOSITION ON THE THERMAL STABILITY OF DIAMIDE STATIONARY PHASES

Phase	Max operating temp.(°C)
N-Docosanoyl-L-leucine *tert*-octylamide	220
N-Docosanoyl-L-valine *tert*-butylamide	190
N-Lauroyl-L-valine *tert*-butylamide	140
N-Lauroyl-L-valine *tert*-octylamide	180

Note: The phases have been used for the gas chromatographic resolution of optical isomers, including those of amino acids and 2-aminoalkan-1-ols.

REFERENCE

1. **Charles, R. and Watabe, K.,** *J. Chromatogr.*, 298, 253, 1984.

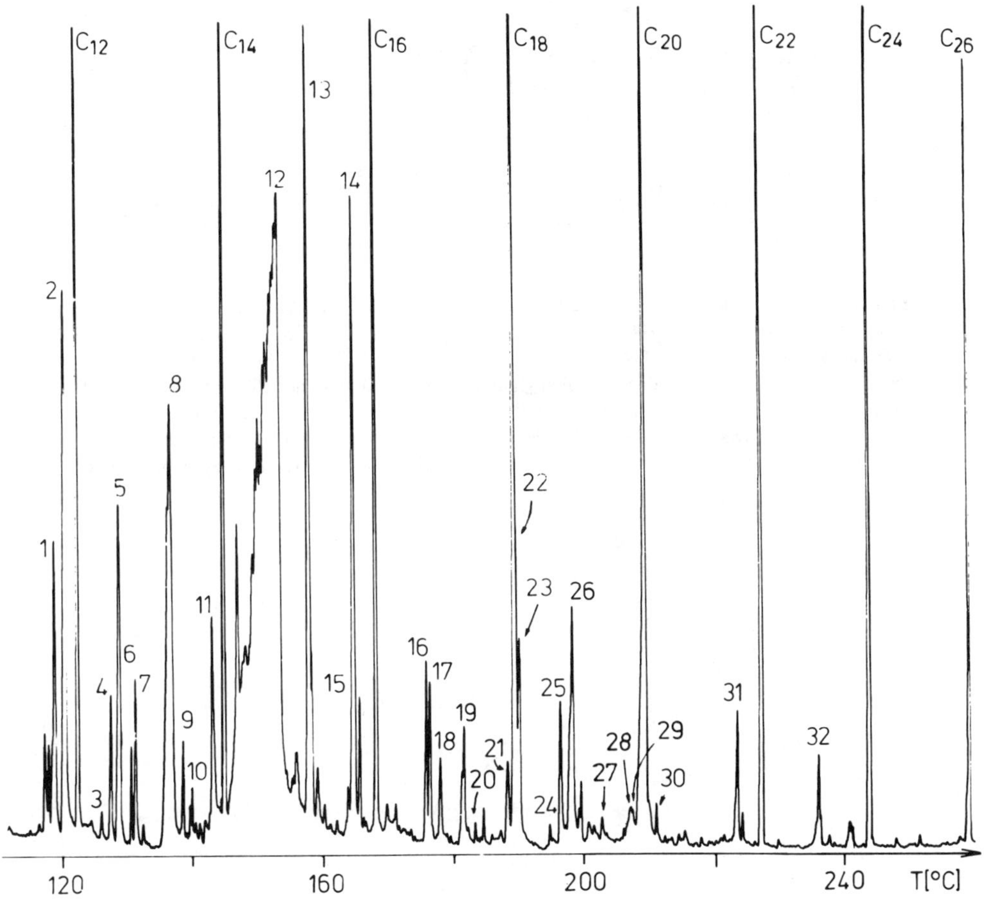

FIGURE GC 1. Urine profile of the *N*-trifluoroacetyl amino acid *n*-butyl esters of a 29-year-old healthy male individual (24-hr urine). Conditions: WCOT capillary column, 50 m × 0.22 mm, carrier gas, hydrogen at a flow rate of 2 mℓ/min, flame ionization detector. Peak identification: (1) alanine, (2) glycine, (3) α-aminoisobutyric acid, (4) threonine, (5) serine, (6) valine, (7) β-aminoisobutyric acid, (8) *N*-methylproline, (9) leucine, (10) isoleucine, (11) nicotinic acid, (12) creatinine, (13) unknown, (14) unknown, (15) unknown, (16) aspartic acid (includes asparagine), (17) phenylalanine, (18) ornithine, (19) unknown, (20) unknown, (21) *N*(*O*)-TFA-tyrosine, (22) glutamic acid (includes glutamine), (23) lysine, (24) *N*-α-methyllysine, (25) 2-aminoadipic acid, (26) 1-methylhistidine + 3-methylhistidine, (27) glycylproline, (28) *N*-TFA-tyrosine, (29) histidine (monoacyl), (30) *N*-α-acetyllysine, (31) prolylhydroxyproline, (32) unknown. Reference: Schneider, K., Neupert, M., Spiteller, G., Henning, H. V., Matthaei, D., and Scheler, F., *J. Chromatogr.*, 345, 19, 1985. (Reproduced from Schneider, K., Neupert, M., Spiteller, G., Henning, H. V., Matthaei, D., and Scheler, F., *J. Chromatogr.*, 345, 19, 1985. With permission.)

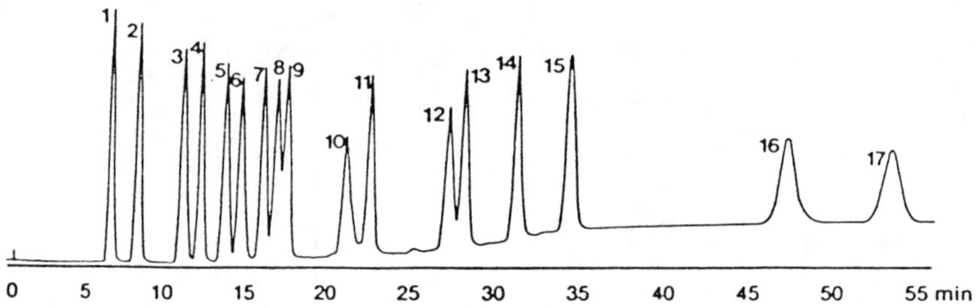

FIGURE GC 2. The separation of aliphatic amines. Packing PT 28% Alltech® 223 + 4% KOH on Gas Chrom®R, 80-100 mesh; column, copper 10 ft × 1/4 in.; gas, He, flow rate 45 mℓ/min; temperature, 75 to 200°C at 4°C/min; detection, FID. Peaks: (1) ethylamine, (2) isopropylamine, (3) *n*-propylamine, (4) diethylamine, (5) *sec*-butylamine, (6) isobutylamine, (7) diisopropylamine, (8) *n*-butylamine, (9) triethylamine, (10) isoamylamine, (11) di-*n*-propylamine, (12) di-*sec*-butylamine, (13) diisobutylamine, (14) tri-*n*-propylamine, (15) di-*n*-butylamine, (16) *N*-(2-methyl-1-butyl)-1-pentylamine, (17) tri-*n*-butylamine. (Reproduced by permission of Alltech Associates Inc.)

Table LC 1
AMINO ACID ANALYSIS USING STANDARD HPLC EQUIPMENT

Packing	P1
Column	
Length (mm)	250
Diameter (mm, I.D.)	3.2
Material	SS
Solvent	S1
Flow rate (mℓ/min)	0.4
Temperature	T1
Detection	D1

Amino acid	**t_r(min)**
Alanine	16.32
Arginine	42.99
Aspartic acid	7.98
Glutamic acid	11.90
Glycine	15.37
Histidine	30.22
Isoleucine	21.33
Leucine	22.01
Lysine	34.69
Methionine	20.11
Norleucine	22.85
Phenylalanine	26.78
Serine	9.92
Threonine	9.35
Tyrosine	25.32
Valine	18.70

Packing: P1 = DC-4A resin

Solvent: S1 = buffer A, 0.2 *N* Na^+, pH 3.25: 3 g of NaOH, 8.9 g of Na_2SO_4, ca. 8.2 mℓ of formic acid, 10 mℓ of 1-propanol; buffer B, 0.2 *N* Na^+, pH 4.25: 8 g of NaOH, ca. 10 mℓ of formic acid; buffer C, 1.1 *N* Na^+, pH 7.9: 28 g of NaOH, 40 g of trisodium citrate, ca. 26.4 mℓ of formic acid; elution was carried out with buffers A, B, and C for 13, 11, and 16 min, respectively

Temperature: T1 = the initial temperature of 38°C was increased to 59°C and then to 75°C after 8 and 26 min, respectively

Detection: D1 = *o*-phthalaldehyde

REFERENCE

1. **Hughes, G. J., Winterhalter, K. H., Boller, E., and Wilson, K. J.,** *J. Chromatogr.*, 235, 417, 1982.

Table LC 2
AMINO ACIDS USING AN AMINO ACID ANALYSIS SYSTEM

Packing	P1
Column	
Length (mm)	120
Diameter (mm, I.D.)	4.6
Solvent	S1
Flow rate (mℓ/min)	0.5
Temperature (°C)	60
Detection	D1

Amino acid	t_r(min)	RSD(%)[a]
Alanine	14.24	0.4
Arginine	30.23	0.3
Aspartic acid	5.46	0.8
Glutamic acid	8.81	0.6
Glycine	13.04	0.4
Histidine	25.55	0.5
Isoleucine	19.80	0.4
Leucine	20.47	0.3
Lysine	26.86	0.3
Methionine	18.49	0.4
Phenylalanine	23.29	0.3
Proline	10.64	0.5
Serine	7.21	0.7
Threonine	6.72	0.7
Tyrosine	22.41	0.3
Valine	16.93	0.3

[a] Relative standard deviation.

Packing: P1 = 6 μm sulfonated ion-exchange resin (Perkin Elmer® high-speed amino acid analysis column)

Solvent: S1 = A step gradient was established using commercial chromatographic buffers. Pierce® Buffelute: A, 8 min; B, 5 min; C, 13 min, step gradient

Detection: D1 = postcolumn reaction with sodium hypochlorite -*o*-phthalaldehyde

REFERENCE

1. **Dong, M. W. and Gant, J. R.,** *J. Chromatogr.*, 327, 17, 1985.

Table LC 3
EFFECT OF CATION SPECIES ON THE ELUTION OF AMINO ACIDS FROM THE SHORT COLUMN OF AN AUTOMATIC ANALYZER

Packing	P1	P1
Column		
Length (cm)	28	28
Diameter (cm)	0.9	0.9
Solvent	S1	S2
Flow rate (mℓ/hr)	70	70
Temperature (°C)	45	45
Detection	D1	D1
Amino acid	**t_r(min)**	
β-Alanine	63.5	67.9
γ-Aminobutyric acid	87.4	100.7
Ammonia	81.1	110.7
Ethanolamine	75.0	100.7
Histidine	117.4	174.6
Lysine	136.2	195.1
Ornithine	97.6	137.0
Tryptophan	181.6	156.5
Tyrosine + phenylalanine	55.5	51.5

Packing: P1 = Beckman® PA-35 resin to a depth of 16 cm in the column

Solvent: S1 = buffers 1 and 2 have the following respective characteristics: pH — 3.70, 4.15; potassium concentration (*M*) — 0.30, 0.40; potassium citrate monohydrate (g/ℓ) — 21.63, 21.63; KCl (g/ℓ) — 7.50, 15.00; caprylic acid (mℓ/ℓ) — 0.1, 0.1; concentrated HCl (mℓ/ℓ) — 11.0, 9.5; benzyl alcohol (mℓ/ℓ) — 10.0, 5.0; the final adjustment of pH is made with HCl or KOH; benzyl alcohol is added after adjustment of the pH

Amino acids are eluted with buffers 1 and 2 for 122 and 90 min with the buffer change between lysine and tryptophan; the resin is regenerated with base for 10 min and equilibrated with buffer 2 for 36 min

S2 = in solvent S2 sodium replaces potassium

Detection: D1 = ninhydrin

REFERENCE

1. **Mortensen, J. V.,** *J. Chromatogr.,* 209, 41, 1981.

Table LC 4
AMINO ACIDS ON AMBERLITE® XAD-4

Packing	P1
Column	
Length (mm)	250
Diameter (mm, I.D.)	8.0
Solvent	S1
Flow rate (mℓ/min)	5.6
Detection	D1

Amino acid	k′
DL-DOPA	3.6
L-Tyrosine	6.4
D-Phenylalanine	20

Packing: P1 = Amberlite® XAD-4, 37—44 μm
Solvent: S1 = 1.25% EtOH (95%)/98.75% H_2O, pH 2, ionic strength, 0.1 *M*
Adjustment of the pH was made with H_3PO_4 buffer or 0.01 *M* HCl
Detection: D1 = absorbance measurement

REFERENCE

1. **Pietrzyk, D. J., Cahill, W. J., Jr., and Stodola, J. D.,** *J. Liquid Chromatogr.*, 5, 443, 1982.

Table LC 5
AMINO ACIDS ON M-72 RESIN

Packing	P1
Column	
Length (cm)	54
Diameter (cm, I.D.)	0.9
Solvent	S1
Flow rate (mℓ/hr)	65
Detection	D1

Amino acid	t_r(min)
Alanine	104.0
Arginine	312.0
Aspartic acid	54.9
Cysteic acid	21.5
Cystine	113.0
Glutamic acid	77.8
Glycine	97.8
Histidine	272.6
Hydroxylysine	250.5
3-Hydroxyproline	48.1
4-Hydroxyproline	52.4
Isoleucine	155.4
Leucine	159.2
Lysine	260.0
Methionine	146.5
Norleucine	164.1

Table LC 5 (continued)
AMINO ACIDS ON M-72 RESIN

Amino acid	t_r(min)
Phenylalanine	185.0
Proline	83.0
Serine	65.6
Threonine	62.2
Tyrosine	175.6
Valine	131.0

Packing: P1 = M-72 resin (in the column of a Multichrom B Beckman® analyzer)
Solvent: S1 = three buffers are employed: 0.2 *N* sodium citrate, pH 3.22 for 120 min, then 0.2 *N* sodium citrate, pH 4.1 for 90 min, and finally 0.2 *N* sodium citrate, pH 5.00 containing 1 *M* NaCl for 95 min
Temperature: T1 = the initial temperature of the column is 39°C, which is increased to 67°C after 40 min
Detection: D1 = ninhydrin

REFERENCE

1. **Szymanowicz, A. G., Poulin, G., Fontaine, N., Werquin, J. P., and Borel, J. P.,** *J. Chromatogr.*, 190, 457, 1980.

Table LC 6
AMINO ACIDS ON AN NH_2 COLUMN

Packing	P1
Column	
Length (mm)	200
Diameter (mm, I.D.)	4.6
Solvent	S1
Temperature (°C)	35
Detection	D1

Amino acid	t_r(min)	RSD (%)[a]
Alanine	14.11	0.23
Arginine	22.98	0.72
Glutamic acid	25.46	0.80
Glycine	15.79	0.21
Histidine	19.82	0.39
Isoleucine	8.85	0.33
Leucine	7.75	0.46
Methionine	9.25	0.36
Phenylalanine	6.85	0.48
Proline	12.02	0.13
Threonine	14.44	0.37
Tryptophan	6.29	0.54
Valine	10.79	0.26

Table LC 6 (continued)
AMINO ACIDS ON AN NH_2 COLUMN

[a] Relative standard deviation.

Packing: P1 = 5 μm LiChrosorb® NH_2
Solvent: S1 = solution A, 0.01 *M* KH_2PO_4, pH 4.3; solution B, acetonitrile-water, 500:70; a linear stepwise solvent gradient was established: 0 min, 95% B; 5 min, 95% B; 20 min, 70% B; 27 min, 50% B; 30 min, 95% B for column conditioning
Detection: D1 = UV at 200 nm

REFERENCE

1. **Schuster, R.,** *Anal. Chem.,* 52, 617, 1980.

Table LC 7
ANION EXCHANGE CHROMATOGRAPHY OF AMINO ACIDS

Packing	Pl	Pl	Pl	Pl
Column				
Length (cm)	50	50	50	50
Diameter (cm, I.D.)	0.4	0.4	0.4	0.4
Material	SS	SS	SS	SS
Solvent	S1	S2	S3	S4
Flow rate (mℓ/min)	0.75	0.75	0.75	0.75
Temperature	Tl	Tl	Tl	Tl
Detection	Dl	Dl	Dl	Dl
Amino acid	**t_r(min)**			
Histidine	7	7	7	7
Phenylalanine	10	10	10	10
Tyrosine	24	28	26	25
Tryptophan	25	28	26	26
Nicotinic acid	30	38	36	32
p-Aminobenzoic acid	45	56	54	42
Hippuric acid	49	65	64	53

Packing: Pl = a strongly basic anion-exchange resin, Diaion® CDR-10, a macroreticular type having a particle size distribution of 5—7 μm (Mitsubishi Chemical Industries, Tokyo)

Solvent: Sl = ammonium perchlorate solution, pH 5.5, using a linear gradient varying in concentration from 0—0.25 *M*

S2 = a concave ammonium perchlorate gradient using a two-chamber gradient generator is employed; the curvature of the gradient can be chosen by the curve coefficient, K, and the total gradient volume, V; the curve coefficient is the ratio of the area of upper phase, V_1, against the area of lower phase, V_2, in the area of rectangle V_t, divided by the gradient

Table LC 7 (continued)
ANION EXCHANGE CHROMATOGRAPHY OF AMINO ACIDS

			curve; then K = V_1/V_2; for S2, K = 3 and V = 20; the gradient is formed by placing 20 mℓ of water in the first chamber and 20 mℓ of 0.25 *M* ammonium perchlorate in the second chamber when K = 3 and V = 20
	S3	=	a gradient is established as for S2, but 15% of ethanol is added to the 0.25 *M* ammonium perchlorate
	S4	=	a gradient is established as for S2, but 15% of acetonitrile is added to the 0.25 *M* ammonium perchlorate
Temperature:	T1	=	the column temperature is raised from 22—70°C over the first 30 min, then maintained at 70°C to the end of the run
Detection:	D1	=	absorbance measurement in the UV

REFERENCE

1. **Seta, K., Washitake, M., Tanaka, I., Takai, N., and Okuyama, T.,** *J. Chromatogr.*, 221, 215, 1980.

Table LC 8
SEPARATION OF THE ION PAIRS OF AMINO ACIDS WITH DODECYL SULFATE

Packing	P1
Column	
Length (cm)	25
Diameter (mm, I.D.)	4.6
Solvent	S1
Flow rate (mℓ/min)	1.0
Temperature (°C)	25
Detection	D1

Amino acid	**k′**
Taurine, hypotaurine, cysteic acid, phosphoenthanolamine	0.4
Glutamine	3.6
Serine	3.8
Aspartic acid	4.3
Glutamic acid	4.6
Glutathione	5.0
Threonine	5.4
Glycine	5.7
Cystine	6.1
α-Alanine	6.2
β-Alanine	7.1
α-Difluoromethylornithine	8.6
4-Aminobutyric acid	9.4
Tyrosine	10.1
Valine	10.6
Methionine	11.1
Norvaline	12.0
Ornithine	13.3

Table LC 8 (continued)
SEPARATION OF THE ION PAIRS OF AMINO ACIDS WITH DODECYL SULFATE

Amino acid	k′
Lysine	16.6
Histidine	17.9
Isoleucine	18.4
Phenylalanine	21.0
Leucine	21.6
4-Aminohex-5-enoic acid	22.4
Homocarnosine	23.8
Carnosine	24.3
Tryptophan	24.9
Arginine	28.0
Putreanine (N^1-[2-carboxyethyl]-1,4-butanediamine)	28.8

Packing: Pl = Beckman Ultrasphere® I.P. (5 μm pellicular material with chemically bonded C_{18} groups); the separation column was protected by a guard column, 7 cm × 2.1 mm I.D., filled with pellicular ODS (C_{18} groups bonded to 37.53-μm particles, Whatman)

Solvent: S1 = aqueous solutions were prepared using tap water distilled over phosphoric acid; gradients were prepared by mixing three solutions: solution A, 0.2 *M* H_3PO_4 containing 10 m*M* sodium dodecyl sulfate; solution B, 0.2 *M* sodium acetate adjusted to pH 4.50 with acetic acid containing 10 m*M* sodium dodecyl sulfate; solution C, methanol; a relatively complicated sequence of solvent mixtures was employed (Table LC 8A)

Table 8A
COMPOSITION OF THE ELUENT FOR THE SEPARATION OF AN AMINO ACID MIXTURE

	Percentage solution		
Elution time (min)	A	B	C
0	63	27	10
10	63	27	10
11	40	60	0
20	40	60	0
21	41	56	3
30	41	56	3
31	35	60	5
57	35	60	5
58	0	75	25

Note: End time = 80 min; equilibration time = 15 min.

Table 8A (continued)
COMPOSITION OF THE ELUENT FOR THE SEPARATION OF AN AMINO ACID MIXTURE

Detection: D1 = *o*-phthalaldehyde-2-mercaptoethanol reagent; fluorescence was continuously recorded, excitation being at 345 nm and emission at 455 nm

REFERENCE

1. **Seiler, N. and Knodgen, B.,** *J. Chromatogr.*, 341, 11, 1985.

Table LC 9
THE EFFECT OF HEPTANE SULFONIC ACID CONCENTRATION ON THE RETENTION OF AMINO ACIDS

Packing	P1	P1	P1	P1	P1
Column					
Length (mm)	250	250	250	250	250
Diameter (mm, I.D.)	4.6	4.6	4.6	4.6	4.6
Solvent	S1	S2	S3	S4	S5
Flow rate (mℓ/min)	1.4	1.4	1.4	1.4	1.4
Temperature	rt	rt	rt	rt	rt
Detection	D1	D1	D1	D1	D1
Amino acid			**k′**		
Aspartic acid	1.0	1.3	1.9	3.0	3.0
Glycine	1.94	2.8	3.7	4.6	4.7
Methionine	7.9	8.4	9.2	11.2	11.1
Serine	1.25	1.9	2.5	3.3	3.2
Tyrosine	5.7	6.0	6.5	7.5	7.2
Valine	6.3	6.9	7.7	9.0	9.1

Packing: P1 = 5 μm silica particles bonded with octadecylsilane (Partisil® PXS-1025 ODS3; Whatman)

Solvent: S1, S2, S3, S4, and S5 = consisting of 0.05 *M* acetate buffer, pH 3.4, 10% acetonitrile with the following respective concentrations of heptane sulfonic acid: 0.001, 0.005, 0.02, 0.2, and 0.5 *M*

Detection: D1 = nickel oxide amperometric detector

REFERENCE

1. **Kafil, J. B. and Huber, C. O.,** *Anal. Chim. Acta,* 175, 275, 1985.

Table LC 10
THE EFFECT ON CAPACITY FACTORS OF ADDING BUTANE BORONIC ACID TO PHOSPHATE BUFFER

Packing	P1	P1
Column		
Length (cm)	15	15
Diameter (mm)	4.6	4.6
Solvent	S1	S2
Detection	D1	D1
Compound	**k′**	
Dopamine	3.35	11.8
Noradrenaline	0.4	2.15
Serotonin	15.0	9.35
Tryptophan	14.15	9.6
Tyrosine	1.55	1.03

Packing: P1 = Hypersil® ODS, 5 μm; a 5 cm × 2.1-mm guard column of pellicular ODS silica (Chrompack®) was also used

Solvent: S1 = 0.1 *M* phosphate buffer, pH 7.5

S2 = 0.1 *M* phosphate buffer, pH 7.5 + 5 m*M* butane boronic acid

Detection: D1 = fluorescence (mercury line excitation, interference filter at 254 nm; emission measured at 365 nm) and electrochemical detection

REFERENCE

1. **Joseph, M. H.,** *J. Chromatogr.*, 342, 370, 1985.

Table LC 11
BASIC AMINO ACIDS AND COLLAGEN CROSS-LINKS

Packing	P1
Column	
Length (cm)	23
Diameter (cm)	0.9
Solvent	S1
Flow rate (mℓ/hr)	70
Temperature	T1
Detection	D1
Amino acid	**t_r(min)**
Tyrosine	20.0
Phenylalanine	21.7
Aldol-histidine	25.9
Hydroxyaldol-histidine	31.9
Dihydroxylysinonorleucine	36.5
Hydroxylysinonorleucine	41.5
Lysinonorleucine	49.5

Table LC 11 (continued)
BASIC AMINO ACIDS AND COLLAGEN CROSS-LINKS

Amino acid	t_r(min)
Hydroxylysine	56.9/58.9[a]
Lysine	67.7
Histidine	78.2
Histidinohydroxymerodesmosine	85.5
Ammonia	95.6
Arginine	117.7

[a] The natural form and allo isomer of hydroxylysine elute as two distinct peaks.

Packing: P1 = Beckman® PA-35 resin filled to a height of 15 cm in the column
Solvent: S1 = a single buffer was used to elute all amino acids and cross-links except arginine; this buffer had a sodium citrate concentration of 0.26 *N* (pH 5.25) and contained 7% *n*-propanol; a second buffer with 0.3 *N* citrate and 0.6 *N* sodium ion concentration was used to decrease the elution time of arginine, the buffer change being timed to coincide with the ammonia peak
Temperature: T1 = 45°C changed to 55°C after 20 min
Detection: D1 = ninhydrin

REFERENCE

1. **Housley, T. J. and Tanzer, M. L.,** *Anal. Biochem.*, 114, 310, 1981.

Table LC 12
CYCLIC IMINO ACIDS

Compound	t_r (min)	Fluorescence[a]	Temp[b]
trans-4-Hydroxyproline	12.2	45	75
2,4-*trans*-4,5-*trans*-Dihydroxypipecolic acid	13.0	33	115
2,4-*trans*-4,5-*cis*-Dihydroxypipecolic acid	13.0	28	115
cis-4-Hydroxyproline	15.3	54	85
trans-4-Acetylamino pipecolic acid	15.8	6	115
trans-4-Hydroxypipecolic acid	16.8	16	130
trans-5-Hydroxypipecolic acid	18.5	29	85
cis-5-Hydroxypipecolic acid	19.6	37	85
Proline	21.6	100	75
Pipecolic acid	33.2	19	120

[a] Percentage fluorescence relative to proline.
[b] Optimum temperature (delay coil A).

Conditions: A modified version of a Dionex® amino acid/peptide analyzer kit equipped with a 0.4 × 12-cm column packed with DC-5A cation exchange resin (Dionex Corp.) was used for the analyses; the analyzer was equipped with

Table LC 12 (continued)
CYCLIC IMINO ACIDS

the Dionex secondary amine kit, with chloramine-T being used as oxidant instead of sodium hypochlorite; the chloramine-T delay coil (delay coil A) was increased in length from 3—5 cm and was immersed in a thermostatically controlled oil bath

Eluant A (Dionex femto-buffer system eluent 1A titrated to pH 3.1 with formic acid) was pumped through the column at a flow rate of 15 mℓ/hr, the column temperature being 45°C; imino acids were then injected into the eluent flow; eluent B (femto-buffer system eluent 18 titrated to pH 3.5 with formic acid) entered the column at 20 min and the column temperature was increased to 65°C; the chloramine T solution was introduced into the postcolumn effluent, which after passage through delay coil A contained oxidized imino acids; the effluent entered a second mixing T where *o*-phthalaldehyde solution was added; fluorophor formation then occurred in a second delay coil (delay coil B) prior to measurement of fluorescence intensity in a fluorometer

REFERENCE

1. **Bleeker, A. B. and Romeo, J. T.,** *Anal. Biochem.*, 121, 295, 1982.

Table LC 13
IODOAMINO ACIDS

Compound	k′	Solvent
Tyrosine	0.55	S1
3-Iodotyrosine	3.55	S1
3,5-Diiodotyrosine	12.8	S1
Thyronine	0.20	S2
3-Monoidothyronine	1.00	S2
3′-Monoiodothyronine	1.40	S2
3,5-Diiodothyronine	1.45	S2
3′,5′-Diiodothyronine	3.35	S2
3,3′-Diiodothyronine	3.35	S2
3,3′,5-Triiodothyronine	5.15	S2
3,3′,5′-Triiodothyronine	6.95	S2
3,3′,5,5′-Tetraiodothyronine	9.05	S2
3-Monoiodothyroacetic acid	0.85	S3
3,5-Diiodothyroacetic acid	2.1	S3
3,3′,5-Triiodothyroacetic acid	4.2	S3
3,3′,5,5′-Tetraiodothyroacetic acid	7.3	S3

Solvent: S1 = 10% methanol/water/5 m*M* H_3PO_4, pH 3.0
S2 = 50% methanol/water/5 m*M* H_2PO_4, pH 3.0
S3 = 60% methanol/water/5 m*M* H_3PO_4, pH 3.0

Conditions: A column of μBondapak®, 10 μm, 30 cm × 4 mm I.D. was employed; chromatography was carried out at about 20°C; detection was performed in the UV

REFERENCE

1. **Hearn, M. T. W. and Hancock, W. S.,** *J. Chromatogr. Sci.*, 18, 288, 1980.

Table LC 14
IODOAMINO ACIDS

Packing	P1
Column	
Length (mm)	300
Diameter (mm)	3
Material	G
Solvent	S1
Flow rate (mℓ/min)	1.2
Detection	D1

Amino acid	$_r$(min)
3,3′-Diiodothyronine	8
3′,5′-Diiodothyronine	8
3,5,3′-Triiodothyronine	13
3,3′,5′-Triiodothyronine	20
Thyroxine	28

Packing: P1 = C_{18} silica gel, 10 μm
Solvent: S1 = methanol/water, 55:45, brought to pH 3 with phosphoric acid
Detection: D1 = UV at 254 nm

REFERENCE

1. **Bianchi, R., Molea, N., Cazzuola, F., Fusani, L., Lotti, M., Bertelli, P., Ferdeghini, M., and Mariani, G.,** *J. Chromatogr.*, 297, 393, 1984.

Table LC 15
PHOSPHOAMINO ACIDS

Packing	P1
Column	
Length (mm)	150
Diameter (mm)	4.6
Solvent	S1
Flow rate (mℓ/min)	1.0
Detection	D1

Compound	k′
Aspartate	1.82
Glutamate	1.82
O-Phosphoserine	5.82
O-Phosphothreonine	5.14
O-Phosphotyrosine	7.46
O-Phosphate	2.67
Chloride	2.19

Packing: P1 = Spherisorb® 5 ODS 2
Solvent: S1 = 0.5 m*M* tetrabutylammonium hydroxide-phthalic acid, pH 6.3
Detection: D1 = UV visualization with *o*-phthalate at 243 nm

REFERENCE

1. **Morrice, N. and Aitken, A.,** *Anal. Biochem.*, 148, 207, 1985.

Table LC 16
PHOSPHOAMINO ACIDS

Packing	P1
Column	
Length (cm)	8
Diameter (cm, I.D.)	0.6
Material	G
Solvent	S1
Flow rate (mℓ/min)	0.4
Temperature (°C)	50
Detection	D1

Phosphoamino acid	t_r(min)
Phosphoserine	19.3 ± 0.1
Phosphothreonine	28.3 ± 0.1
Phosphotyrosine	64.1 ± 0.2

Packing: P1 = Durrum® DC6A resin
Solvent: S1 = solvents used for elution, washing, regeneration, and equilibration, respectively, were (A) 0.01 *M* trifluoroacetic acid, (B) 1.0 *M* Na^+ (34.8 g NaCl and 39.2 g trisodium citrate per liter), (C) 0.2 *M* NaOH, (D) 1 *M* trifluoracetic acid; addition of 1 mℓ of phenol per liter prevented bacterial growth in A and B

Table LC 16 (continued)
PHOSPHOAMINO ACIDS

Detection: D1 = *o*-phthalaldehyde

REFERENCE

1. **Capony, J.-P. and Demaille, J. G.,** *Anal. Biochem.,* 128, 206, 1983.

Table LC 17
N-METHYLATED AMINO ACIDS

Packing	P1
Column	
Length (cm)	10
Diameter (cm)	0.5
Solvent	S1
Flow rate (mℓ/min)	3.0
Detection	D1

Amino acid	**k′**
1-*N*-Methylhistidine	2.30
3-*N*-Methylhistidine	2.86
3-Hydroxy-6-*N*-trimethyllysine	3.30
6-*N*-Methyllysine	4.33
6-*N*-Dimethyllysine	4.77
6-*N*-Trimethyllysine	5.13
N^G-Methylarginine	7.48
N^G,N'^G-Dimethylarginine	7.56
6-*N*-Triethyllysine	11.11

Packing: P1 = 10 μm nominal particle diameter Radial-Pak® C_{18} (Waters)

Solvent: S1 = 30% acetonitrile in water containing 5×10^{-2} *M* sodium dodecyl sulfate and 5×10^{-2} *M* NaH_2PO_4; to 700 mℓ of water are added 6.9 g of NaH_2PO_4 and 14.4 g of sodium dodecyl sulfate; the solution is adjusted to pH 2.75 with concentrated H_2SO_4, filtered through a cellulose acetate membrane, and 300 mℓ of filtered acetonitrile is added with thorough magnetic stirring

Detection: D1 = reaction with *o*-phthalaldehyde and fluorometric detection

REFERENCE

1. **Davis, A. T., Ingalls, S. T., and Hoppel, C. L.,** *J. Chromatogr.,* 306, 79, 1984.

Table LC 18
N-SUBSTITUTED AND METHYLATED *N*-SUBSTITUTED AMINO ACIDS

Solvent (% methanol)[a]	*N*-Substituted amino acid	k′	Methylated *N*-substituted amino acid	k′
45	Boc-Ala	0.92	Boc-MeAla	2.02
52	Z-Asp	0.38	Z-MeAsp	0.59
58	Z-Leu	2.16	Z-MeLeu	5.01
13[b]	Boc-Lys	2.48	Boc-Lys(Me_2)	3.84
65	Tos-Lys(Bz)	0.33	Tos-MeLys(Bz)	0.60
70	Z-Phe	0.47	Z-MePhe	0.83
50	Z-Thr	0.74	Z-Thr(Me)	1.63
58	Boc-Thr(Bzl)	3.53	Boc-MeThr(Bzl)	6.95
50	Boc-Val	1.66	Boc-MeVal	3.95
50	Bz-Val	0.98	Bz-MeVal	1.66

[a] The remainder of the solvent was water, 0.01% of acetic acid being added.

[b] Methanol plus 87% 0.01 *M* sodium acetate, pH 4.0.

Conditions: μBondapak®-C_{18}, 10 μm, column 30 × 3.9 mm, I.D.; flow rate 1.0 mℓ/min Absorbance was measured at 208 and 215 nm for benzoyl derivatives

REFERENCE

1. **Steinauer, R., Chen, F. M. F., and Benoitin, N. L.,** *Int. J. Pept. Protein Res.,* 26, 109, 1985.

Table LC 19
S-ADENOSYLMETHIONINE AND HOMOLOGS, AMINO ACIDS, AND POLYAMINES

Compound	t_r(min)	Detection limit (pmol)
UV Detection		
S-Adenosyl-D-cysteine	10.9	2.0
S-Adenosylethionine	17.93, 18.08[a]	0.8[a]
S-Adenosyl-L-homocysteine	15.1	0.8
S-Adenosylmethionine	17.29	1.1
Tryptophan	21.73	4.0
Tyrosine	8.65	50
Fluorescence Detection		
Putrescine	16.95	3.5
Spermidine	24.91	3.5
Spermine	28.07	5.0
N-Acetylputrescine	4.89	7.0
N[1]-Acetylspermidine	19.66	5.0
N[8]-Acetylspermidine	20.33	3.4
Ethionine	15.85	6.0
Methionine	7.71	8.5
Tryptophan	22.11	8.0
Tyrosine	9.06	11.0

Table LC 19 (continued)
S-ADENOSYLMETHIONINE AND HOMOLOGS, AMINO ACIDS, AND POLYAMINES

[a] S-Adenosylethionine gives two peaks of equal height; the detection limit is given for the sum of the two peaks.

Column: An Ultrasphere ion pair column (5-μm particle size 250 × 4.6 mm I.D.), fitted with a guard column (70 × 2 mm I.D.) filled with Co:Pell ODS, 30—38 μm

Temperature: The two columns are kept in a jacket thermostated at 40°C

Solvent: Solvent A is a mixture of 980 mℓ of 0.1 *M* NaH_2PO_4 and 20 mℓ of acetonitrile containing octanesulfonic acid at 8 × 10^{-3} *M*; the pH is adjusted to 2.65 with about 10.5 mℓ of 3 *M* H_3PO_4; solvent B is a mixture of 740 mℓ of 0.15 *M* NaH_2PO_4 and 260 mℓ of acetonitrile and contains octanesulfonic acid at 8 × 10^{-3} *M*; the pH is adjusted to 3.25 with about 10.5 mℓ of 3 *M* H_3PO_4; a linear gradient is used, starting with 85% of solvent A and 15% of solvent B and leading to 100% of solvent B in 30 min; EDTA, 1 × 10^{-4} *M*, is added to solvent A to compensate for the difference in absorption between the two solvents

Detection: UV at 254 and/or 280 nm and fluorescence measurement after reaction with *o*-phthalaldehyde

REFERENCE

1. **Wagner, J., Claverie, N., and Danzin, C.,** *Anal. Biochem.*, 140, 108, 1984.

Table LC 20
METHYL HISTIDINES AND OTHER COMPOUNDS

Packing	P1
Column	
Length (cm)	13.0
Diameter (cm)	0.28
Solvent	S1
Flow rate (mℓ/hr)	7.5
Temperature (°C)	64
Detection	D1

Compound	**t_r(min)**
Ammonia	33.3
Arginine	62.5
Carnosine	56.9
Creatinine	33.6
Histidine	41.3
Hydroxylysine	29.5
Lysine	36.0
1-Methylhistidine	41.6
3-Methylhistidine	46.7
Ornithine	35.6
Tryptophan	23.9

Table LC 20 (continued)
METHYL HISTIDINES AND OTHER COMPOUNDS

Packing: P1 = AA-10 resin (Beckman), resin bed height 4.5 cm, used in conjunction with model 121 M amino acid analyzer (Beckman)
Solvent: S1 = buffer, pH 3.950 ± 0.010, lithium 0.500 mol/ℓ, citrate 0.067 mol/ℓ
Detection: D1 = ninhydrin

REFERENCE

1. **Zunic, G., Stanimirovic, S., and Savic, J.,** *J. Chromatogr.,* 311, 69, 1984.

Table LC 21
THE SEPARATION OF AMINO ACID ENANTIOMERS ON A REVERSED-PHASE PACKING COATED WITH *N*-ALKYL-L-HYDROXYPROLINE

Packing	P1
Column	
Length (mm)	100
Diameter (mm)	4.2
Solvent	S1
Flow rate (mℓ/min)	2
Temperature (K)	293
Detection	D1

Amino acid	k'_L	k'_D
Alanine	0.58	0.91
Allo-hydroxyproline		1.24
Allo-isoleucine		18.92
Aminobutyric acid	1.00	2.49
Arginine	2.97	3.67
Asparagine	0.66	0.66
Aspartic acid	0.08	0.10
Citrulline	0.82	1.06
Cysteic acid	0	0
DOPA	1.94	4.51
Ethionine	6.57	18.15
Glycine	0.63	0.63
Glutamic acid	0.13	0.17
Glutamine	0.68	0.85
Histidine	0.97	0.57
Hydroxyproline	0.65	
Isoleucine	3.18	
Leucine	5.01	17.82
Lysine	2.24	
Methionine	2.54	5.06
Methionine sulfone	0.79	0.98
Norleucine	6.25	25.05
Norvaline	2.33	7.94
Ornithine	3.00	3.00
Phenylalanine	10.33	38.30
Proline	0.81	6.10

Table LC 21 (continued)
THE SEPARATION OF AMINO ACID ENANTIOMERS ON A REVERSED-PHASE PACKING COATED WITH *N*-ALKYL-L-HYDROXYPROLINE

Amino acid	k'_L	k'_D
Serine	0.73	0.73
Threonine	0.83	0.85
Tyrosine	2.51	7.01
Tryptophan	21.52	54.22
Valine	1.63	7.42

Note: The L-enantiomer elutes before the D-enantiomer with the exception of histidine, where the reverse is the case.

Packing: P1 = LiChrosorb® RP-18, particle diameter 5 μm; coating of the support is achieved by forcing 2 mℓ of a solution of 100 mg of *N*-hexadecyl-L-hydroxyproline in methanol through the column, which is then washed with 2—4 mℓ of a concentrated solution of Cu(II) acetate in methanol/water, 15:85

Solvent: S1 = 10^{-4} *M* Cu(II) acetate in methanol/water, 15:85

Detection: D1 = UV at 254 nm

REFERENCE

1. **Davankov, V. A., Bochkov, A. S., Kurganov, A. A., Roumeliotis, P., and Unger, K. K.,** *Chromatographia,* 13, 677, 1980.

Table LC 22
THE SEPARATION OF AMINO ACID ENANTIOMERS ON A REVERSED-PHASE PACKING COATED WITH *N*-HEXADECYL-L-HYDROXYPROLINE AS A FUNCTION OF COPPER(II) ACETATE CONCENTRATION

Packing	P1	P1	P1
Column			
Length (mm)	100	100	100
Diameter (mm)	4.2	4.2	4.2
Solvent	S1	S2	S3
Flow rate (mℓ/min)	2	2	2
Temperature (K)	293	293	293
Detection	D1	D1	D1

Amino acid	k'_L	k'_D	k'_L	k'_D	k'_L	k'_D
Ala	0.76	0.86	1.11	1.26	1.43	1.63
Abu	1.06	1.67	1.58	2.40	1.99	2.97
Asp	0.42	0.42	0.39	0.39	0.41	0.41
His	1.03	1.03	1.53	1.53	2.21	2.21
Met	2.72	4.39	3.92	6.24	4.93	7.85

Table LC 22 (continued)
THE SEPARATION OF AMINO ACID ENANTIOMERS ON A REVERSED-PHASE PACKING COATED WITH *N*-HEXADECYL-L-HYDROXYPROLINE AS A FUNCTION OF COPPER(II) ACETATE CONCENTRATION

Amino acid	k'_L	k'_D	k'_L	k'_D	k'_L	k'_D
Nleu	6.21	15.7	8.31	21.3	10.2	25.6
Phe	12.5	27.1	17.2	37.6	21.3	46.9
Pro	1.08	3.61	1.51	4.81	2.20	6.08
Thr	0.86	0.86	1.32	1.32	1.81	1.81
Trp	25.1	41.4	34.5	58.1	43.7	72.4
Tyr	3.32	5.53	4.57	7.58	6.11	9.63
Val	1.86	4.47	2.62	6.14	3.29	7.52

Packing: P1 = LiChrosorb® RP-18, particle diameter 5 μm; coating of the support is achieved by forcing 2 mℓ of a solution of 100 mg of *N*-hexadecyl-L-hydroxyproline in methanol through the column, which is then washed with 2—4 mℓ of a concentrated solution of Cu(II) acetate in methanol/water, 15:85

Solvent: S1 = 2×10^{-4} *M* Cu(II) acetate in ethanol/water, 15:85

S2 = 10^{-4} *M* Cu(II) acetate in ethanol/water, 15:85

S3 = 5×10^{-5} *M* Cu(II) acetate in ethanol/water, 15:85

Detection: D1 = UV at 254 nm

REFERENCE

1. **Davankov, V. A., Bachkov, A. S., Kurganov, A. A., Roumeliotis, P., and Unger, K. K.,** *Chromatographia,* 13, 677, 1980.

Table LC 23
THE SEPARATION OF AMINO ACID ENANTIOMERS ON A REVERSED-PHASE PACKING COATED WITH *N*-DECYL-L-HYDROXYPROLINE AS A FUNCTION OF COPPER(II) ACETATE CONCENTRATION

Packing	P1	P1	P1
Column			
Length (mm)	100	100	100
Diameter (mm)	4.2	4.2	4.2
Solvent	S1	S2	S3
Flow rate (mℓ/min)	2	2	2
Temperature (K)	293	293	293
Detection	D1	D1	D1

Amino acid	k'_L	k'_D	k'_L	k'_D	k'_L	k'_D
Ala	0.94	1.13	1.40	1.64	1.89	2.21
Abu	1.26	2.08	1.87	2.94	2.55	3.94
Asp	0.81	0.81	0.87	0.87	0.74	0.74
His	1.17	1.17	2.13	2.13	2.53	2.53
Met	2.89	5.06	4.47	7.27	6.19	10.0
Nleu	6.47	18.6	9.30	24.5	11.5	31.3
Phe	13.5	34.1	18.2	42.4	23.1	54.7
Pro	1.21	4.30	1.85	5.98	2.98	9.57
Thr	1.06	1.06	1.62	1.62	2.62	2.62
Trp	30.5	54.7	34.7	60.8	43.7	77.5
Tyr	3.32	6.08	4.74	8.25	6.53	11.8
Val	1.89	5.27	2.89	7.32	3.74	9.40

Table LC 23 (continued)
THE SEPARATION OF AMINO ACID ENANTIOMERS ON A REVERSED-PHASE PACKING COATED WITH *N*-DECYL-L-HYDROXYPROLINE AS A FUNCTION OF COPPER(II) ACETATE CONCENTRATION

Packing: P1 = LiChrosorb® RP-18, particle diameter 5 μm; coating of the support is achieved by forcing 2 mℓ of the solution of 100 mg of *N*-decyl-L-hydroxyproline in methanol-water, 20:80 through the column, which is then washed with 2—4 mℓ of a concentrated solution of Cu(II) acetate in methanol/water, 15:85

Solvent: S1 = 2×10^{-4} *M* Cu(II) acetate in ethanol/water, 15:85
S2 = 10^{-4} *M* Cu(II) acetate in ethanol/water, 15:85
S3 = 5×10^{-5} *M* Cu(II) acetate in ethanol/water, 15:85

Detection: D1 = UV at 254 nm

REFERENCE

1. **Davankov, V. A., Bochkov, A. S., Kurganov, A. A., Roumeliotis, P., and Unger, K. K.,** *Chromatographia,* 13, 677, 1980.

Table LC 24
THE SEPARATION OF AMINO ACID ENANTIOMERS ON A REVERSED-PHASE PACKING COATED WITH *N*-HEPTYL-L-HYDROXYPROLINE AS A FUNCTION OF COPPER(II) ACETATE CONCENTRATION

Packing	P1	P1	P1
Column			
Length (mm)	100	100	100
Diameter (mm)	4.2	4.2	4.2
Solvent	S1	S2	S3
Flow rate (mℓ/min)	2	2	2
Temperature (K)	293	293	293
Detection	D1	D1	D1

Amino acid	k'_L	k'_D	k'_L	k'_D	k'_L	k'_D
Ala	0.51	1.02	0.45	0.87	1.17	2.00
Abu	0.91	2.72	0.83	2.29	1.83	5.17
Asp	0.13	0.13	0.15	0.15		
His	0.63	0.94	0.72	1.38	1.11	1.87
Met	2.96	7.77	3.55	8.64	6.15	15.6
Nleu	5.72	28.8	5.96	30.8	12.0	59.4
Phe	12.8	50.1	13.5	52.0	25.8	97.3
Pro	0.62	6.77	0.69	6.55	1.49	13.6
Thr	0.94	0.94	0.96	0.96	1.75	1.75
Trp	30.5	30.5	34.1	85.9	56.4	132.0
Tyr	2.87	10.0	3.26	9.47	5.26	16.5
Val	1.43	7.64	1.40	7.94	2.90	16.6

Packing: P1 = LiChrosorb® RP-18, particle diameter 5 μm; coating of the support is achieved by forcing 2 mℓ of a solution of 100 mg of *N*-heptyl-L-hydroxyproline in methanol-water, 40:60 through the column, which is then washed with 2—4 mℓ of a concentrated solution of Cu(II) acetate in methanol/water, 15:85

Solvent: S1 = 2×10^{-4} *M* Cu(II) acetate in ethanol/water, 15:85
S2 = 10^{-4} *M* Cu(II) acetate in ethanol/water, 15:85

Table LC 24 (continued)
THE SEPARATION OF AMINO ACID ENANTIOMERS ON A REVERSED-PHASE PACKING COATED WITH *N*-HEPTYL-L-HYDROXYPROLINE AS A FUNCTION OF COPPER(II) ACETATE CONCENTRATION

S3 = 5×10^{-5} *M* Cu(II) acetate in ethanol/water, 15:85

Detection: D1 = UV at 254 nm

REFERENCE

1. **Davankov, V. A., Bochkov, A. S., Kurganov, A. A., Roumeliotis, P., and Unger, K. K.,** *Chromatographia,* 13, 677, 1980.

Table LC 25
THE EFFECT OF pH AND AMMONIUM ACETATE CONCENTRATION ON THE RETENTION OF AMINO ACID ENANTIOMERS ON A REVERSED-PHASE PACKING COATED WITH *N*-HEPTYL-L-HYDROXYPROLINE

Packing	P1	P1	P1	P1
Column				
Length (mm)	100	100	100	100
Diameter (mm)	4.2	4.2	4.2	4.2
Solvent	S1	S2	S3	S4
Flow rate (mℓ/min)	2	2	2	2
Temperature (K)	293	293	293	293
Detection	D1	D1	D1	D1

Amino acid	k'_L	k'_D	k'_L	k'_D	k'_L	k'_D	k'_L	k'_D
Ala	0.94	1.47	0.83	1.41	4.79	8.53	0.60	1.28
Abu	1.47	3.66	0.98	3.42	6.13	21.3	1.11	3.89
Arg	5.04	6.36	1.47	2.53	4.74	8.70	1.81	2.47
Asp	0.37	0.37	0.36	0.36	2.06	2.06	0.21	0.21
His	1.01	1.40	0.43	0.85	3.85	9.06	0.79	2.89
Met	4.87	11.1	3.91	9.96	22.1	63.7	4.74	13.3
Nleu	9.42	43.3	8.02	40.0	39.8		9.42	47.3
Phe	18.9	75.0	16.0	67.5	80.7		16.2	70.0
Pro	1.16	10.0	0.68	9.94	4.39	65.0	0.89	14.6
Thr	1.40	1.40	0.79	0.79	5.85	5.85	0.91	0.91
Tyr	5.00	14.7	3.51	11.9	14.4	61.1	4.02	14.6
Trp	44.7	116.0	34.4	99.4	84.5		31.5	96.4
Val	2.58	11.7	1.66	9.81	0.64	72.8	1.87	12.9

Packing: P1 = LiChrosorb® RP-18, particle diameter 5 μm; the support is coated by forcing 2 mℓ of a solution of *N*-heptyl-L-hydroxyproline in methanol/water, 40:60, through the column; the column is then washed with 2—4 mℓ of a concentrated solution of Cu(II) acetate in methanol/water, 15:85

Solvent: S1 = 10^{-4} *M* Cu(II) acetate pH 5.5 in methanol/water, 15:85

S2 = 10^{-4} Cu(II) acetate + 4×10^{-3} *M* ammonium acetate, pH 5.5 in methanol/water, 15:85

S3 = 10^{-4} Cu(II) acetate + 4×10^{-3} *M* ammonium acetate, pH 7.2, in methanol/water, 15:85

S4 = 10^{-4} *M* Cu(II) acetate + 4×10^{-2} *M* ammonium acetate, pH 7.2 in methanol/water, 15:85

Detection: D1 = UV at 254 nm

REFERENCE

1. **Davankov, V. A., Bochkov, A. S., Kurganov, A. A., Roumeliotis, P., and Unger, K. K.,** *Chromatographia,* 13, 677, 1980.

Table LC 26
THE EFFECT OF ELUENT COMPOSITION ON THE RETENTION OF AMINO ACID ENANTIOMERS ON A REVERSED-PHASE PACKING COATED WITH *N*-HEXADECYL-L-HYDROXYPROLINE

Packing	P1		P1		P1		P1	
Column								
Length (mm)	100		100		100		100	
Diameter (mm)	4.2		4.2		4.2		4.2	
Solvent	S1		S2		S3		S4	
Flow rate (mℓ/min)	2		2		2		2	
Temperature (K)	293		293		293		293	
Detection	D1		D1		D1		D1	
Amino acid	**k'_L**	**k'_D**	**k'_L**	**k'_D**	**k'_L**	**k'_D**	**k'_L**	**k'_D**
Abu	0.94	0.94	0.96	1.16	0.68	0.90	0.78	1.17
Asp	0.33	0.33	0.43	0.43	0	0	0	0
His	1.45	1.45	1.32	1.32	1.55	1.55	0.90	0.90
Met	1.21	1.21	1.38	1.80	1.58	2.19	2.20	3.27
Nleu	1.60	2.16	2.31	2.92	3.29	7.20	4.91	10.9
Phe	2.03	2.60	3.28	5.03	4.63	8.03	8.57	16.1
Pro	0.80	1.11	0.92	1.45	0.61	1.28	0.73	1.78
Thr	0.84	0.84	0.73	0.73	0.82	0.82	0.83	0.83
Trp	2.76	3.13	5.85	7.74	10.6	14.9	21.4	32.4
Tyr	1.21	1.21	1.32	1.60	1.42	1.88	2.29	3.32
Val	0.87	1.10	0.95	1.46	0.90	1.70	0.98	2.59

Packing: P1 = LiChrosorb® RP-18, particle diameter 5 μm; the support is coated by forcing 2 mℓ of a solution of 100 mg of *N*-hexadecyl-L-hydroxyproline in methanol through the column; the column is then washed with 2—4 mℓ of a concentrated solution of Cu(II) acetate in methanol/water, 15:85

Solvent: S1 = 10^{-4} *M* Cu(II) acetate in acetonitrile/water, 30:70
S2 = 10^{-4} *M* Cu(II) acetate in acetonitrile/water, 20:80
S3 = 10^{-4} *M* Cu(II) acetate in acetonitrile/water, 15:85
S4 = 10^{-4} *M* Cu(II) acetate in acetonitrile/water, 10:90

Detection: D1 = UV at 254 nm

REFERENCE

1. **Davankov, V. A., Bochkov, A. S., Kurganov, A. A., Roumeliotis, P., and Unger, K. K.,** *Chromatographia,* 13, 677, 1980.

Table LC 27
THE EFFECT OF ELUENT COMPOSITION ON THE RETENTION OF AMINO ACID ENANTIOMERS ON A REVERSED-PHASE PACKING COATED WITH *N*-DECYL-L-HYDROXYPROLINE

Packing	P1	P1	P1	P1
Column				
Length(mm)	100	100	100	100
Diameter (mm)	4.2	4.2	4.2	4.2
Solvent	S1	S2	S3	S4
Flow rate (mℓ/min)	2	2	2	2
Temperature (K)	293	293	293	293
Detection	D1	D1	D1	D1

Amino acid	k'_L	k'_D	k'_L	k'_D	k'_L	k'_D	k'_L	k'_D
Abu	1.09	1.09	1.20	1.45	1.48	1.91	1.69	2.38
Asp	1.76	1.76	1.55	1.55	1.34	1.34	0.92	0.92
His	1.05	1.05	1.35	1.35	1.56	1.56	1.47	1.47
Met	1.41	1.41	1.90	2.43	2.75	3.76	3.81	5.72
Nleu	1.72	2.35	3.00	5.30	4.73	9.33	7.45	16.1
Phe	2.10	2.76	4.37	7.13	7.46	13.1	13.3	26.0
Pro	0.82	1.02	1.10	1.68	1.33	2.35	1.54	3.53
Thr	0.87	0.87	1.16	1.16	1.27	1.27	1.49	1.49
Trp	2.78	3.20	6.99	9.65	14.4	21.0	27.6	40.7
Tyr	1.29	1.29	1.84	2.16	2.57	3.29	3.59	5.39
Val	1.03	1.27	1.44	2.26	1.63	2.80	2.45	5.12

Packing: P1 = LiChrosorb® RP-18, particle diameter 5 μm; the support is coated by forcing 2 mℓ of a solution of 100 mg of *N*-decyl-L-hydroxyproline in methanol/water, 20:80, through the column; the column is then washed with 2—4 mℓ of a concentrated solution of Cu(II) acetate in methanol/water, 15:85

Solvent: S1 = 10^{-4} *M* Cu(II) acetate in acetonitrile/water, 30:70
S2 = 10^{-4} *M* Cu(II) acetate in acetonitrile/water, 20:80
S3 = 10^{-4} *M* Cu(II) acetate in acetonitrile/water, 15:85
S4 = 10^{-4} *M* Cu(II) acetate in acetonitrile/water, 10:90

Detection: D1 = UV at 254 nm

REFERENCE

1. **Davankov, V. A., Bochkov, A. S., Kurganov, A. A., Roumeliotis, P., and Unger, K. K.,** *Chromatographia,* 13, 677, 1980.

Table LC 28
THE SEPARATION OF AMINO ACID ENANTIOMERS IN DIMETHYLSULFOXIDE-WATER ON REVERSED-PHASE PACKINGS COATED WITH *N*-ALKYL-L-HYDROXYPROLINE

Packing	P1	P2
Column		
Length (mm)	100	100
Diameter (mm)	4.2	4.2
Solvent	S1	S1
Flow rate (mℓ/min)	2	2
Temperature (K)	293	293
Detection	D1	D1

Amino acid	k'_L	k'_D	k'_L	k'_D
Aminobutyric acid	1.30	2.00	2.11	3.04
Aspartic acid	0.41	0.41	1.06	1.06
Histidine	1.54	1.54	2.00	2.00
Methionine	4.07	6.85	4.87	8.01
Norleucine	7.46	17.6	10.6	25.4
Phenylalanine	17.0	34.2	24.6	49.6
Proline	1.27	4.04	2.04	6.72
Threonine	1.28	1.28	1.53	1.53
Tyrosine	5.78	9.67	8.63	13.9
Tryptophan	43.7	68.1	60.7	93.7
Valine	2.27	5.34	2.70	6.55

Packing: P1 = LiChrosorb® RP-18, particle diameter 5 μm; coating of the support is achieved by forcing 2 mℓ of a solution of 100 mg of *N*-hexadecyl-L-hydroxyproline in methanol through the column, which is then washed with 2—4 mℓ of a concentrated solution of Cu(II) acetate in methanol/water, 15:85

P2 = LiChrosorb® RP-18, particle diameter 5 μm; coating of the support is achieved by forcing 2 mℓ of a solution of *N*-decyl-L-hydroxyproline in methanol/water, 20:80 through the column, which is then washed with 2—4 mℓ of a concentrated solution of Cu(II) acetate in methanol/water, 15:85

Solvent: S1 = 10^{-4} *M* Cu(II) acetate in dimethylsulfoxide/water, 15:85

Detection: D1 = UV at 254 nm

REFERENCE

1. **Davankov, V. A., Bochkov, A. S., Kurganov, A. A., Roumeliotis, P., and Unger, K. K.,** *Chromatographia,* 13, 677, 1980.

Table LC 29
THE EFFECT OF TEMPERATURE ON THE RETENTION OF AMINO ACID ENANTIOMERS ON A REVERSED-PHASE PACKING COATED WITH *N*-HEXADECYL-L-PROLINE

Packing	P1	P1	P1
Column			
Length (mm)	100	100	100
Diameter (mm)	4.2	4.2	4.2
Solvent	S1	S1	S1
Flow rate (mℓ/min)	2	2	2
Temperature	T1	T2	T3
Detection	D1	D1	D1

Amino acid	k'_L	k'_D	k'_L	k'_D	k'_L	k'_D
Abu	1.58	2.40	1.59	2.45	1.56	2.34
Asp	0.39	0.39	0.38	0.38	0.18	0.18
His	1.53	1.53	1.56	1.56	1.30	1.30
Met	3.92	6.24	3.74	6.09	3.56	5.52
Nleu	8.31	21.3	8.21	21.1	6.76	16.9
Phe	17.2	37.6	15.4	34.8	11.5	25.3
Pro	1.51	4.18	1.59	4.92	1.22	3.60
Thr	1.32	1.32	1.34	1.34	1.08	1.08
Tyr	4.57	7.58	4.19	6.86	2.88	4.50
Trp	34.5	58.1	27.6	46.9	18.4	31.4
Val	2.62	6.14	2.66	6.31	2.20	5.10

Packing: P1 = LiChrosorb® RP-18, particle diameter 5 μm; coating of the support is achieved by forcing 2 mℓ of a solution of 100 mg of *N*-hexadecyl-L-hydroxyproline in methanol through the column, which is then washed with 2—4 mℓ of a concentrated solution of Cu(II) acetate in methanol/water, 15:85

Solvent: S1 = 10^{-4} *M* Cu(II) acetate in methanol/water, 15:85

Temperature: T1 = 293 K
T2 = 308 K
T3 = 323K

Detection: D1 = UV at 254 nm

REFERENCE

1. **Davankov, V. A., Bochkov, A. S., Kurganov, A. A., Roumeliotis, P., and Unger, K. K.,** *Chromatographia,* 13, 677, 1980.

Table LC 30
THE EFFECT OF TEMPERATURE ON THE RETENTION OF AMINO ACID ENANTIOMERS ON A REVERSED-PHASE PACKING COATED WITH *N*-DECYL-L-HYDROXYPROLINE

Packing	P1	P1	P1
Column			
Length (mm)	100	100	100
Diameter (mm)	4.2	4.2	4.2
Solvent	S1	S1	S1
Flow rate (mℓ/min)	2	2	2
Temperature	T1	T2	T3
Detection	D1	D1	D1

Amino acid	k'_L	k'_D	k'_L	k'_D	k'_L	k'_D
Abu	1.87	2.94	1.85	2.91	1.89	2.86
Asp	0.87	0.87	0.66	0.66	0.74	0.74
His	2.13	2.13	2.00	2.00	1.79	1.79
Met	4.47	7.27	4.15	6.79	3.98	6.53
Nleu	9.30	24.5	8.85	23.5		
Phe	18.2	42.4				
Pro	1.85	5.98	1.81	5.74	1.76	5.51
Thr	1.62	1.62	1.53	1.53	1.59	1.59
Tyr	4.74	8.25	4.11	7.04	3.55	5.92
Trp	34.7	60.8				
Val	2.89	7.32	2.76	6.98	2.87	7.08

Packing: P1 = LiChrosorb® RP-18, particle diameter 5 μm; coating of the support is achieved by forcing 2 mℓ of a solution of 100 mg of *N*-decyl-L-hydroxyproline in methanol-water, 20:80 through the column, which is then washed with 2—4 mℓ of a concentrated solution of Cu(II) acetate in methanol/water, 15:85

Solvent: S1 = 10^{-4} *M* Cu(II) acetate in methanol/water, 15:85

Temperature: T1 = 293 K

T2 = 308 K

T3 = 323 K

Detection: D1 = UV at 254 nm

REFERENCE

1. **Davankov, V. A., Bochkov, A. S., Kurganov, A. A., Roumeliotis, P., and Unger, K. K.,** *Chromatographia,* 13, 677, 1980.

Table LC 31
THE EFFECT OF TEMPERATURE ON THE RETENTION OF AMINO ACID ENANTIOMERS ON A REVERSED-PHASE PACKING COATED WITH *N*-HEPTYL-L-HYDROXYPROLINE

Packing	P1	P1	P1
Column			
Length (mm)	100	100	100
Diameter (mm)	4.2	4.2	4.2
Solvent	S1	S1	S1
Flow rate (mℓ/min)	2	2	2
Temperature	T1	T2	T3
Detection	D1	D1	D1

Amino acid	k'_L	k'_D	k'_L	k'_D	k'_L	k'_D
Abu	1.38	3.36	0.87	2.36	0.91	2.45
Asp	0.17	0.17	0.11	0.11	0.17	0.17
His	0.89	1.23	0.30	0.72	0.34	0.74
Met	3.79	8.70	2.60	6.26	2.55	5.98
Nleu	8.49	34.7	5.83	25.8	6.02	24.5
Phe	15.6	54.7	10.1	38.8	8.94	34.8
Pro	1.12	8.66	0.79	5.89	0.70	5.62
Thr	1.13	1.13	0.75	0.75	0.83	0.83
Tyr	3.68	10.1	2.36	6.55	2.08	5.28
Trp	30.3	69.2	19.1	46.7	14.6	
Val	2.32	10.0	1.53	7.30	1.70	7.57

Packing: P1 = LiChrosorb® RP-18, particle diameter 5 μm; coating of the support is achieved by forcing 2 mℓ of a solution of 100 mg of *N*-heptyl-L-hydroxyproline in methanol/water, 40:60 through the column, which is then washed with 2—4 mℓ of a concentrated solution of Cu(II) acetate in methanol/water, 15:85

Solvent: S1 = 10^{-4} *M* Cu(II) acetate in methanol/water, 15:85

Temperature: T1 = 293 K
T2 = 308 K
T3 = 323 K

Detection: D1 = UV at 254 nm

REFERENCE

1. **Dovankov, V. A., Bochkov, A. S., Kurganov, A. A., Roumeliotis, P., and Unger, K. K.,** *Chromatographia,* 13, 677, 1980.

Table LC 32
AMINO ACID RACEMATES ON RESINS CONTAINING L-PROLINE OR L-HYDROXYPROLINE GROUPS

Packing	P1		P1		P2		P2	
Column								
Length (cm)	20		20		20		20	
Diameter (cm, I.D.)	0.2		0.2		0.4		0.4	
Material	G		G		SS		SS	
Solvent	S1		S2		S3		S1	
Flow rate (mℓ/hr)	5		5		6.4		6.4	
Detection	D1		D1		D1		D1	
Amino acid	k'_L	k'_D	k'_L	k'_D	k'_L	k'_D	k'_L	k'_D
Alanine	0.6	1.4	2.1	5.5	1.62	1.62	0.47	0.47
Arginine	2.4	1.8						
Asparagine			6.1	4.4				
Cysteine	7.8	5.0						
Histidine	6.6	1.4			22.8	11.9	2.93	1.05
Leucine		7.0	7.0		4.87	4.71	2.17	2.07
Lysine			6.4	4.6				
Methionine	2.4	2.1					2.07	1.53
Phenylalanine	10.8	6.0			19.6	16.1	19.8	14.6
Proline	0.6	1.4	2.1	5.5			0.53	0.93
Threonine	1.1	0.8	5.8	4.2	0.67	0.51	0.63	0.39
Tryptophan	58.0	27.6						
Tyrosine	13.0	4.6			1.0	0.72	11.0	6.38
Valine	1.0	0.8	4.1	3.7	2.21	1.85	1.17	0.83

Packing: P1 = polymethacrylate-type resin was synthesized by reaction of the glycidyl groups of the polymer matrix with L-hydroxyproline
P2 = similar to P1 but with L-proline in place of L-hydroxyproline
The structures of the resins are shown below (P1 = I, P2 = II)
Solvent: S1 = 0.1 M $(NH_4)_2CO_3$, pH 9.5—10^{-4} M Cu(II)
S2 = 0.02 M $(NH_4)_2CO_3$, pH 9.5—2 × 10^{-5} M Cu(II)
S3 = 0.1 M aqueous $NH_3/10^{-5}$ M Cu(II)
Detection: D1 = UV at 250 nm

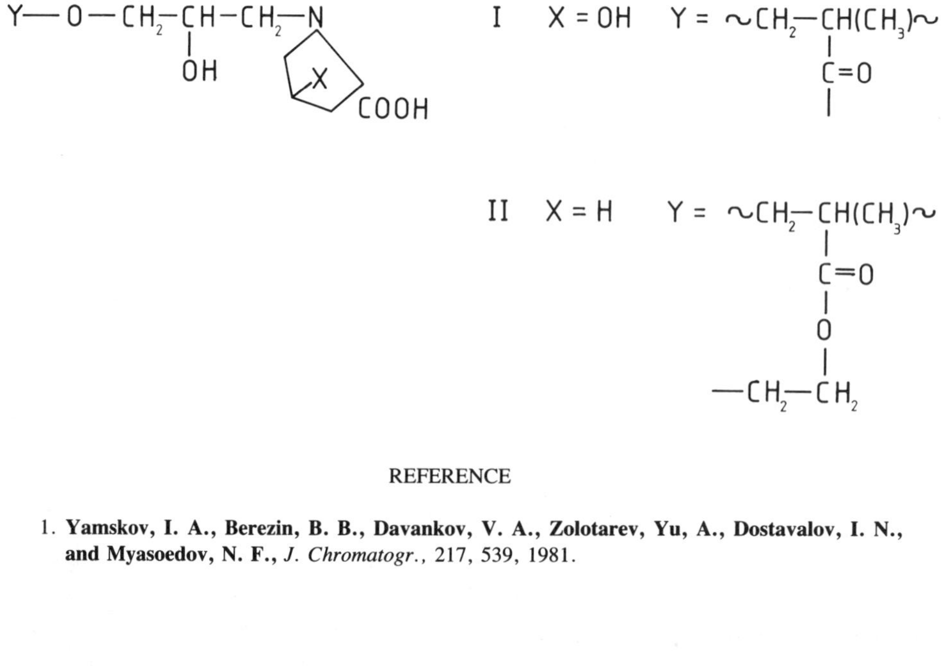

REFERENCE

1. **Yamskov, I. A., Berezin, B. B., Davankov, V. A., Zolotarev, Yu, A., Dostavalov, I. N., and Myasoedov, N. F.,** *J. Chromatogr.*, 217, 539, 1981.

Table LC 33
THE EFFECT OF pH AND IONIC STRENGTH ON THE ENANTIOMERIC SEPARATION OF AMINO ACIDS

Packing	P1	P1
Column		
Length (mm)	250	250
Diameter (mm, I.D.)	4.6	4.6
Solvent	S1	S2
Flow rate (mℓ/min)	2	2
Temperature (°C)	35	35
Detection	D1	D1

Amino acid	k'_L	k'_D	k'_L	k'_D
Aminobutyric acid	7.20	11.60	2.30	5.61
Arginine	4.77	4.77	0.63	0.63
Aspartic acid	0.86	0.63	0.50	0.50
Histidine	0.50	0.50	0.20	0.20
Methionine	34.00	43.00	11.45	15.11
Proline	4.66	6.70	2.00	3.00
Threonine	1.03	1.33	0.35	0.35
Tyrosine	31.78	44.21	9.17	13.29
Valine	19.33	35.33	7.41	9.26

Packing: P1 = Zorbax® C_8, 10 μm, coated with *N*-decyl-L-histidine by passing a 1% ethanolic solution of the compound through the column followed by an aqueous solution of Cu(II) acetate

Solvent: S1 = 10^{-4} *M* Cu(II) acetate, pH 6.18

S2 = 10^{-4} *M* Cu(II) acetate, pH 4.85 + 0.1 *M* NH_4Cl

Detection: D1 = UV at 254 nm

REFERENCE

1. **Davankov, V. A., Bochkov, A. S., and Belov, Yu. P.,** *J. Chromatogr.*, 218, 541, 1981.

Table LC 34
THE EFFECT OF COPPER ACETATE CONCENTRATION ON THE ENANTIOMERIC SEPARATION OF AMINO ACIDS

Packing	P1	P1	P1
Column			
Length (mm)	250	250	250
Diameter (mm, I.D.)	4.6	4.6	4.6
Solvent	S1	S2	S3
Flow rate (mℓ/min)	2	2	2
Temperature (°C)	35	35	35
Detection	D1	D1	D1

Amino acid	k'_L	k'_D	k'_L	k'_D	k'_L	k'_D
Ala	2.17	2.17	1.80	1.80	1.50	1.50
allo-Hyp	1.17	1.17	0.90	0.90	0.67	0.67
allo-Thr	1.96	1.96	1.90	1.90	1.50	1.50
Aaa	1.95	1.95	1.60	1.60	1.47	1.47
Abu	3.40	3.40	2.76	2.76	2.57	2.57
Arg	8.20	8.20	5.70	5.70	4.17	4.17
Asp	1.69	1.24	1.50	1.13	1.36	0.96
Cit	1.40	1.40	1.20	1.20	1.18	1.18
Glu	1.67	1.67	1.36	1.36	1.17	1.17
His	5.50	5.50	1.88	1.88	1.72	1.72
Ival	6.50	6.95	5.60	6.00	3.83	4.25
Lys	6.67	6.67	3.80	3.80	3.50	3.50
Met	11.40	12.06	8.00	8.50	7.25	7.76
Orn	0.83	0.83	0.71	0.71	0.54	0.54
Pro	2.40	3.05	1.82	2.37	1.38	1.84
Ser	2.00	2.00	1.40	1.40	1.31	1.31
Thr	2.00	2.24	1.60	1.80	1.33	1.52
Tyr	9.33	11.00	6.50	7.75	6.33	7.66
Val	7.60	9.35	5.30	6.70	4.67	6.02

Abbreviations: Aaa = α-aminoadipic acid; Abu = aminobutyric acid; Ival = isovaline

Packing: P1 = Zorbax® C_8, 10 μm, coated with *N*-decyl-L-histidine by passing a 1% ethanolic solution of the compound through the column followed by an aqueous solution of Cu(II) acetate

Solvent: S1 = 0.5 × 10^{-4} *M* Cu(II) acetate in ethanol/water, 15:85
S2 = 10^{-4} *M* Cu(II) acetate in ethanol/water, 15:85
S3 = 2 × 10^{-4} *M* Cu(II) acetate in ethanol/water, 15:85

Detection: D1 = UV at 254 nm

REFERENCE

1. **Davankov, V. A., Bochkov, A. S., and Belov, Yu. P.,** *J. Chromatogr.*, 218, 547, 1984.

Table LC 35
THE EFFECT OF ELUENT COMPOSITION OF THE ENANTIOMERIC SEPARATION OF AMINO ACIDS

Packing	P1	P1	P1	P1
Column				
Length (mm)	250	250	250	250
Diameter (mm, I.D.)	4.6	4.6	4.6	4.6
Solvent	S1	S2	S3	S4
Flow rate (mℓ/min)	2	2	2	2
Temperature (°C)	35	35	35	35
Detection	D1	D1	D1	D1

Amino acid	k'_L	k'_D	k'_L	k'_D	k'_L	k'_D	k'_L	k'_D
Ala	1.86	1.86	1.80	1.80	1.71	1.71	2.48	3.35
allo-Thr	2.29	2.29	1.90	1.90	1.88	1.88	3.31	4.00
Aaa	1.47	1.47	1.60	1.60	2.00	2.00	5.46	6.43
Abu	2.79	2.79	2.76	2.76	3.25	3.88	7.20	11.60
Arg	5.73	5.73	5.70	5.70	7.50	7.50	4.77	4.77
Asn	0.86	0.86	0.82	0.82	0.82	0.82	0.63	0.63
Asp	1.71	1.29	1.50	1.13	1.47	1.09	0.86	0.63
Cit	1.29	1.29	1.20	1.20	1.83	1.83	2.00	2.00
Glu	1.50	1.50	1.36	1.36	0.93	0.93	2.39	2.39
His	1.67	1.67	1.88	1.88	1.88	1.88	0.50	0.50
Ival	1.71	1.71	5.60	6.00	6.88	7.63	19.57	24.33
Lys	3.07	3.07	3.80	3.80	7.33	7.33	3.29	3.29
Met	4.00	4.24	8.00	8.50	10.25	11.38	34.00	43.00
Pro	1.71	2.21	1.82	2.37	2.00	2.66	4.66	6.70
Ser	1.57	1.57	1.40	1.40	1.15	1.33	1.23	1.54
Thr	1.69	1.92	1.60	1.80	1.50	1.75	1.03	1.33
Tyr	4.86	4.86	6.50	7.75	7.33	9.67	31.78	44.22
Val	3.43	3.43	5.30	6.70	8.00	10.33	19.33	35.33

Abbreviations: Aaa = α-aminoadipic acid; Abu = aminobutyric acid; Ival = isovaline

Packing: P1 = Zorbax® C_8, 10 μm coated with *N*-decyl-L-histidine by passing a 1% ethanolic solution of the compound through the column followed by an aqueous solution of Cu(II) acetate

Solvent: S1 = 10^{-4} *M* Cu(II) acetate in water/ethanol, 70:30
S2 = 10^{-4} *M* Cu(II) acetate in water/ethanol, 85:15
S3 = 10^{-4} *M* Cu(II) acetate in water/ethanol, 92.5:7.5
S4 = 10^{-4} *M* copper acetate in water

Detection: D1 = UV at 254 nm

REFERENCE

1. **Davankov, V. A., Bochkov, A. S., and Belov, Yu. P.,** *J. Chromatogr.*, 218, 547, 1981.

Table LC 36
THE EFFECT OF ELUENT COMPOSITION ON THE ENANTIOMERIC RESOLUTION OF AMINO ACIDS

Packing	P1		P2		P1		P2	
Column								
Length (mm)	250		250		250		250	
Diameter (mm, I.D.)	4.6		4.6		4.6		4.6	
Solvent	S1		S1		S2		S2	
Flow rate (mℓ/min)	2		2		2		2	
Temperature (°C)	35		35		35		35	
Detection	D1		D1		D1		D1	
Amino acid	**k'_L**	**k'_D**	**k'_L**	**k'_D**	**k'_L**	**k'_D**	**k'_L**	**k'_D**
Ala	0.96	0.96	3.01	3.01	1.83	1.83	2.37	2.37
allo-Hyp	0.71	0.71	1.00	1.00	0.82	0.82	0.93	0.93
allo-Thr	0.44	0.44	1.67	1.67	2.11	2.11	3.21	3.21
Aaa	0.52	0.52	2.25	2.25	1.13	1.13	1.98	1.98
Abu	3.24	3.24	3.67	3.67	4.52	4.52	6.31	6.31
Asn	0.43	0.43	1.40	1.40	0.61	0.61	0.75	0.75
Asp	0.78	0.44	1.06	0.66	0.83	0.51	1.21	0.81
Cit	0.79	0.79	2.40	2.40	1.75	1.75	3.02	3.02
Glu	0.57	0.57	0.88	0.88	0.81	0.81	0.96	0.96
Ival	4.75	5.75	11.37	13.67	9.00	10.71	14.17	17.15
Lys	2.00	2.00	5.57	5.57	5.72	5.72	7.36	7.36
Met	5.25	6.63	13.51	16.48	12.33	14.54	17.81	20.48
Pro	1.00	1.51	3.20	4.80	3.51	4.86	4.62	6.05
Ser	0.74	0.74	1.75	1.75	1.12	1.12	1.23	1.23
Thr	0.59	0.59	1.75	1.75	1.66	1.91	2.31	2.59
Tyr	4.67	5.88	6.68	8.41	12.33	15.33	15.02	17.72
Val	3.58	5.84	10.60	16.85	8.27	10.67	13.92	16.98

Abbreviations: Aaa = α-aminoadipic acid; Abu = aminobutyric acid; Ival = isovaline

Packing: P1 = Zorbax®C_8, 10 μm coated with *N*-decyl-L-histidine by passing a 1% ethanolic solution of the compound through the column followed by an aqueous solution of Cu(II) acetate

P2 = Zorbax® ODS, 10 μm treated similarly to P1

Solvent: S1 = 10^{-4} *M* Cu(II) acetate in water/acetonitrile, 85:15

S2 = 10^{-4} *M* Cu(II) acetate in water/methanol, 85:15

Detection: D1 = UV at 254 nm

REFERENCE

1. **Davankov, V. A., Bochkov, A. S., and Belov, Yu. P.,** *J. Chromatogr.*, 218, 547, 1981.

Table LC 37
THE EFFECT OF COLUMN TEMPERATURE ON THE ENANTIOMERIC RESOLUTION OF AMINO ACIDS

Packing	P1	P1	P1
Column			
Length (mm, I.D.)	250	250	250
Diameter (mm, I.E.)	4.6	4.6	4.6
Solvent	S1	S1	S1
Flow rate (mℓ/min)	2	2	2
Temperature (°C)	18	35	45
Detection	D1	D1	D1

Amino acid	k'_L	k'_D	k'_L	k'_D	k'_L	k'_D
Ala	1.85	1.85	1.80	1.80	1.38	1.38
allo-Hyp	0.95	0.95	0.90	0.90	0.63	0.63
allo-Thr	2.27	2.27	1.90	1.90	1.36	1.36
Aaa	1.80	1.80	1.60	1.60	1.36	1.36
Abu	3.13	3.13	2.76	2.76	2.70	2.70
Arg	6.23	6.23	5.70	5.70	5.21	5.21
Asn	0.88	0.88	0.82	0.82	0.50	0.50
Asp	2.07	1.50	1.50	1.13	1.39	0.98
Cit	1.33	1.33	1.20	1.20	0.88	0.88
Glu	1.63	1.63	1.36	1.36	1.00	1.00
His	2.40	2.40	1.88	1.88	1.77	1.77
Ival	6.33	6.64	5.60	6.00	4.00	4.36
Lys	4.47	4.47	3.80	3.80	2.88	2.88
Met	10.80	11.34	8.00	8.50	6.43	6.94
Orn	0.87	0.87	0.71	0.71	0.50	0.50
Pro	2.01	2.54	1.82	2.37	1.72	2.27
Ser	1.45	1.45	1.40	1.40	1.33	1.33
Thr	1.71	1.94	1.60	1.80	1.29	1.48
Tyr	11.93	14.55	6.50	7.75	5.28	6.50
Val	6.21	7.50	5.30	6.70	3.60	4.64

Abbreviations: Aaa = α-aminoadipic acid; Abu = aminobutyric acid; Ival = isovaline

Packing: P1 = Zorbax® C_8, 10 μm, coated with *N*-decyl-L-histidine by passing a 1% ethanolic solution of the compound through the column followed by an aqueous solution of Cu(II) acetate

Solvent: S1 = 10^{-4} *M* Cu(II) acetate in water/ethanol, 85:15

Detection: D1 = UV at 254 nm

REFERENCE

1. **Davankov, V. A., Bochkov, A. S., and Belov, Yu. P.,** *J. Chromatogr.*, 218, 547, 1981.

Table LC 38
ENANTIOMERIC RESOLUTION OF STRONGLY RETAINED AMINO ACIDS

Packing	P1		P1		P1		P1		P1		P1	
Column												
Length (mm)	80		80		80		80		80		80	
Diameter (mm, I.D.)	4.6		4.6		4.6		4.6		4.6		4.6	
Solvent	S1		S1		S2		S3		S4		S5	
Flow rate	na		na		na		na		na		na	
Temperature (°C)	18		35		18		18		18		18	
Detection	D1		D1		D1		D1		D1		D1	
Amino acid	k'_L	k'_D	k'_L	k'_D	k'_L	k'_D	k'_L	k'_D	k'_L	k'_D	k'_L	k'_D
Leucine	5.90	9.00	4.33	7.67	3.17	3.83	6.91	10.50	6.71	9.13	4.12	6.22
Norleucine	8.25	13.78	6.33	10.30	5.29	6.29	10.39	18.80	11.31	17.98	6.11	10.88
Phenylalanine	25.00	40.00	13.70	21.65	14.30	17.73	43.60	75.43	29.06	45.04	21.10	36.29
Phenylserine	4.75	10.74	3.67	8.00	2.91	5.00	5.64	13.48	5.24	11.16	4.01	9.38
Tryptophan	40.50	51.00	33.00	42.60	17.10	18.63	85.90	113.38	48.20	59.77	35.30	46.24

Packing: P1 = LiChrosorb® RP-18, 5 μm; the packing was coated with *N*-decyl-L-histidine by passing a 1% ethanolic solution of the compound through the column followed by an aqueous solution of Cu(II) acetate

Solvent: S1 = 10^{-4} *M* Cu(II) acetate in water/ethanol, 85:15
S2 = 10^{-4} *M* Cu(II) acetate in water/ethanol, 70:30
S3 = 10^{-4} *M* Cu(II) acetate in water/ethanol, 92.5:7.5
S4 = 5×10^{-5} *M* Cu(II) acetate in water/ethanol, 85:15
S5 = 2×10^{-4} *M* Cu(II) acetate in water/ethanol, 85:15

Detection: D1 = UV at 254 nm

REFERENCE

1. **Davankov, V. A., Bochkov, A. S., and Belov, Yu. P.,** *J. Chromatogr.*, 218, 547, 1981.

Table LC 39
ENANTIOMERIC AMINO ACIDS

Packing	P1
Column	
Length (mm)	200
Diameter (mm, I.D.)	1.3
Material	G
Solvent	S1
Flow rate (mℓ/hr)	1—8
Temperature	r t
Detection	D1

Amino acid	k'_D	k'_L
Alanine	0.35	0.35
Aspartic acid	0.18	0.18
N,N-Dibenzylserine	3.90	3.90
Dihydroxyphenylalanine	2.40	2.25
Isovaline	0.42	0.42
Leucine	1.10	1.10
Methionine	1.67	1.67
Norleucine	1.70	1.70
Norvaline	0.80	0.80
Phenylalanine	5.96	5.75
Phenylglycine	2.00	2.00
Phenylserine	7.33	6.81
Serine	0.37	0.37
Tryptophan	10.30	9.00
Tyrosine	1.87	1.67
Valine	0.85	0.85

Packing: P1 = LiChrosorb® RP-18, 5 μm
Solvent: S1 = 5 × 10^{-4} *M* solution of $Cu_2(tmpn)_2(OH)_2(ClO_4)_2$ in acetonitrile/water, 1:5; tmpn = *N,N,N′,N′*-tetramethyl-(R)-propanediamine-1,2; the complex was obtained by mixing ethanolic solutions of (R)-tmpn and Cu(II) perchlorate and crystallization from aqueous ethanol
Detection: D1 = UV at 254 nm

REFERENCE

1. **Kurganov, A. A. and Davankov, V. A.,** *J. Chromatogr.*, 218, 559, 1981.

Table LC 40
THE EFFECT OF ELUENT COMPOSITION ON THE ENANTIOMERIC SEPARATION OF AMINO ACIDS

Packing	P1		P1		P1		P1	
Column								
Length (mm)	100		100		100		100	
Diameter (mm, I.D.)	1.3		1.3		1.3		1.3	
Material	G		G		G		G	
Solvent	S1		S2		S3		S4	
Flow rate (mℓ/hr)	1—8		1—8		1—8		1—8	
Detection	D1		D1		D1		D1	
Amino acid	**k'_L**	**k'_D**	**k'_L**	**k'_D**	**k'_L**	**k'_D**	**k'_L**	**k'_D**
Alanine	16.8	16.8	15.8	15.8	13.2	13.2	8.05	8.05
Aspartic acid	7.44	10.5	3.40	4.30	1.28	1.50	0.55	0.55
Aminobutyric acid	12.1	11.2	8.77	8.33	7.33	7.33	6.77	6.77
Glutamic acid	15.1	17.4	6.66	6.66	1.72	1.72	0.55	0.55
Isoleucine	8.05	7.39	7.50	7.00	6.75	6.50	4.83	4.83
Isovaline	13.7	13.7			9.83	9.83	5.89	5.89
Leucine	6.78	6.78			5.83	5.83	4.61	4.61
Methionine	7.89	7.22	6.11	5.77	5.43	5.17	4.94	4.94
Norleucine	7.23	7.23	6.25	6.25	5.00	5.00	4.05	4.05
Norvaline	8.16	7.61			6.72	6.72	5.00	5.00
Phenylalanine	6.11	5.00	5.67	4.89	5.14	4.66	4.77	4.38
Phenylglycine	7.17	7.17			4.66	4.66	4.28	4.28
Phenylserine	6.10	5.15	4.89	4.55	4.28	3.89	3.94	3.78
Proline	16.1	14.4	11.0	10.2	9.30	9.30	6.11	6.11
Serine	37.3	35.4	25.7	24.7	18.3	18.3	9.61	9.61
Threonine	18.6	17.6	12.8	12.4	10.4	10.4	8.17	8.17
Tyrosine	8.55	6.94	9.11	7.67	7.17	6.22	6.05	5.39
Tryptophan	6.80	5.40	6.33	5.33	5.50	4.85	4.88	4.33
Valine	8.83	8.05	7.87	7.37	7.00	6.80	5.78	5.78

Packing: P1 = LiChrosorb® Si 100, 10 μm

Solvent: S1 = 5×10^{-4} *M* $Cu_2(tmpn)_2(OH)_2(ClO_4)_2$ and 5×10^{-3} *M* (R)-tmpn in acetonitrile/water, 90:10

S2 = as S1 but acetonitrile/water, 85:15

S3 = as S1 but acetonitrile/water, 75:25

S4 = as S1 but acetonitrile/water, 50:50

tmpn = *N,N,N′,N′*-tetramethyl-(R)-propanediamine-1,2; the complex was obtained by mixing ethanolic solutions of (R)-tmpn and Cu(II) perchlorate and crystallization from aqueous ethanol

Detection: D1 = UV at 254 nm

REFERENCE

1. **Kurganov, A. A. and Davankov, V. A.,** *J. Chromatogr.*, 218, 559, 1981.

Table LC 41
THE EFFECT OF pH ON THE RETENTION OF AMINO ACID ENANTIOMERS USING PACKING CONTAINING [Cu(II)3-(L-PROLYL) PROPYL]$^+$

Packing	P1	P1
Column		
Length (mm)	250	250
Diameter (mm, I.D.)	4	4
Solvent	S1	S2
Temperature (K)	323	323
Detection	D1	D1

Amino acid	k'_L	k'_D	k'_L	k'_D
Asp	0.30	0.31	0.58	0.58
Glu	0.22	0.25	0.49	0.56
His	1.97	1.67	3.13	2.55
Ala	0.45	0.47	0.83	0.90
Asn	0.81	0.79	1.08	1.08
Gln			0.84	0.49
Ser	0.58	0.47	1.06	0.88
Pro	0.58	0.88	1.06	1.89
Cit	0.64	0.64	1.12	1.12
Thr	0.84	0.72	1.28	1.18
Val	0.74	0.94	1.69	1.86
Lys	1.51	1.50	2.95	3.06
Nval	0.85	0.85	1.76	1.76
Tyr	1.16	1.04	2.26	2.01
Met	1.05	1.09	1.94	2.04
Arg	2.00	2.10	3.79	4.02
Ile	0.89	1.09	1.76	2.11
Leu	0.92	1.05	1.77	1.94
Nleu	0.97	0.97	1.95	1.95
Eth	1.27	1.27	2.32	2.32
Phe	1.29	1.30	2.43	2.60
Trp	2.67	2.55	5.12	4.94

Packing: P1 = [Cu(II)3-(L-prolyl)propyl]$^+$ on LiChrosorb® Si 100, particle diameter 10 μm

Solvent: S1 = 10^{-4} *M* Cu(II) acetate, 0.01 *M* ammonium acetate, pH 4.75, pH being adjusted with acetic acid

S2 = 10^{-4} *M* Cu(II) acetate, 0.01 *M* ammonium acetate, pH 5.0, the pH being adjusted with acetic acid

Detection: D1 = UV at 254 nm

REFERENCE

1. **Roumeliotis, P., Unger, K. K., Kurganov, A. A., and Davankov, V. A.,** *J. Chromatogr.*, 255, 51, 1983.

Table LC 42
THE EFFECT OF pH ON THE RETENTION OF AMINO ACID ENANTIOMERS USING PACKING CONTAINING [Cu(II)3-(L-HYDROXYPROLYL)PROPYL]$^+$

Packing	P1	P1	P1
Column			
Length (mm)	250	250	250
Diameter (mm, I.D.)	4	4	4
Solvent	S1	S2	S3
Temperature (K)	323	323	323
Detection	D1	D1	D1

Amino acid	k'_L	k'_D	k'_L	k'_D	k'_L	k'_D
Asp	0.50	0.42	0.93	0.76	2.09	1.96
Glu	0.43	0.47	0.96	1.07	2.71	3.25
His	2.62	2.01	5.65	4.32	25.14	25.14
Ala	0.42	0.47	1.94	2.08	6.70	7.21
Asn	0.99	0.85	2.50	2.15	5.14	5.54
Gln	0.62	0.47	2.01	1.69	2.83	2.79
Ser	0.67	0.60	2.01	2.00	4.74	4.75
Pro	0.57	1.08	2.51	3.97	8.00	10.85
Cit	1.36	1.36	3.14	3.14	6.87	6.87
Thr	1.01	0.82	2.58	2.32	6.33	6.08
Val	0.87	0.94	2.74	3.08	8.53	9.72
Dopa	1.11	1.26	6.53	6.53		
Lys	1.78	2.07	8.18	9.31	40.24	42.19
Nval	0.85	0.85	4.62	4.62	10.83	11.19
Tyr	1.13	0.78	2.96	2.14	7.12	6.05
Met	1.10	1.10	3.01	3.11	8.55	8.93
Arg	2.62	2.61	10.61	11.03	46.08	48.76
Ile	1.07	1.12	3.11	3.56	8.93	10.07
Leu	1.07	1.10	2.83	3.21	8.51	9.31
Nleu	0.96	0.96	4.72	4.72	10.83	11.62
Eth	1.32	1.32	5.05	5.05	11.76	11.76
Phe	1.53	1.31	3.82	3.44	9.25	9.18
Trp	2.60	1.78	6.14	4.33	12.20	10.25

Packing: P1 = [Cu(II)3-(L-hydroxyprolyl)propyl]$^+$ on LiChrosorb® Si 100, particle diameter 10 μm
Solvent: S1 = 10^{-4} *M* Cu(II) acetate, pH 4.85
S2 = 10^{-4} *M* Cu(II) acetate, pH 5.0
S3 = 10^{-4} *M* Cu(II) acetate, pH 5.35 pH values were adjusted with acetic acid
Detection: D1 = UV at 254 nm

REFERENCE

1. **Roumeliotis, P., Unger, K. K., Kurganov, A. A., and Davankov, V. A.,** *J. Chromatogr.*, 255, 51, 1983.

Table LC 43
THE EFFECT OF pH ON THE RETENTION OF AMINO ACID ENANTIOMERS USING PACKING CONTAINING [Cu(II)3-(L-HYDROXYPROLYL)PROPYL]$^+$ AND 3-(IODOPROPYL) GROUPS

Packing	P1	P1
Column		
Length (mm)	250	250
Diameter (mm, I.D.)	4	4
Solvent	S1	S2
Temperature (K)	323	323
Detection	D1	D1

Amino acid	k'_L	k'_D	k'_L	k'_D
Asp	1.45	1.32	1.96	1.76
Glu	1.12	1.18		
His	1.52	2.71	2.02	3.74
Ala	0.69	0.88	1.37	1.82
Asn	1.54	1.42	1.87	1.70
Gln			2.14	2.34
Ser	1.25	0.93	1.91	1.43
Pro	1.12	2.00	2.38	4.64
Thr	1.73	1.40	2.45	1.90
Val	1.89	1.54	3.35	3.15
Lys	0.93	1.05	2.00	2.29
Tyr	3.95	2.73	6.21	3.94
Met	2.70	2.63	4.15	4.49
Arg	1.59	1.73	3.10	3.45
Ile	2.79	2.48	4.18	4.96
Leu	2.25	2.22	4.20	4.02
Phe	5.73	3.90	8.11	6.16
Trp	11.37	6.49	17.68	10.80

Packing: P1 = [Cu(II)3-(L-hydroxyprolyl)propyl]$^+$ with remaining excess 3-(iodopropyl) groups on LiChrosorb® Si 100

Solvent: S1 = 10^{-4} *M* Cu(II) acetate, 0.01 *M* ammonium acetate, pH 5.0, the pH being adjusted with acetic acid

S2 = 10^{-4} *M* Cu(II) acetate, 0.01 *M* ammonium acetate, pH 5.5, the pH being adjusted with acetic acid

Detection: D1 = UV at 254 nm

REFERENCE

1. **Roumeliotis, P., Unger, K. K., Kurganov, A. A., and Davankov, V. A.,** *J. Chromatogr.*, 255, 53, 1983.

Table LC 44
THE EFFECT OF AMMONIUM ACETATE CONCENTRATION ON THE RETENTION OF AMINO ACID ENANTIOMERS USING PACKING CONTAINING [Cu(II)3-(L-PROLYL)PROPYL]$^+$

Packing	P1		P1		P1	
Column						
Length (mm)	250		250		250	
Diameter (mm, I.D.)	4		4		4	
Solvent	S1		S2		S3	
Temperature (K)	323		323		323	
Detection	D1		D1		D1	
Amino acid						
	k'_L	**k'_D**	**k'_L**	**k'_D**	**k'_L**	**k'_D**
Asp	2.19	2.12	0.58	0.58	0.20	0.20
Glu	2.51	2.91	0.49	0.56	0.06	0.06
His	8.62	8.62	3.13	2.55	1.08	0.78
Ala	4.37	4.60	0.83	0.90	0.11	0.14
Asn	4.27	4.16	1.08	1.08	0.32	0.32
Gln	4.29	4.22	0.84	0.49	0.17	0.06
Ser	4.21	4.15	1.06	0.88	0.23	0.24
Pro	5.39	6.86	1.06	1.89	0.16	0.28
Cit			1.12	1.12		
Thr	4.65	4.51	1.28	1.18	0.33	0.29
Val	5.54	6.01	1.69	1.86	0.24	0.31
Lys	14.4	16.3	2.95	3.06	0.31	0.31
Nval			1.76	1.76		
Tyr	5.62	5.44	2.26	2.01	0.49	0.45
Met	5.70	5.94	1.94	2.04	0.38	0.41
Arg	18.46	18.70	3.79	4.02	0.46	0.46
Ile	5.34	6.11	1.76	2.11	0.29	0.38
Leu	5.61	5.92	1.77	1.94	0.31	0.36
Nleu	7.57	7.57	1.99	1.99		
Eth	7.69	7.69	2.32	2.32		
Phe	6.50	6.55	2.43	2.60	0.55	0.60
Trp	10.67	10.32	5.12	4.94	1.30	1.27

Packing: P1 = [Cu(II)3-(L-prolyl)]propyl$^+$ on LiChrosorb® Si 100, particle diameter 10 μm

Solvent: S1 = 10^{-4} *M* Cu(II) acetate, 0.001 *M* ammonium acetate, pH 5.0

S2 = 10^{-4} *M* Cu(II) acetate, 0.01 *M* ammonium acetate, pH 5.0

S3 = 10^{-4} *M* Cu(II) acetate, 0.1 *M* ammonium acetate, pH 5.0

The pH values were adjusted with acetic acid

Detection: D1 = UV at 254 nm

REFERENCE

1. **Roumeliotis, P., Unger, K. K., Kurganov, A. A., and Davankov, V. A.,** *J. Chromatogr.*, 255, 51, 1983.

Table LC 45
THE EFFECT OF AMMONIUM ACETATE CONCENTRATION ON THE RETENTION OF AMINO ACID ENANTIOMERS USING PACKING CONTAINING [Cu(II)3-(L-HYDROXYPROLYL)PROPYL]$^+$

Packing	P1	P1	P1	P1
Column				
Length (mm)	250	250	250	250
Diameter (mm, I.D.)	4	4	4	4
Solvent	S1	S2	S3	S4
Temperature (K)	323	323	323	323
Detection	D1	D1	D1	D1

Amino acid	k'_L	k'_D	k'_L	k'_D	k'_L	k'_D	k'_L	k'_D
Asp	0.93	0.76	0.60	0.50	0.36	0.33	0.47	0.47
Glu	0.96	1.07	0.67	0.74	0.19	0.22	0.28	0.33
His	5.65	4.32	4.25	3.47	3.43	5.39	2.08	1.18
Ala	1.94	2.08	1.63	1.74	0.58	0.64	0.36	0.39
Asn	2.50	2.15	2.18	1.76	1.49	1.32	0.83	0.74
Gln	2.01	1.69	1.67	0.71	0.89	0.24	0.54	0.35
Ser	2.01	2.00	1.61	1.21	1.07	0.81	0.60	0.47
Pro	2.51	3.97	1.82	3.14	0.76	1.42	0.51	0.96
Cit	3.14	3.14	1.83	1.83	1.07	1.07	0.61	0.61
Thr	2.58	2.32	1.99	1.87	1.51	1.33	0.83	0.72
Val	2.74	3.08	2.39	2.44	1.25	1.28	0.77	0.82
Dopa	6.53	6.53	4.00	4.00	2.08	2.08	1.57	1.57
Lys	8.18	9.31	6.61	6.75	3.05	3.12	0.90	0.96
Nval	4.62	4.62	2.60	2.60	1.17	1.17	0.68	0.68
Tyr	2.96	2.14	2.25	1.63	1.68	1.12	1.03	0.69
Met	3.01	3.11	2.49	2.54	1.55	1.49	0.90	0.89
Arg	10.61	11.03	8.07	8.69	3.87	4.12	1.15	1.24
Ile	3.11	3.56	2.50	2.74	1.37	1.42	0.89	0.93
Leu	2.83	3.21	2.36	2.57	1.21	1.40	0.76	0.88
Nleu	4.72	4.72	2.74	2.74	1.36	1.36	0.80	0.80
Eth	5.05	5.05	2.82	2.82	2.00	2.00	1.12	1.12
Phe	3.82	3.44	2.87	2.63	2.08	1.69	1.28	1.08
Trp	6.14	4.33	4.29	3.07	3.57	2.28	2.18	1.42

Packing: P1 = [Cu(II)3-(L-hydroxyprolyl)propyl]$^+$ on LiChrosorb® Si 100, particle diameter 10 μm

Solvent: S1 = 10^{-4} *M* Cu(II) acetate, pH 5.0

S2 = 10^{-4} *M* Cu(II) acetate, 0.001 *M* ammonium acetate, pH 5.0

S3 = 10^{-4} *M* Cu(II) acetate, 0.01 *M* ammonium acetate, pH 5.0

S4 = 10^{-4} *M* Cu(II) acetate, 0.1 *M* ammonium acetate, pH 5.0

The pH values were adjusted with acetic acid

Detection: D1 = UV at 254 nm

REFERENCE

1. **Roumeliotis, P., Unger, K. K., Kurganov, A. A., and Davankov, V. A.,** *J. Chromatogr.,* 255, 51, 1983.

Table LC 46
THE EFFECT OF METHANOL CONCENTRATION ON THE RETENTION OF AMINO ACID ENANTIOMERS USING PACKING CONTAINING [Cu(II)3-(L-PROLYL)PROPYL]$^+$

Packing	P1	P1	P1
Column			
Length (mm)	250	250	250
Diameter (mm, I.D.)	4	4	4
Solvent	S1	S2	S3
Temperature (K)	323	323	323
Detection	D1	D1	D1

Amino acid	k'_L	k'_D	k'_L	k'_D	k'_L	k'_D
Asp	0.58	0.58	1.05	1.05	2.12	2.13
Glu	0.49	0.56	1.00	1.10	2.00	2.16
His	3.13	2.55	3.81	3.10	6.38	5.10
Ala	0.83	0.90	1.23	1.32	2.61	2.82
Asn	1.08	1.08	2.08	1.96	3.30	3.17
Gln	0.84	0.49			2.64	1.91
Ser	1.06	0.88	1.95	1.96	3.09	2.64
Pro	1.06	1.89	1.90	2.70	2.96	4.18
Cit	1.12	1.12			3.19	3.19
Thr	1.28	1.18	2.06	1.96	3.21	2.87
Val	1.69	1.86	2.05	2.36	2.91	3.47
Lys	2.95	3.06	4.14	4.20	7.88	7.75
Nval	1.76	1.76	2.22	2.22	3.02	3.02
Tyr	2.26	2.01	2.96	2.55	4.69	3.74
Met	1.94	2.04	2.36	2.43	3.57	3.57
Arg	3.79	4.02	5.03	5.33	8.95	9.79
Ile	1.76	2.11	2.14	2.52	2.93	3.42
Leu	1.77	1.94	2.13	2.31	2.97	3.21
Nleu	1.99	1.99	2.33	2.33	3.01	3.01
Eth	2.32	2.32	2.60	2.60	3.66	3.66
Phe	2.43	2.60	2.86	2.95	4.02	3.93
Trp	5.12	4.94	5.78	5.32	8.00	6.68

Packing: P1 = [Cu(II)3-(L-hydroxyprolyl)propyl]$^+$ 3-(iodopropyl) groups on LiChrosorb® Si 100, particle diameter 10 μm

Solvent: S1 = 10^{-4} *M* Cu(II) acetate, 0.01 *M* ammonium acetate adjusted to pH 5.0

S2 = 10^{-4} *M* Cu(II) acetate, 0.01 *M* ammonium acetate, adjusted to pH 5.0, 10% methanol

S3 = 10^{-4} *M* Cu(II) acetate, 0.01 *M* ammonium acetate, adjusted to pH 5.0, 30% methanol

Detection: D1 = UV at 254 nm

REFERENCE

1. **Roumeliotis, P., Unger, K. K., Kurganov, A. A., and Davankov, V. A.,** *J. Chromatogr.*, 255, 51, 1983.

Table LC 47
THE EFFECT OF METHANOL CONCENTRATION ON THE RETENTION OF AMINO ACID ENANTIOMERS USING PACKING CONTAINING [Cu(II)3-(L-HYDROXYPROPYL)PROPYL]$^+$

Packing	P1	P1	P1
Column			
Length(mm)	250	250	250
Diameter (mm, I.D.)	4	4	4
Solvent	S1	S2	S3
Temperature (K)	323	323	323
Detection	D1	D1	D1

Amino acid	**k'_L**	**k'_D**	**k'_L**	**k'_D**	**k'_L**	**k'_D**
Asp	0.36	0.33	0.79	0.71	0.96	0.83
Glu	0.19	0.22	0.75	0.92	0.93	1.08
His	3.44	5.39	3.75	2.56	3.82	2.61
Ala	0.58	0.64	1.35	1.48	1.47	1.67
Asn	1.49	1.32	1.74	1.57	1.90	1.69
Gln	0.89	0.24	1.49	0.89	1.65	1.12
Ser	1.07	0.81	1.58	1.29	1.75	1.39
Pro	0.76	1.42	1.55	2.96	1.65	3.42
Cit	1.07	1.07	1.85	1.85	2.20	2.20
Thr	1.51	1.33	1.75	1.53	1.86	1.62
Val	1.25	1.28	1.83	2.05	1.90	2.22
Dopa	2.08	2.08	3.11	3.11	4.02	4.02
Lys	3.05	3.12	5.17	5.53	6.44	6.99
Nval	1.17	1.17	1.77	1.77	1.94	1.94
Tyr	1.68	1.12	2.30	1.57	2.37	1.57
Met	1.55	1.49	1.86	1.83	1.94	1.93
Arg	3.87	4.12	5.79	6.33	6.74	7.36
Ile	1.37	1.42	1.86	2.05	1.87	2.15
Leu	1.21	1.40	1.62	1.94	1.67	2.03
Nleu	1.36	1.36	1.85	1.85	1.96	1.96
Eth	2.00	2.00	2.08	2.08		
Phe	2.08	1.69	2.27	1.96	2.27	1.90
Tryp	3.57	2.28	3.87	2.44	3.87	2.33

Packing: P1 = [Cu(II)3-(hydroxyprolyl)propyl]$^+$ on LiChrosorb® Si 100, particle diameter 10 μm

Solvent: S1 = 10^{-4} *M* Cu(II) acetate, 0.01 *M* ammonium acetate adjusted to pH 5.0

S2 = 10^{-4} *M* Cu(II) acetate, 0.01 *M* ammonium acetate adjusted to pH 5.0, 10% methanol

S3 = 10^{-4} *M* Cu(II) acetate, 0.01 *M* ammonium acetate adjusted to pH 5.0, 30% methanol

Detection: D1 = UV at 254 nm

REFERENCE

1. **Roumeliotis, P., Unger, K. K., Kurganov, A. A., and Davankov, V. A.,** *J. Chromatogr.*, 255, 51, 1982.

Table LC 48
THE EFFECT OF METHANOL CONCENTRATION ON THE RETENTION OF AMINO ACID ENANTIOMERS USING PACKING CONTAINING [Cu(II)3-(L-HYDROXYPROLYL) PROPYL]$^+$ AND 3-IODOPROPYL) GROUPS

Packing	P1	P1	P1
Column			
Length (mm)	250	250	250
Diameter (mm, I.D.)	4	4	4
Solvent	S1	S2	S3
Temperature (K)	323	323	323
Detection	D1	D1	D1

Amino acid	k'_L	k'_D	k'_L	k'_D	k'_L	k'_D
Asp	1.45	1.32	2.13	1.96	6.16	5.59
Glu	1.12	1.18	1.64	1.70	4.86	5.82
His	1.52	2.71	1.78	3.03	3.81	6.03
Ala	0.69	0.88	0.96	1.21	2.59	3.34
Asn	1.54	1.42	2.03	1.88	4.54	4.14
Ser	1.25	0.93	1.74	1.30	4.14	3.21
Pro	1.12	2.00	1.27	2.36	3.06	6.50
Thr	1.73	1.40	2.21	1.79	4.75	3.86
Val	1.89	1.54	2.10	1.71	4.41	3.83
Lys	0.93	1.05	1.19	1.34	2.96	3.00
Tyr	3.95	2.73	4.42	2.74	7.60	5.14
Met	2.70	2.63	2.87	2.75	5.49	5.52
Arg	1.59	1.73	1.82	2.01	3.72	4.15
Ile	2.79	2.48	2.90	2.56	5.48	5.16
Leu	2.25	2.22	2.32	2.35	4.64	4.94
Phe	5.73	3.90	5.77	3.90	9.63	6.49
Trp	11.37	6.49	10.50	5.95	14.88	8.85

Packing: P1 = [Cu(II)3-(L-prolyl) propyl $^+$ on LiChrosorb® Si 100, particle diameter, 10 μm

Solvent: S1 = 10^{-4} *M* Cu(II) acetate, 0.01 *M* ammonium acetate, adjusted to pH 5.0

S2 = 10^{-4} *M* Cu(II) acetate, 0.01 *M* ammonium acetate, adjusted to pH 5.0, 10% methanol

S3 = 10^{-4} *M* Cu(II) acetate, 0.01 *M* ammonium acetate, adjusted to pH 5.0, 30% methanol

Detection: D1 = UV at 254 nm

REFERENCE

1. **Roumeliotis, P., Unger, K. K., Kurganov, A. A., and Davankov, V. A.,** *J. Chromatogr.*, 255, 51, 1983.

Table LC 49
THE EFFECT OF ORGANIC MODIFIERS ON THE RETENTION OF AMINO ACID ENANTIOMERS USING PACKING CONTAINING [Cu(II)3-(L-PROLYL)PROPYL]$^+$

Packing	P1	P1	P1
Column			
Length (mm)	250	250	250
Diameter (mm, I.D.)	4	4	4
Solvent	S1	S2	S3
Temperature (K)	323	323	323
Detection	D1	D1	D1

Amino acid	k'_L	k'_D	k'_L	k'_D	k'_L	k'_D
Asp	1.05	1.05	0.95	0.95	0.86	0.86
Glu	1.00	1.10	0.80	0.95	0.72	0.83
His	3.81	3.10	4.08	3.35	3.62	2.88
Ala	1.23	1.32	1.35	1.60	1.17	1.30
Asn	2.08	1.96	2.07	2.00	1.96	1.87
Gln			1.35	0.83	1.15	0.74
Ser	1.95	1.96	2.00	1.95	1.84	1.79
Pro	1.90	2.70	1.77	2.49	1.34	2.21
Cit			1.97	1.97	1.74	1.74
Thr	2.06	1.96	2.12	2.00	1.94	1.80
Val	2.05	2.36	2.03	2.34	1.78	2.00
Dopa			1.31	1.31		
Lys	4.14	4.20	4.65	4.61	3.80	3.74
Nval	2.22	2.22	2.16	2.16	1.90	1.90
Tyr	2.06	2.55	2.51	2.51	2.86	2.38
Met	2.36	2.40	2.39	2.44	2.14	2.17
Arg	5.03	5.33	5.65	5.64	4.70	4.96
Ile	2.14	2.52	2.09	2.44	1.86	2.10
Leu	2.13	2.31	2.13	2.28	1.83	1.96
Nleu	2.33	2.33	2.24	2.24	1.95	1.95
Eth	2.60	2.60	2.56	2.56	2.26	2.26
Phe	2.86	2.95	2.77	2.83	2.48	2.53
Trp	5.78	5.32	5.31	4.74	5.17	4.57

Packing: P1 = [Cu(II)3-(L-prolyl)propyl]$^+$ on LiChrosorb® Si 100, particle diameter 10 μm

Solvent: S1 = 10^{-4} *M* Cu(II) acetate, 0.01 *M* ammonium acetate, pH 5.0/methanol, 90:10

S2 = 10^{-4} *M* Cu(II) acetate, 0.01 *M* ammonium acetate, pH 5.0/acetonitrile, 90:10

S3 = 10^{-4} *M* Cu(II) acetate, 0.01 *M* ammonium acetate, pH 5.0/tetrahydrofuran, 90:10

Detection: D1 = UV at 254 nm

REFERENCE

1. **Roumeliotis, P., Unger, K. K., Kurganov, A. A., and Davankov, V. A.,** *J. Chromatogr.*, 255, 51, 1983.

Table LC 50
THE EFFECT OF ORGANIC MODIFIERS ON THE RETENTION OF AMINO ACID ENANTIOMERS USING PACKING CONTAINING [Cu(II)3-(L-HYDROXYPROLYL)PROPYL]$^+$

Packing	P1		P1		P1	
Column						
Length (mm)	250		250		250	
Diameter (mm, I.D.)	4		4		4	
Solvent	S1		S2		S3	
Temperature (K)	323		323		323	
Detection	D1		D1		D1	
Amino acid	k'_L	k'_D	k'_L	k'_D	k'_L	k'_D
His	3.82	2.13	4.18	2.55	3.69	2.04
Ala	1.35	1.61	1.74	2.04	1.41	1.71
Ser	1.46	1.21	1.86	1.59	1.54	1.28
Pro	1.63	4.24	1.91	4.68	1.52	3.91
Thr	1.65	1.39	1.97	1.71		
Val	2.18	2.53	2.50	2.78		
Nval	1.83	2.21	2.04	2.33	1.64	1.93
Tyr	2.86	1.88	3.20	2.07	3.37	1.92
Leu	2.27	2.71	2.48	2.87	1.83	2.14
Nleu	2.30	2.69	2.37	2.72	1.81	2.10
Phe	3.40	3.03	3.40	2.99	3.40	2.44
Trp	6.81	4.24	6.40	3.85	5.80	3.35

Packing: P1 = [Cu(II)3-(L-hydroxyprolyl)propyl]$^+$ on LiChrosorb® Si 100, particle diameter 10 μm

Solvent: S1 = 10^{-4} *M* Cu(II) acetate, 0.01 *M* ammonium acetate, pH 5.0/methanol, 90:10

S2 = 10^{-4} *M* Cu(II) acetate, 0.01 *M* ammonium acetate, pH 5.0/acetonitrile, 90:10

S3 = 10^{-4} *M* Cu(II) acetate, 0.01 *M* ammonium acetate, pH 5.0/tetrahydrofuran, 90:10

Detection: D1 = UV at 254 nm

REFERENCE

1. **Roumeliotis, P., Unger, K. K., Kurganov, A. A., and Davankov, V. A.,** *J. Chromatogr.,* 255, 51, 1983.

Table LC 51
THE EFFECT OF TEMPERATURE ON THE RETENTION OF AMINO ACID ENANTIOMERS

Packing	P1	P1
Column		
Length (mm)	250	250
Diameter (mm, I.D.)	4	4
Solvent	S1	S1
Temperature (K)	298	323
Detection	D1	D1

Amino acid	**k'_L**	**k'_D**	**k'_L**	**k'_D**
Asp	1.21	1.12	0.36	0.33
Glu	0.87	1.10	0.19	0.22
His	3.43	2.29	3.44	5.39
Ala	0.74	0.85	0.58	0.64
Asn	1.38	1.30	1.49	1.32
Gln	1.11	1.11	0.89	0.24
Ser	1.13	0.96	1.07	0.81
Pro	1.17	2.21	0.76	1.42
Cit			1.07	1.07
Thr	1.44	1.28	1.51	1.33
Val	1.50	1.62	1.25	1.28
Dopa			2.08	2.08
Lys	2.53	2.68	3.05	3.12
Nval			1.17	1.17
Tyr	1.99	1.37	1.68	1.12
Met	1.96	1.97	1.55	1.49
Arg	3.44	3.71	3.87	4.12
Ile	1.71	1.87	1.37	1.42
Leu	1.65	1.85	1.21	1.40
Nleu			1.36	1.36
Eth			2.00	2.00
Phe	2.39	2.13	2.08	1.69
Trp	4.62	3.30	3.57	2.28

Packing: P1 = [Cu(II)3-(L-hydroxyprolyl)propyl]$^+$ on LiChrosorb® Si 100, particle diameter 10 μm

Solvent: S1 = 10^{-4} *M* Cu(II) acetate, 0.01 *M* ammonium acetate, pH 5.0

Detection: D1 = UV at 254 nm

REFERENCE

1. **Roumeliotis, P., Unger, K. K., Kurganov, A. A., and Davankov, V. A.,** *J. Chromatogr.*, 255, 51, 1983.

Table LC 52
AMINO ACID ENANTIOMERS ON THE NICKEL FORM OF A POLYSTYRENE SORBENT CONTAINING METHIONINE SULFOXIDE

Packing	P1		P2	
Column				
Length (cm)	20		20	
Diameter (cm, I.D.)	0.4		0.4	
Material	SS		SS	
Solvent	S1		S1	
Flow rate (mℓ/hr)	7		7	
Detection	D1		D1	
Amino acid	k'_D	k'_L	k'_D	k'_L
Alanine	1.12	1.12	1.71	1.71
Aspartic acid	0.35	0.43	0.47	0.50
	2.95[a]	4.54[a]	5.80[a]	7.35[a]
Histidine	18.0	40.0	35.0	43.0
Leucine	5.53	5.00	6.86	6.40
Phenylalanine	30.0	24.7	57.6	48.4
Proline	1.57	1.14	2.33	2.17
Threonine	2.67	2.33	4.57	3.81
Tyrosine	27.3	22.7	28.7	24.0
Valine	2.14	2.14	2.95	2.86

[a] Water as eluent.

Packing: P1 = sorbent containing L-methionine-*d*-sulfoxide groups in a macronet polystyrene matrix

P2 = a similar sorbent containing L-methionine-1-sulfoxide groups The polymers were loaded with Ni(II) ions using a 0.05 *M* solution of nickel chloride in 0.5 *M* NH_4OH; the columns contained about 2 g of dry sorbent, particle size 50 μm, with a swelling capacity in water of approximately 100%

Solvent: S1 = 0.1 *M* $(NH_4)_2CO_3$, pH 9.5 containing 2×10^{-6} *M* nickel ions

Detection: D1 = UV at 206 nm

REFERENCE

1. **Berezin, B. B., Yamskov, I. A., and Davankov, V. A.,** *J. Chromatogr.*, 261, 301, 1983.

Table LC 53
AMINO ACID ENANTIOMERS ON THE COPPER FORM OF A POLYSTYRENE SORBENT CONTAINING METHIONINE SULFOXIDE

Packing	P1	P2
Column		
Length (cm)	20	20
Diameter (cm, I.D.)	0.4	0.4
Material	SS	SS
Solvent	S1	S1
Flow rate (mℓ/hr)	7	7
Detection	D1	D1

Amino acid	k'_D	k'_L	k'_D	k'_L
Alanine	2.92	2.73	2.73	2.29
Aspartic acid	7.00[a]	5.50[a]	7.50[a]	6.00[a]
Histidine	10.0[b]	15.5[b]	11.0[b]	20.0[b]
Leucine	25.1	19.3	24.2	20.0
Methionine	12.7	11.6	18.3	14.3
Phenylalanine	40.0	27.1	37.6	26.8
Proline	5.66[c]	4.63[c]	13.0	7.83
	10.8[d]	8.00[d]	21.4[e]	12.0[e]
	15.6	10.2		
	24.4[e]	14.2[e]		
Threonine	1.02	0.82	0.85	0.67
Tyrosine	3.64	2.54	5.36	4.40
Valine	9.09	7.27	7.22	5.60

[a] Water as eluent.
[b] 1.0 *M* NH_4OH + 5 × 10^{-5} *M* copper ions.
[c] 0.3 *M* NH_4OH + 2 × 10^{-5} *M* copper ions.
[d] 0.2 *M* NN_4OH + 2 × 10^{-5} *M* copper ions.
[e] 0.05 *M* NH_4OH + 2 × 10^{-5} *M* copper ions.

Packing: P1 = sorbent containing L-methionine-*d*-sulfoxide groups in a macronet polystyrene matrix

P2 = a similar sorbent containing L-methionine-1-sulfoxide groups; the polymers were loaded with Cu(II) ions during a 0.1 *M* solution of Cu(II) chloride in 0.5 *M* NH_4OH; the columns contained about 2 g of dry sorbent, particle size 50 μm, with a swelling capacity in water of approximately 100%

Solvent: S1 = 0.1 *M* NH_4OH containing 2 × 10^{-5} *M* copper ions

Detection: D1 = UV at 206 nm

REFERENCE

1. **Berezin, B. B., Yamskov, I. A., and Davankov, V. A.,** *J. Chromatogr.*, 261, 301, 1983.

Table LC 54
THE EFFECT OF pH ON THE RETENTION OF AMINO ACID ENANTIOMERS USING AS PACKING [Cu(II) (L-HYDROXYPROLYL)METHYL]$^+$ ON LICHROSORB Si 60

Packing	P1	P1	P1
Column			
Length (mm)	100	100	100
Diameter (mm)	4.6	4.6	4.6
Solvent	S1	S2	S3
Temperature (K)	323	323	323
Detection	D1	D1	D1

Amino acid	**k'_L**	**k'_D**	**k'_L**	**k'_D**	**k'_L**	**k'_D**
Asp	0.44	0.44	1.10	1.03	0.76	0.71
Glu	0.39	0.42	1.02	1.13	0.77	0.82
His	1.62	1.08	3.80	2.33	4.62	3.08
Ala	0.47	0.51	1.40	1.47	1.79	1.86
Asn	0.81	0.77	2.05	1.76	1.71	1.71
Ser	0.57	0.55	1.72	1.46	1.66	1.54
Pro	0.79	0.90	1.93	2.51	3.42	4.64
Thr	0.79	0.85	2.14	1.94	2.69	2.48
Val	0.80	0.80	2.10	2.12	3.30	3.47
Lys	0.99	0.92	2.77	2.61	6.57	6.23
Tyr	0.75	0.62	2.30	1.52	2.42	1.59
Met	0.91	0.97	2.87	2.82	3.32	3.38
Arg	1.21	1.21	3.40	3.44	7.90	8.00
Ile	0.88	0.85	2.45	2.51	3.67	3.87
Leu	0.97	0.89	2.83	3.02	3.77	3.75
Phe	1.10	1.02	3.35	2.87	3.43	3.28
Trp	1.58	1.10	4.69	2.69	4.33	3.32

Packing: P1 = [Cu(L-hydroxyprolyl)methyl]$^+$ on LiChrosorb® Si 60, particle diameter 10 μm

Solvent: S1 = 10^{-4} *M* copper acetate + 10^{-2} *M* ammonium acetate, pH 4.5

S2 = 10^{-4} *M* copper acetate + 10^{-2} *M* ammonium acetate, pH 5.0

S3 = 10^{-4} *M* copper acetate + 10^{-2} *M* ammonium acetate, pH 5.5

The pH values were adjusted with acetic acid

Detection: D1 = UV at 254 nm

REFERENCE

1. **Roumeliotis, P., Kurganov, A. A., and Davankov, V. A.,** *J. Chromatogr.*, 266, 439, 1983.

Table LC 55
THE EFFECT OF pH ON THE RETENTION OF AMINO ACID ENANTIOMERS USING AS PACKING [Cu(II) (L-HYDROXYPROLYL)METHYL]$^+$ ON LICHROSORB Si 100

Packing	P1	P1	P1
Column			
Length (mm)	100	100	100
Diameter (mm)	4.6	4.6	4.6
Solvent	S1	S2	S3
Temperature (K)	323	323	323
Detection	D1	D1	D1

Amino acid	k'_L	k'_D	k'_L	k'_D	k'_L	k'_D
Asp	0.13	0.14	0.05	0.05	0.20	0.20
Glu	0.10	0.25	0.05	0.13	0.20	0.20
His	1.70	1.27	1.97	1.78	3.10	2.87
Ala	0.48	0.58	0.50	0.58	1.13	1.13
Asn	0.52	0.80	0.41	0.57	1.08	1.15
Ser	0.65	0.65	0.45	0.45	1.08	1.08
Pro	0.63	0.90	1.05	1.47	2.28	2.35
Thr	0.73	1.00	0.57	0.65	1.22	1.30
Val	0.80	0.90	1.42	1.18	2.65	1.58
Lys	1.42	1.48	2.82	2.72		
Tyr	1.37	0.95	1.01	0.80	1.55	1.33
Met	1.23	1.40	1.53	1.53	2.21	2.56
Arg	2.05	1.78	3.53	3.37		
Ile	1.23	1.23			3.15	3.15
Leu	1.30	1.03	1.67	1.45	2.60	2.43
Phe	1.32	1.32	1.58	1.78	2.00	2.35
Trp	2.42	1.58			2.22	1.95

Packing: P1 = [Cu(L-hydroxyprolyl)methyl]$^+$ on LiChrosorb® Si 100, particle diameter 10 μm

Solvent: S1 = 10^{-4} *M* copper acetate + 10^{-2} *M* ammonium acetate, pH 5.0

S2 = 10^{-4} *M* copper acetate + 10^{-2} *M* ammonium acetate, pH 5.5

S3 = 10^{-4} *M* copper acetate + 10^{-2} *M* ammonium acetate, pH 6.0

The pH values were adjusted with acetic acid

Detection: D1 = UV at 254 nm

REFERENCE

1. **Roumeliotis, P., Kurganov, A. A., and Davankov, V. A.,** *J. Chromatogr.*, 266, 439, 1983.

Table LC 56
THE EFFECT OF pH ON THE RETENTION OF AMINO ACID ENANTIOMERS USING AS PACKING [Cu(II) (L-HYDROXYPROLYL) *n*- OCTYL]$^+$ ON LICHROSORB Si 100

Packing	P1	P1	P1
Column			
Length (mm)	100	100	100
Diameter (mm)	4.6	4.6	4.6
Solvent	S1	S2	S3
Temperature (k)	323	323	323
Detection	D1	D1	D1

Amino acid	k'_L	k'_D	k'_L	k'_D	k'_L	k'_D
Asp	1.48	1.55	2.01	2.13	2.92	3.18
Glu	1.02	1.02	1.57	1.57	2.78	2.78
His	1.46	1.22	2.22	1.92	3.75	1.82
Ala	0.17	0.24	0.40	0.58	0.85	1.12
Asn	0.55	0.50	0.89	0.82	1.83	1.68
Ser	0.46	0.42	0.82	0.74	1.92	1.78
Pro	0.56	1.10	1.07	2.06	2.85	5.25
Thr	0.71	0.71	1.14	1.16	2.18	2.17
Val	1.61	1.53	2.81	2.71	5.98	5.83
Lys	0.09	0.12	0.07	0.02	0.42	0.32
Tyr	7.39	5.94	12.25	9.72	24.30	19.20
Met	2.75	2.71	4.05	4.56	8.91	8.72
Arg	0.05	0.09	0.31	0.39	0.97	1.28
Ile	3.15	3.15	5.86	5.94	12.13	12.27
Leu	2.78	2.96	4.80	5.14	10.20	10.90
Phe	10.43	9.42	17.60	15.45	38.35	31.02
Trp	21.56	14.19				

Packing: Pl = [Cu(L-hydroxyprolyl)*n*-octyl]$^+$ on LiChrosorb® Si 100, particle diameter 10 μm.

Solvent: S1 = 10^{-4} *M* copper acetate + 10^{-2} *M* ammonium acetate, pH 4.7

S2 = 10^{-4} *M* copper acetate + 10^{-2} *M* ammonium acetate, pH 5.0

S3 = 10^{-4} *M* copper acetate + 10^{-2} *M* ammonium acetate, pH 5.4

The pH values were adjusted with acetic acid

Detection: D1 = UV at 254 nm

REFERENCE

1. **Roumeliotis, P., Kurganov, A. A., and Davankov, V. A.,** *J. Chromatogr.*, 266, 439, 1983.

Table LC 57
THE EFFECT OF METHANOL CONCENTRATION ON THE RETENTION OF AMINO ACID ENANTIOMERS USING AS PACKING [Cu(II) (L-HYDROXYPROLYL)METHYL]$^+$ ON LICHROSORB® Si60

Packing	P1	P1	P1	P1
Column				
Length (mm)	100	100	100	100
Diameter (mm)	4.6	4.6	4.6	4.6
Solvent	S1	S2	S3	S4
Temperature (k)	323	323	323	323
Detection	D1	D1	D1	D1

Amino acid	k'_L	k'_D	k'_L	k'_D	k'_L	k'_D	k'_L	k'_D
Asp	0.44	0.44	1.00	1.00	1.13	1.13	2.64	2.52
Glu	0.39	0.42	0.88	0.96	0.74	0.84	2.04	2.33
His	1.62	1.08	3.42	1.85	3.58	1.71	4.42	2.02
Ala	0.47	0.51	1.18	1.26	0.79	0.87	2.10	2.34
Asn	0.81	0.77	1.94	1.80	1.74	1.62	4.03	3.40
Ser	0.57	0.55	1.52	1.41	1.17	1.11	3.00	2.65
Pro	0.79	0.90	1.46	2.07	0.97	1.22	2.33	3.92
Thr	0.79	0.85	1.91	1.97	1.53	1.68	3.41	3.51
Val	0.80	0.80	1.66	1.68	1.04	1.05	2.49	2.52
Lys	0.99	0.92	2.59	2.35	1.47	1.32	4.90	4.15
Tyr	0.75	0.62	1.88	1.44	1.30	1.06	3.13	2.43
Met	0.91	0.97	1.87	2.03	1.18	1.39	2.66	3.00
Arg	1.21	1.21	2.97	3.02	1.54	1.58	4.48	4.64
Ile	0.88	0.85	1.66	1.63	0.98	0.94	2.20	2.13
Leu	0.97	0.89	1.89	1.74	1.12	1.01	2.56	2.30
Phe	1.10	1.02	2.30	1.97	1.41	1.26	3.04	2.50
Trp	1.58	1.10	3.77	2.40	2.42	1.57	5.27	3.47

Packing: P1 = [Cu(-hydroxyprolyl)methyl]$^+$ on LiChrosorb® Si 60, particle diameter 10 μm

Solvent: S1 = 10^{-4} *M* copper acetate + 10^{-2} *M* ammonium acetate, pH 5.0

S2 = methanol/10^{-4} *M* copper acteate + 10^{-2} *M* ammonium acetate, pH 5.0, 10:90

S3 = methanol/10^{-4}*M* copper acetate + 10^{-2} *M* ammonium acetate, pH 5.0, 30:70

S4 = methanol/10^{-4} *M* copper acetate + 10^{-2} *M* ammonium acetate, pH 5.0, 50:50

Detection: D1 = UV at 254 mm

REFERENCE

1. **Roumeliotis, P., Kurganov, A. A., and Davankov, V. A.,** *J. Chromatogr.*, 266, 439, 1983.

Table LC 58
THE EFFECT OF METHANOL CONCENTRATION ON THE RETENTION OF AMINO ACID ENANTIOMERS USING AS PACKING [Cu(II) (L-HYDROXYPROLYL)METHYL]$^+$ ON LICHROSORB® Si 100

Packing	P1		P1		P1		P1	
Column								
Length (mm)	100		100		100		100	
Diameter (mm)	4.6		4.6		4.6		4.6	
Solvent	S1		S2		S3		S4	
Temperature (K)	323		323		323		323	
Detection	D1		D1		D1		D1	
Amino acid	k'_L	k'_D	k'_L	k'_D	k'_L	k'_D	k'_L	k'_D
Asp	0.13	0.14	0.10	0.10	0.25	0.30	0.55	0.62
Glu	0.10	0.25	0.10	0.10	0.20	0.52	0.58	0.82
His	1.70	1.27	1.26	0.95	1.60	1.07	1.70	1.13
Ala	0.48	0.58	0.32	0.42	0.53	0.63	0.80	1.00
Asn	0.52	0.80	0.42	0.60	0.60	1.05	0.93	1.33
Ser	0.65	0.65	0.45	0.45	0.77	0.67	0.97	1.00
Pro	0.63	0.90	0.42	0.62	0.47	0.75	0.85	1.78
Thr	0.73	1.00	0.57	0.75	0.72	1.08	1.18	1.62
Val	0.80	0.90	0.52	0.48	0.75	0.93	1.05	1.00
Lys	1.42	1.48	1.08	1.13	2.30	2.47	2.60	2.72
Tyr	1.37	0.95	0.92	0.73	1.68	1.38	1.80	1.57
Met	1.23	1.40	0.77	0.77	1.13	1.45	1.37	1.47
Arg	2.05	1.78	1.35	1.35	3.58	2.78	2.68	2.67
Ile	1.23	1.23	0.67	0.70	1.17	1.30	1.23	1.35
Leu	1.30	1.03	0.70	0.57	1.15	0.87	1.30	1.07
Phe	1.32	1.32	0.77	0.70	1.10	1.23	1.00	1.12
Trp	2.42	1.58	1.33	0.97	2.32	1.65	2.00	1.55

Packing: P1 = [Cu(L-hydroxyprolyl)methyl]$^+$ on LiChrosorb® Si 100, particle diameter 10 μm

Solvent: S1 = 10^{-4} *M* copper acetate + 10^{-2} *M* ammonium acetate, pH 5.0

S2 = methanol/10^{-4} *M* copper acetate + 10^{-2} *M* ammonium acetate, pH 5.0, 10:90

S3 = methanol/10^{-4} *M* copper acetate + 10^{-2} *M* ammonium acetate, pH 5.0, 30:70

S4 = methanol/10^{-4} *M* copper acetate + 10^{-2} *M* ammonium acetate, pH 5.0, 50:50

Detection: D1 = UV at 254 nm

REFERENCE

1. **Roumeliotis, P., Kurganov, A. A., and Davankov, V. A.,** *J. Chromatogr.*, 266, 439, 1983.

Table LC 59
THE EFFECT OF METHANOL CONCENTRATION ON THE RETENTION OF AMINO ACID ENANTIOMERS USING AS PACKING [Cu(II) (L-HYDROXY-PROLYL) *n*-OCTYL]$^+$ ON LICHROSORB® Si 100

Packing	P1	P1	P1	P1	P1
Column					
Length (mm)	100	100	100	100	100
Diameter (mm)	4.6	4.6	4.6	4.6	4.6
Solvent	S1	S2	S3	S4	S5
Temperature (K)	323	323	323	323	323
Detection	D1	D1	D1	D1	D1

Amino acid	k'_L	k'_D	k'_L	k'_D	k'_L	k'_D	k'_L	k'_D	k'_L	k'_D
Asp	2.01	2.13	3.05	3.27	6.32	6.45	10.70	10.58	21.82	21.22
Glu	1.57	1.57	2.43	2.30	4.62	4.65	6.42	6.47	8.72	8.34
His	2.22	1.92	2.35	3.10	2.43	5.72	2.77	7.85	5.60	10.82
Ala	0.40	0.58	0.73	0.83			2.00	1.68	2.94	3.50
Asn	0.89	0.82	1.12	1.05	1.55	1.38	2.12	1.82	4.76	3.94
Ser	0.82	0.74	1.23	1.07	1.90	1.60	2.55	2.38	5.44	4.96
Pro	1.07	2.06	1.35	2.68	1.82	3.68	1.97	4.18	3.20	6.80
Thr	1.14	1.16	1.37	1.42	1.72	1.82	2.92	2.62	5.42	4.96
Val	2.81	2.71	3.05	2.93	3.31	3.27	3.38	3.28	3.60	3.60
Lys	0.07	0.02	0.12	0.05	0.37	0.25	0.47	0.35	1.48	1.24
Tyr	12.25	9.72	12.15	9.55	9.87	7.58	7.48	5.72	8.32	6.64
Met	4.65	4.56	4.92	4.63	4.92	4.87	4.45	4.37	5.62	5.52
Arg	0.31	0.39	0.30	0.37	0.53	0.67	0.55	0.68	1.54	1.76
Ile	5.86	5.94	5.95	5.97			4.55	4.63	4.32	4.62
Leu	4.80	5.14	4.73	5.05	4.40	5.10	3.77	4.10	3.88	4.26
Phe	17.60	15.45	16.55	13.78	12.33	10.22	7.82	6.57	6.42	5.76
Trp					41.13	30.82	22.17	16.92	15.62	12.32

Packing:	P1	=	[Cu(L-hydroxyprolyl *n*-octyl]$^+$ on LiChrosorb® Si 100, particle diameter 10 μm
Solvent:	S1	=	10^{-4} *M* copper acetate + 10^{-2} *M* ammonium acetate, pH 5.0
	S2	=	methanol/10^{-4} *M* copper acetate + 10^{-2} *M* ammonium acetate, pH 5.0, 10:90
	S3	=	methanol/10^{-4} *M* copper acetate + 10^{-2} *M* ammonium acetate, pH 5.0, 30:70
	S4	=	methanol/10^{-4} *M* copper acetate + 10^{-2} *M* ammonium acetate, pH 5.0, 50:50
	S5	=	methanol/10^{-4} *M* copper acetate + 10^{-2} *M* ammonium acetate, pH 5.0, 75:25
Detection:	D1	=	UV at 254 nm

REFERENCE

1. **Roumeliotis, P., Kurganov, A. A., and Davankov, V. A.,** *J. Chromatogr.*, 266, 439, 1983.

Table LC 60
THE EFFECT OF THE ORGANIC SOLVENT ON THE RETENTION OF AMINO ACID ENANTIOMERS USING AS PACKING [Cu (II) (L-HYDROXYPROLYL) METHYL]$^+$ ON LICHROSORB® Si 60

Packing	P1	P1	P1
Column			
Length (mm)	100	100	100
Diameter (mm)	4.6	4.6	4.6
Solvent	S1	S2	S3
Temperature (K)	323	323	323
Detection	D1	D1	D1

Amino acid	k'_L	k'_D	k'_L	k'_D	k'_L	k'_D
Asp	1.52	1.44	3.40	3.36	2.64	2.52
Glu	1.25	1.46	2.92	3.27	2.04	2.33
His	2.65	1.42	5.86	3.46	4.42	2.02
Ala	1.47	1.61	2.68	2.95	2.10	2.34
Asn	2.37	2.07	4.87	4.47	4.03	3.40
Ser	1.83	1.73	3.59	3.32	3.00	2.65
Pro	1.74	2.81	2.61	4.25	2.33	3.92
Thr	2.08	2.32	4.00	4.42	3.41	3.51
Val	1.71	1.77	3.01	3.08	2.49	2.52
Lys	2.50	2.27			4.90	4.15
Tyr	2.05	1.72	3.12	2.64	3.13	2.43
Met	1.67	2.02	2.90	3.45	2.66	3.00
Arg	2.60	2.80			4.42	4.64
Ile	1.52	1.48	2.58	2.49	2.20	2.13
Leu	1.75	1.56	2.92	2.62	2.56	2.30
Phe	1.96	1.86	2.87	2.69	3.04	2.65
Trp	3.05	2.39	4.28	3.02	5.27	3.47

Packing: P1 = [Cu(L-hydroxyprolyl)methyl]$^+$ on LiChrosorb® Si 60, particle diameter 10 μm

Solvent: S1 = tetrahydrofuran/10^{-4} *M* copper acetate + 10^{-2} *M* ammonium acetate, pH 5.0, 50:50

S2 = acetonitrile/10^{-4} *M* copper acetate + 10^{-2} *M* ammonium acetate, pH 5.0, 50:50

S3 = methanol/10^{-4} *M* copper acetate + 10^{-2} *M* ammonium acetate, pH 5.0, 50:50

Detection: D1 = UV at 254 nm

REFERENCE

1. **Roumeliotis, P., Kurganov, A. A., and Davankov, V. A.,** *J. Chromatogr.*, 266, 439, 1983.

Table LC 61
THE EFFECT OF THE ORGANIC SOLVENT ON THE RETENTION OF AMINO ACID ENANTIOMERS USING AS PACKING [Cu(II) (L-HYDROXYPROLYL)METHYL]$^+$ ON LICHROSORB® Si 100

Packing	P1	P1	P1
Column			
Length (mm)	100	100	100
Diameter (mm)	4.6	4.6	4.6
Solvent	S1	S2	S3
Temperature (K)	323	323	323
Detection	D1	D1	D1

Amino acid	k'_L	k'_D	k'_L	k'_D	k'_L	k'_D
Asp			0.47	0.55	0.55	0.62
Glu			0.55	0.75	0.58	0.82
His	0.52	0.37	1.83	1.37	1.70	1.13
Ala	0.20	0.33	0.86	1.18	0.80	1.00
Asn	0.10	0.30	0.76	1.27	0.93	1.33
Ser	0.25	0.25	0.98	0.95	0.97	1.00
Pro	0.23	0.52	0.82	1.52	0.85	1.78
Thr	0.28	0.52	1.05	1.55	1.18	1.62
Val	0.43	0.37	1.05	0.97	1.05	1.00
Lys	0.63	0.62	3.15	3.23	2.60	2.72
Tyr	1.08	0.98	1.65	1.43	1.80	1.57
Met	0.60	0.68	1.27	1.40	1.37	1.47
Arg	0.78	0.75	3.13	3.15	2.68	2.67
Ile	0.60	0.72	1.20	1.38	1.23	1.35
Leu	0.70	0.48	1.28	1.03	1.30	1.07
Phe	0.40	0.63	0.68	0.90	1.00	1.12
Trp	1.32	0.95	1.33	1.18	2.00	1.55

Packing: P1 = [Cu(L-hydroxyprolyl)methyl]$^+$ on LiChrosorb® Si 100, particle diameter 10 μm

Solvent: S1 = tetrahydrofuran/10^{-4} *M* copper acetate + 10^{-2} *M* ammonium acetate, pH 5.0, 50:50

S2 = acetonitrile/10^{-4} *M* copper acetate + 10^{-2} *M* ammonium acetate, pH 5.0, 50:50

S3 = methanol/10^{-4} *M* copper acetate + 10^{-2} *M* ammonium acetate, pH 5.0, 50:50

Detection: D1 = UV at 254 nm

REFERENCE

1. **Roumeliotis, P., Kurganov, A. A., and Davankov, V. A.,** *J. Chromatogr.*, 266, 439, 1983.

Table LC 62
THE EFFECT OF THE ORGANIC SOLVENT ON THE RETENTION OF AMINO ACID ENANTIOMERS USING AS PACKING [Cu(II) (L-HYDROXYPROLYL) *n*-OCTYL]$^+$ ON LICHROSORB® Si 100

Packing	P1		P1		P1	
Column						
Length (mm)	100		100		100	
Diameter (mm)	4.6		4.6		4.6	
Solvent	S1		S2		S3	
Temperature (K)	323		323		323	
Detection	D1		D1		D1	
Amino acid	k'_L	k'_D	k'_L	k'_D	k'_L	k'_D
Asp	4.90	5.30	6.85	7.37	10.70	10.58
Glu	2.45	2.43	5.05	5.17	6.42	5.47
His	3.23	2.98	3.22	6.28	2.77	7.85
Ala	0.87	0.87	1.67	1.95	2.00	1.68
Asn	1.80	1.53	1.85	1.50	2.12	1.82
Ser	1.87	1.87	2.47	2.27	2.55	2.38
Pro	0.93	1.78	1.92	3.83	1.97	4.18
Thr	2.15	1.92	2.58	2.38	2.92	2.62
Val	1.82	1.58	3.08	3.20	3.38	3.28
Lys	0.58	0.47	0.70	0.57	0.47	0.35
Tyr	2.85	2.03	5.67	4.37	7.48	5.72
Met	2.20	2.03	3.58	3.38	4.45	4.37
Arg	0.67	0.77	0.87	0.97	0.55	0.68
Ile	2.33	2.02	4.30	4.47	4.55	4.63
Leu	1.60	1.65	3.28	3.68	3.77	4.10
Phe	3.02	2.25	5.45	4.70	7.82	6.57
Trp	5.00	3.63	11.70	9.25	22.17	16.92

Packing: P1 = [Cu(L-hydroxyprolyl) *n*-octyl]$^+$ on LiChrosorb® Si 100, particle diameter 10 μm
Solvent: S1 = tetrahydrofuran/10^{-4} *M* copper acetate + 10^{-2} *M* ammonium acetate, pH 5.0, 50:50
S2 = acetonitrile/10^{-4} *M* copper acetate + 10^{-2} *M* ammonium acetate, pH 5.0, 50:50
S3 = methanol/10^{-4} *M* copper acetate + 10^{-2} *M* ammonium acetate, pH 5.0, 50:50
Detection: D1 = UV at 254 nm

REFERENCE

1. **Roumeliotis, P., Kurganov, A. A., and Davankov, V. A.,** *J. Chromatogr.*, 266, 439, 1983.

Table LC 63
THE EFFECT OF COLUMN TEMPERATURE ON THE RETENTION OF AMINO ACID ENANTIOMERS USING AS PACKING [Cu (II) (L- HYDROXYPROLYL) METHYL]$^+$ ON LICHROSORB® Si 60

Packing	P1	P1	P1
Column			
Length (mm)	100	100	100
Diameter (mm)	4.6	4.6	4.6
Solvent	S1	S1	S1
Temperature (K)	298	323	348
Detection	D1	D1	D1

Amino acid	k'_L	k'_D	k'_L	k'_D	k'_L	k'_D
Asp	1.92	2.10	1.10	1.03	0.86	0.81
Glu	1.71	1.84	1.02	1.13	0.84	0.83
His	4.13	2.34	3.80	2.33	4.01	2.52
Ala	1.26	1.39	1.40	1.47	1.46	1.53
Asn	2.39	2.18	2.05	1.76	1.47	1.50
Ser	1.92	1.61	1.72	1.46	1.63	1.32
Pro	1.99	2.67	1.93	2.51	1.70	2.27
Thr	2.57	2.18	2.14	1.94	2.10	1.94
Val	3.30	2.08	2.10	2.12	2.17	2.32
Lys	2.01	2.01	2.77	2.61	3.72	3.47
Tyr	2.64	1.74	2.30	1.52	2.13	1.51
Met	3.07	3.32	2.87	2.82	2.38	2.51
Arg	2.75	2.87	3.40	3.44	4.26	4.26
Ile	2.87	3.04	2.45	2.51	2.40	2.43
Leu	2.92	3.47	2.83	3.02	2.62	2.51
Phe	4.17	3.66	3.35	2.87	2.92	2.57
Trp	6.67	3.36	4.69	2.69	3.70	2.66

Packing: P1 = [Cu(L-hydroxyprolyl)methyl]$^+$ on LiChrosorb® Si 60, particle diameter 10 μm

Solvent: S1 = 10^{-4} *M* copper acetate + 10^{-2} *M* ammonium acetate, pH 5.0

Detection: D1 = UV at 254 nm

REFERENCE

1. **Roumeliotis, P., Kurganov, A. A., and Davankov, V. A.,** *J. Chromatogr.*, 266, 493, 1983.

Table LC 64
THE EFFECT OF COLUMN TEMPERATURE ON THE RETENTION OF AMINO ACID ENANTIOMERS USING AS PACKING [Cu (II) (L- HYDROXYPROLYL) METHYL]$^+$ ON LICHROSORB® Si 100

Packing	P1	P1	P1
Column			
Length (mm)	100	100	100
Diameter (mm)	4.6	4.6	4.6
Solvent	S1	S1	S1
Temperature (K)	298	323	348
Detection	D1	D1	D1

Amino acid	k'_L	k'_D	k'_L	k'_D	k'_L	k'_D
Asp	0.17	0.17	0.05	0.05	0.05	0.05
Glu	0.17	0.22	0.05	0.13	0.18	0.15
His	1.83	1.45	1.97	1.78	2.15	1.98
Ala	0.48	0.60	0.50	0.58	0.60	0.65
Asn	1.10	1.10	0.41	0.57	0.33	0.60
Ser	0.45	0.45	0.45	0.45	0.50	0.50
Pro	1.07	1.57	1.05	1.47	1.15	1.28
Thr	1.06	1.13	0.57	0.65	0.62	0.68
Val			1.42	1.18	1.58	1.23
Lys	2.57	2.75	2.82	2.72	3.70	3.60
Tyr	1.07	0.68	1.01	0.80	0.83	0.72
Met	1.33	1.48	1.53	1.53	1.23	1.63
Arg	4.48	4.58	3.53	3.37	4.28	4.12
Ile	1.48	1.57			1.72	1.70
Leu	1.43	1.25	1.67	1.45	1.67	1.28
Phe			1.58	1.78	1.55	1.55
Trp	2.17	1.28			1.53	0.98

Packing: P1 = [Cu(L-hydroxyprolyl)methyl]$^+$ on LiChrosorb® Si 100, particle diameter 10 μm

Solvent: S1 = 10^{-4} copper acetate + 10^{-2} *M* ammonium acetate, pH 5.5

Detection: D1 = UV at 254 nm

REFERENCE

1. **Roumeliotis, P., Kurganov, A. A., and Davankov, V. A.,** *J. Chromatogr.,* 266, 439, 1983.

Table LC 65
THE EFFECT OF COLUMN TEMPERATURE ON THE RETENTION OF AMINO ACID ENANTIOMERS USING AS PACKING [Cu (II) (L-HYDROXYPROLYL) *n*-OCTYL]$^+$ ON LICHROSORB® Si 100

Packing	P1		P1		P1	
Column						
Length (mm)	100		100		100	
Diameter (mm)	4.6		4.6		4.6	
Solvent	S1		S1		S1	
Temperature (K)	298		323		348	
Detection	D1		D1		D1	
Amino acid	**k'_L**	**k'_D**	**k'_L**	**k'_D**	**k'_L**	**k'_D**
Asp	1.75	1.85	2.01	2.13	1.90	2.02
Glu	1.25	1.33	1.57	1.57	1.57	1.57
His	2.43	1.47	2.22	1.92	2.77	1.85
Ala	0.33	0.42	0.40	0.58	0.63	0.83
Asn	0.85	0.80	0.89	0.82	1.03	1.03
Ser	0.70	0.60	0.82	0.74	1.08	0.98
Pro	0.90	1.80	1.07	2.06	1.08	2.22
Thr	1.10	1.07	1.14	1.16	1.65	1.62
Val	2.17	1.97	2.81	2.71	3.65	3.58
Lys	0.03	0.03			0.28	0.25
Tyr	11.78	10.37	12.25	9.72	11.98	9.90
Met	4.32	4.25	4.00	4.56	5.32	5.25
Arg	0.35	0.43	0.31	0.39	0.70	0.83
Ile	4.65	4.30	5.86	5.94	7.22	7.32
Leu	3.77	3.97	4.80	5.14	6.00	6.53
Phe	17.70	14.33	17.60	15.45	18.53	16.60
Trp					65.66	55.13

Packing: P1 = [Cu(L-hydroxyprolyl)*n*-octyl]$^+$ on LiChrosorb® Si 100, particle diameter 10 μm

Solvent: S1 = 10^{-4} *M* copper acetate + 10^{-2} *M* ammonium acetate, pH 5.0

Detection: D1 = UV at 254 nm

REFERENCE

1. **Roumeliotis, P., Kurganov, A. A., and Davankov, V. A.,** *J. Chromatogr.*, 266, 439, 1983.

Table LC 66
PHENYLTHIOCARBAMYL DERIVATIVES OF AMINO ACIDS

Packing	P1
Column	
Length (cm)	15
Diameter (mm)	3.9
Solvent	S1
Flow rate (mℓ/min)	1.0
Detection	D1

Amino acid	t_r
Alanine	0.21
Arginine	0.23
Aspartic acid	0.50
Cystine	0.14
Glutamic acid	0.45
Glycine	0.37
Histidine	0.19
Isoleucine	0.07
Leucine	0.08
Lysine	0.06
Methionine	0.22
Phenylalanine	0.11
Proline	0.16
Serine	0.41
Threonine	0.13
Tyrosine	0.37
Valine	0.22

Packing: P1 = Pico-Tag® analysis column (Waters)
Solvent: S1 = The solvent system consists of two solutions: (a) 0.14 *M* sodium acetate containing 0.5 mℓ/ℓ of triethylamine and titrated to pH 6.35 with glacial acetic acid and (b) 60% acetonitrile in water The gradient used consists of 10% B traversing to 51% B in 10 min using a convex curve
Detection: D1 = UV at 254 nm

REFERENCE

1. **Bidlingmeyer, B. A., Cohen, S. A., and Tarvin, T. L.,** *J. Chromatogr.*, 336, 93, 1984.

Table LC 67
PHENYLTHIOCARBAMYL AMINO ACIDS

Packing	P1
Column	
Length (mm)	250
Diameter (mm, I.D.)	4.6
Solvent	S1
Flow rate (mℓ/min)	1.0
Temperature (°C)	50
Detection	D1

Amino acid	**t_r(min)**
Alanine	19.7
Arginine	20.3
Aspartic acid	7.3
Cysteic acid	5.7
Cystine	32.2
Glutamic acid	8.4
Glycine	14.1
Histidine	17.6
Isoleucine	34.6
Leucine	35.2
Lysine	39.2
Methionine	30.7
Phenylalanine	37.1
Proline	21.8
Serine	13.1
Threonine	19.0
Tyrosine	27.7
Valine	29.8

Packing: P1 = LiChrospher® RP-18, 5 μm
Solvent: S1 = solvent A = 10 m*M* potassium phosphate, pH 6.5; solvent B = 70% acetonitrile in the same buffer (final concentration) The following gradient was established: 0—15 min, 9% B; 15—35 min, 36% B; 35—38 min, 100% B, this concentration is maintained for 3 min, 41—44 min, 0% B, washing for 5 min
Detection: D1 = UV at 254 nm

REFERENCE

1. **Scholze, H.,** *J. Chromatogr.*, 350, 453, 1985.

Table LC 68
PHENYLTHIOHYDANTOIN AMINO ACIDS

Packing	P1	P2	P3
Column			
Length (mm)		250	500
Diameter (mm)		4.6	1.0
Solvent	S1	S2	S3
Flow rate	1.5 mℓ/min	1.5 mℓ/min	7.5 μℓ/min
Temperature (°C)	62	35	rt
Detection	D1	D1	D1
Reference	1	2	3
Phenylthiohydantoin of		**t_r(min)**	
Alanine	3.88	7.82	9.32
α-Aminobutyric acid	5.28		
α-Aminoisobutyric acid	5.23		
Arginine	8.85[a]	10.44	16.24
Asparagine	2.14	4.10	5.88
Aspartic acid	1.24	1.75	
Aspartic acid methyl ester	4.31		9.90
Carboxymethylcysteine	1.30		
Carboxymethylcysteine methyl ester	6.10		
Cysteic acid	1.22[b]	1.44	
Glutamic acid	1.40	3.22	
Glutamic acid methyl ester	5.19		11.06
Glutamine	2.33	5.02	6.12
Glycine	2.92	6.09	7.69
Histidine	3.30[a]	6.76	10.17
Hydroxylysine	6.77		
Hydroxyproline	3.32, 3.75[c]		
Isoleucine	12.06	12.95	21.86
Leucine	13.16	13.31	23.71
Lysine		12.71	20.68
Methionine	7.30	11.01	14.67
Methionine sulfone	3.22		
Methylcysteine	5.72		
Norleucine	13.98	13.54	24.88
Norvaline	8.38		
Phenylalanine	11.51	12.65	21.11
Phenylthiocarbamyl aminoethyl cysteine	12.67		
Phenylthiocarbamyl lysine	12.59		
Proline	8.08	11.32	15.85
Serine	2.35	4.74[d]	6.36
Threonine	2.61	5.26[e]	6.75
Tryptophan	9.58	12.08	18.25
Tyrosine	4.19	8.71	9.70
Valine	7.60	11.12	14.93

[a] Retention time on a new column, retention time not constant.
[b] Na salt.
[c] Elutes in two peaks.
[d] By-products from the degradation of PTH-serine had retention times of 6.46, 11.40, and 13.21 min.
[e] A by-product from the degradation of PTH-threonine had a retention time of 9.64 min.

Packing:	P1	=	Hibar® 250-4, LiChrosorb® RP-18, 5 μm (Merck)
	P2	=	Altex Ultrasphere® ODS, 5 μm
	P3	=	LiChrosorb® RP-18, 7 μm
Solvent:	S1	=	0.01 *M* sodium acetate, pH 5.2/acetonitrile/dichloroethane, 68.5:31.5:0.5

Table LC 68 (continued)
PHENYLTHIOHYDANTOIN AMINO ACIDS

	S2	=	solvent A, 10 m*M* sodium acetate buffer, pH 4.85; the buffer is adjusted to pH 4.85 (20°C) with dilute acetic acid and degassed under vacuum immediately before use; solvent B, acetonitrile An elution gradient is set up according to the following schedule: at time 0 the solvent S2 contains 26% B lasting for 1 min, at time 1 min the solvent S2 contains 32% B lasting for 0.5 min, at time 1.5 min the solvent S2 contains 54% B lasting for 7.5 min, and at time 14 min the solvent S2 contains 90% B lasting for 2 min
	S3	=	acetonitrile/0.01 *M* sodium acetate, pH 4.5, 42:58
Detection:	D1	=	UV at 254 nm

REFERENCES

1. **Lottspeich, F.,** *Hoppe-Seylers Z. Physiol. Chem.*, 361, 1829, 1980.
2. **Pucci, P., Sannia, G., and Marino, G.,** *J. Chromatogr.*, 270, 371, 1983.
3. **Silver, M. R., Trosper, T. D., Gould, M. R., Dickinson, J. E., and Desotelle, G. A.,** *J. Liquid Chromatogr.*, 7, 559, 1984.

Table LC 69
PHENYLTHIOHYDANTOIN AMINO ACIDS

Packing	P1
Column[a]	
Length (mm)	250
Diameter (mm)	4.6
Solvent	S1
Flow rate (mℓ/min)	1.0
Temperature (°C)	50
Detection	D1

Amino acid	**t_r(min)**
Alanine	5.78 ± 0.027
Arginine	7.73 ± 0.019
Asparagine	3.36 ± 0.021
Aspartic acid	2.34 ± 0.010
Cysteic acid	2.09 ± 0.017
Glutamic acid	3.07 ± 0.021
Glutamine	3.58 ± 0.024
Glycine	4.63 ± 0.027
Histidine	5.18 ± 0.033
Isoleucine	10.48 ± 0.040
Leucine	10.93 ± 0.043
Lysine	9.60 ± 0.044
Methionine	8.14 ± 0.032
Norleucine[b]	11.20 ± 0.046
Phenylalanine	9.88 ± 0.039
Proline	8.69 ± 0.031
Serine	3.73 ± 0.023
Threonine	4.11 ± 0.025
Tryptophan	8.93 ± 0.040
Tyrosine	5.98 ± 0.031
Valine	8.39 ± 0.031

[a] Preceded by an Altex® PTH-amino acid (ODS) precolumn, 4.6 × 45 mm.

[b] Internal standard.

Table LC 69 (continued)
PHENYLTHIOHYDANTOIN AMINO ACIDS

Packing:	P1	=	Altex Ultrasphere® ODS:PTH column, 5 μm
Solvent:	S1	=	buffers A and B were prepared from water, acetonitrile, and a 1 *M* solution of acetic acid adjusted to pH 4.5 with concentrated NH_4OH, the final ammonium acetate concentration in both being 17 m*M*; buffer A contained 10% acetonitrile and buffer B contained 90% acetonitrile; the buffers were degassed at the beginning of the day and were not used when more than 48 hr old Initially, the solvent consisted of 70% buffer A and 30% buffer B; 1 min after injection, the concentration of buffer B was increased linearly to 50% over 3 min; after being held at this level for 8 min, the concentration of buffer B was reduced to 30% over 1 min and following a reequilibration time of 8 min another sample was injected
Detection:	D1	=	UV at 254 nm

REFERENCE

1. **Meuth, J. L. and Fox, J. L.,** *Anal. Biochem.,* 154, 478, 1986.

Table LC 70
PHENYLTHIOHYDANTOIN AMINO ACIDS

Packing	P1
Column	
Length (mm)	250
Diameter (mm)	4.6
Material	SS
Solvent	S1
Flow rate (mℓ/min)	1.55
Temperature (°C)	37
Detection	D1

Amino acid	**t_r(min)**
Alanine	7.25
Arginine	10.21
Asparagine	4.06
Aspartic acid	1.98
S-Carboxamidomethylcysteine	5.95
S-Carboxymethylcysteine	2.46
cysteine	
Glutamic acid	2.73
Glutamine	5.11
Glycine	5.61
Histidine	6.73
Hydroxyproline	7.65
Isoleucine	14.99
Leucine	11.75
Methionine	11.75
Phenylalanine	15.44

Table LC 70 (continued)
PHENYLTHIOHYDANTOIN AMINO ACIDS

Amino acid	t_r(min)
N^{ϵ}-Phenylthiocarbamoyl lysine	16.64
Proline	12.14
Serine	4.43
Threonine	4.80
Tryptophan	15.28
Tyrosine	9.11
Valine	11.60

Packing: P1 = Spherisorb® S5-ODS (5-μm particles, bonded with octadecylsilane)
Solvent: S1 = solution A = 0.02 *M* lithium acetate buffer, pH 5.2; solution B = 80% acetonitrile + 20% 0.1 *M* lithium acetate buffer, pH 5.2; a gradient was established as follows: 0—2 min, 25% B; 2—2.8 min, 25—32% B; 2.8—13 min, 32—45% B; 13—13.5 min, 45—55% B; 13.5—16.5 min, 55—60% B; 16—16.5 min, 60—95% B
Detection: D1 = UV at 254 nm

REFERENCE

1. **Moser, P. W. and Rickli, E. R.,** *J. Chromatogr.*, 176, 451, 1979.

Table LC 71
PHENYLTHIOHYDANTOIN AMINO ACIDS

Packing	P1
Column[a]	
Length (cm)	15
Diameter (mm, I.D.)	4.6
Solvent	S1
Flow rate (mℓ/min)	0.75
Temperature (°C)	33
Detection	D1

Amino acid	t_r(min)[b]
Cysteic acid	−4.3
2-Fluorobenzoic acid[c]	0.0
Asparagine	0.7
Aspartic acid	1.0
Serine	1.4
Threonine	1.75
Glutamine	2.1
Glycine	2.4
Histidine	2.75
S-Carboxymethylcysteine	2.9
Glutamic acid	3.4

Table LC 71 (continued)
PHENYLTHIOHYDANTOIN AMINO ACIDS

Amino acid	t_r(min)[b]
Methionine sulfone	3.6
First 4-hydroxyproline	3.85
Arginine	4.2
Alanine	4.4
Second 4-hydroxyproline	4.95
Tyrosine	5.5
S-Methylcysteine	6.75
Proline	7.2
Methionine	7.55
Valine	7.9
Tryptophan	8.7
Lysine	9.1
Phenylalanine	9.3
First isoleucine	9.65
Second isoleucine	9.75
Leucine	9.95
Norleucine	10.35
α-Naphthylacetic acid[c]	11.5

[a] Plus a guard column of Vydac® A, 2.3 cm × 4.6 mm I.D.
[b] Relative to 2-fluorobenzoic acid.
[c] Internal standard.

Packing: P1 = Ultrasphere® ODS, 5 μm

Solvent: S1 = buffer A consists of 10% acetonitrile, 42 m*M* trifluoroacetic acid, 21 m*M* acetic acid, pH 3.90 (920 g of water, 86.43 g of acetonitrile, 5.28 g of trifluoroacetic acid, and 1.39 g of glacial acetic acid adjusted at room temperature to pH 3.90 with x mℓ of 2 *N* NaOH in HPLC grade water); the solution is made up to a final volume of 1100 mℓ by addition of y g of water (y = 65 − x); buffer B consists of 22% acetonitrile, 34% 2-propanol, 20% buffer A, and 24% water (43.2 g of acetonitrile, 66.8 g of 2-propanol, 49.0 g of buffer A, and 60.0 g of water); pH adjustment is not necessary; the final volume is slightly less than 250 mℓ, owing to solvent mixture contraction; a gradient is established as follows: (A + B = 100%): 0—0.2 min, 0% B (isocratic); 0.2—1.7 min, 0—27% B (linear); 1.7—4.2 min, 27% B (isocratic); 4.2—14.2 min, 27—95% B (linear); 14.2—17.2 min, 95% B (isocratic); 17.2—17.7 min, 95—0% B (linear); 17.7—38.0 min, 0% B (isocratic)

Detection: D1 = UV at 254 nm

REFERENCE

1. **Kolbe, H. V. J., Lu, R. C., and Wohlrab, H.,** *J. Chromatogr.*, 327, 1, 1985.

Table LC 72
PHENYLTHIOHYDANTOIN AMINO ACIDS ON AN OCTADECYL-BONDED SILICA

Packing	P1
Column	
Length (cm)	20
Diameter (cm, I.E.)	0.46
Material	SS
Solvent	S1
Flow rate (mℓ/min)	1.5
Temperature (°C)	41
Detection	D1

Amino acid	t_r(min)	RSD (%)
Alanine	10.24	1.75
Arginine	12.14	0.80
Asparagine	5.14	0.77
Aspartic acid	3.79	0.94
S-Carboxymethylcysteine	13.06	0.78
Cysteic acid	1.59	1.51
Glutamic acid	9.00	3.04
Glutamine	8.25	2.40
Glycine	7.60	0.90
Histidine	10.78	1.38
Isoleucine	17.39	0.46
Leucine	17.69	0.47
Methionine	14.11	0.65
Norleucine	18.50	0.47
Phenylalanine	17.18	0.45
N-Phenylthiocarbamyl lysine	16.83	0.46
Proline	13.60	0.74
Serine	6.08	0.74
Threonine	7.10	0.74
Tyrosine	11.37	1.13
Tryptophan	15.79	0.48
Valine	14.54	0.63

Packing: P1 = PepRPC® (Pharmacia), an octadecyl-bonded silica of particle size 5 μm

Solvent: S1 = solvent A, 20 m*M* sodium acetate buffer, pH 4.1, with the addition of 1% hexanesulfonate and 20% methanol; solvent B, 1-propanol — the percentage of B is increased linearly from 0—18% over 13 min and is maintained at this value for 5 min, than reduced to over 3 min

Detection: D1 = UV at 254 nm

REFERENCE

1. **Johansson, B.-L. and Isaksson, K.,** *J. Chromatogr.*, 356, 383, 1986.

Table LC 73
PHENYLTHIOHYDANTOIN AMINO ACIDS ON A CONVENTIONAL AND MICROBORE COLUMN

Packing	P1	P1
Column		
Length (cm)	15	15
Diameter (mm, I.D.)	4.6	1.0
Solvent	S1	S2
Flow rate (mℓ/min)	1.0	0.06
Temperature (°C)	30	30
Detection	D1	D1
Amino acid	**t_r(min)**	
Arginine	15.27	17.82
Glutamic acid	3.81	9.66
Lysine	25.70	32.58

Packing: P1 = PTHAA-4 column, a column with maximum selectivity for the separation of PTH amino acids, the stationary phase having both CN and alkyl character

Solvent: S1 = solvent A, 0.01 *M* KH_2PO_4, pH 5.0; solvent B, methanol; solvent C, acetonitrile; gradient conditions: t = 0, A = 92%, C = 8%; t = 4 min, A = 85%, B = 15%; t = 8.0 min, A = 85%, B = 10%, C = 5%; t = 8.1 min, A = 85%, C = 15%; t = 12.0 min, A = 70%, C = 30%; t = 30 min, A = 40%, B = 60%

S2 = solvent A, 0.01 *M* sodium acetate, pH 5.0; solvent B, methanol; solvent C, acetonitrile; elution conditions developed with the conventional column were transferred to the microbore column; a preformed ternary gradient was applied so that methanol was the principal strong solvent in the beginning of the chromatogram and acetonitrile at the end

Detection: D1 = UV at 270 nm

REFERENCE

1. **Cunico, R. L., Simpson, R., Correia, L., and Wehr, C. T.,** *J. Chromatogr.,* 336, 105, 1984.

Table LC 74
CHROMATOGRAPHY OF PHENYLTHIOHYDANTOIN AMINO ACIDS USING COLUMNS OF DIFFERNT INTERNAL DIAMETERS

Packing	P1	P1	P1	P1
Column				
Length (cm)	25	25	25	25
Diameter (mm)	4.5	3.0	2.1	1.5
Material	SS	SS	SS	SS
Solvent	S1	S1	S1	S1
Flow rate (mℓ/min)	3.0	1.3	0.9	0.4
Temperature	rt	rt	rt	rt
Detection	D1	D1	D1	D1
Amino acid		**t_r(min)**		
Alanine	13.23	10.98	12.51	11.93
Asparagine	5.01	3.54	2.94	2.72
Aspartic acid	3.93	3.11	2.46	2.23
Glutamic acid	9.14	7.27	7.72	7.08
Glutamine	7.32	5.80	4.60	4.23
Glycine	7.82	6.18	5.15	4.67
Isoleucine + phenylalanine	20.25	21.37	21.12	22.40
Methionine + valine	19.07	20.41	20.24	21.40
Proline	17.04	18.18	17.86	17.60
Serine	5.76	3.94	3.36	3.13
Threonine	6.22	4.88	4.07	3.76
Tryptophan	19.66	20.78	20.24[a]	21.70[a]
Tyrosine	14.80	16.09	15.40	14.60

[a] Shoulder.

Packing: P1 = LiChrosorb® RP-18, 10 μm

Solvent: S1 = solvent A, 0.01 *M* sodium acetate, pH 4.3; solvent B, 50% acetonitrile + 50% A; solvent B was used isocratically for 10 min (8 min when 1.5-mm column was used), a gradient to 92% B over 16 min then being used

Detection: D1 = UV at 254 nm

REFERENCE

1. **Godtfredsen, S. E. and Oliver, R. W. A.,** *Carlsberg Res. Commun.*, 45, 35, 1980.

Table LC 75
PHENYLTHIOHYDANTOIN AMINO ACIDS ON A DIMETHYL *n*-OCTYL COLUMN

Packing	P1	P1	P1	P1	P1	P1	P1
Column							
Length (cm)	8.0	8.0	8.0	8.0	8.0	8.0	8.0
Diameter (mm, I.D.)	0.62	0.62	0.62	0.62	0.62	0.62	0.62
Solvent	S1	S2	S3	S4	S5	S6	S7
Flow rate (mℓ/min)	2.0	2.0	2.0	2.0	2.0	2.0	2.0
Temperature (°C)	35	35	35	35	35	35	35
Detection	D1	D1	D1	D1	D1	D1	D1
PTH amino acid				**t_r(min)**			
Asn	2.85	3.57	3.29	3.78	2.97	2.97	3.03
Ser	3.39	3.96	4.27	4.41	3.81	3.64	3.87
Asp	3.53	4.72	5.00	4.79	4.67	4.45	4.49
His	3.29	3.92	2.66	4.08	2.69	2.63	2.77
Thr	3.89	4.60	4.64	4.96	4.17	4.10	4.17
Gly	4.30	5.47	5.18	5.57	4.65	4.47	5.32
Gln	4.47	4.92	4.43	5.50	4.01	4.13	4.18
Glu	6.28	6.76	8.69	7.45	6.98	7.17	6.91
Arg	5.03	5.95	4.86	6.29	4.58	4.51	4.72
CMC	6.61	7.25	9.75	7.85	7.58	7.84	7.46
Ala	7.22	7.98	8.28	8.48	7.35	7.23	7.31
Tyr	9.21	10.20	14.40	10.93	11.52	11.77	12.26
Val	12.92	12.77	15.12	13.87	12.95	13.33	13.01
Met	12.67	12.90	15.85	13.87	13.16	13.73	13.20
Pro	12.33	13.00	11.61	13.49	10.99	10.92	11.09
Trp	14.14	15.13	20.03	16.20	16.47	17.54	16.58
Ile	15.77	15.34	18.83	16.63	16.11	16.60	16.95
Leu	16.34	15.98	19.66	17.16	16.75	17.48	16.75
Phe	15.76	15.47	18.48	16.82	15.88	16.91	16.27
Lys	16.21	16.30	23.69	17.79	17.64	18.63	17.74

Abbreviation: CMC = carboxymethyl cysteine

Packing: P1 = the column was prepared using Zorbax® silica, mean particle size 3.04 μm, mean pore diameter 75 Å; a monofunctional silane was used to make the dimethyl-*n*-octyl (C_8) packing with a coverage of 3.19 μmol/m² as determined by elemental carbon analysis; the packing was then endcapped with trimethylsilyl groups

Solvent: The components (initial % [final%]) of the mobile phase are shown
The gradient time was 15 min, the final percentage being maintained for 15 min
S1 = water 78.0 (45.0), methanol 22.0 (55.0)
S2 = water 85.0 (55.0), acetonitrile 15.0 (45.0)
S3 = water 88.0 (66.0), tetrahydrofuran 12.0 (34.0)
S4 = water 81.5 (50.0), methanol 11.0 (27.5), acetonitrile 7.5 (22.5)
S5 = water 86.5 (60.5), acetonitrile 7.5 (22.5), tetrahydrofuran 6.0 (17.0)
S6 = water 83.0 (55.5), methanol 11.0 (27.5), tetrahydrofuran 6.0 (17.0)
S7 = water 83.7 (55.4), methanol 7.3 (18.3), acetonitrile 5.0 (15.0), tetrahydrofuran 4.0 (11.3)

Detection: D1 = UV absorbance

Note: The data given in Tables LC 75—77 form part of a study to devise a general method of optimizing gradient-elution separations as applied to the separation of PTH amino acids. The data should be compared with those of a previously reported optimized isocratic separation. (Glajch, J. L., Gluckman, J. C., Charikofsky, J. G., Minor, J. M., and Kirkland, J. J., *J. Chromatogr.*, 318, 23, 1985.)

Table LC 75 (continued)
PHENYLTHIOHYDANTOIN AMINO ACIDS ON A DIMETHYL *n*-OCTYL COLUMN

REFERENCE

1. **Glajch, J. L. and Kirkland, J. J.,** *J. Chromatogr. Sci.*, 25, 4, 1987.

Table LC 76
PHENYLTHIOHYDANTOIN AMINO ACIDS ON A DIMETHYLCYANOPROPYL COLUMN

Packing	P1	P1	P1	P1	P1	P1	P1
Column							
Length (cm)	8.0	8.0	8.0	8.0	8.0	8.0	8.0
Diameter (mm, I.D.)	0.62	0.62	0.62	0.62	0.62	0.62	0.62
Solvent	S1	S2	S3	S4	S5	S6	S7
Flow rate (mℓ/min)	2.0	2.0	2.0	2.0	2.0	2.0	2.0
Temperature (°C)	35	35	35	35	35	35	35
Detection	D1	D1	D1	D1	D1	D1	D1
PTH amino acid				**t_r(min)**			
Asn	2.46	2.50	3.13	2.43	2.70	2.67	2.59
Ser	2.58	2.67	3.67	2.55	3.06	2.94	2.85
Asp	2.67	2.86	4.81	2.69	3.49	3.28	3.32
His	3.40	3.76	4.06	3.58	3.77	3.73	3.78
Thr	2.69	2.89	3.94	2.69	3.31	3.12	3.04
Gly	3.14	3.42	4.53	3.13	3.80	3.59	3.49
Gln	2.91	3.19	4.06	3.05	3.42	3.40	3.30
Glu	4.02	4.13	7.08	3.87	5.12	4.88	4.74
Arg	5.47	5.61	7.55	5.24	6.19	5.94	5.91
CMC	4.23	4.23	8.30	4.12	5.43	5.22	5.11
Ala	4.38	4.77	6.89	4.33	5.54	5.12	4.98
Tyr	7.85	7.89	13.88	7.49	10.00	10.01	9.18
Val	7.60	8.18	12.82	7.40	9.69	9.10	8.69
Met	8.87	9.01	14.30	8.53	10.65	10.52	9.83
Pro	8.48	8.69	11.07	8.05	9.27	8.85	8.60
Trp	12.83	12.67	19.47	12.18	14.91	15.45	14.16
Ile	9.99	10.59	16.73	9.72	12.45	12.09	11.40
Leu	10.64	11.23	17.82	10.38	13.32	12.95	12.10
Phe	11.51	11.83	17.51	11.07	13.62	13.60	12.75
Lys	15.41	14.15	22.53	14.03	16.44	17.55	15.79

Abbreviation: CMC = carboxymethyl cysteine

Packing: P1 = the column was prepared using Zorbax® silica, mean particle size 3.04 μm, mean pore diameter 75 Å; a monofunctional silane was used to make the dimethylcyanopropyl (CN) packing with a coverage of 3.23 μmol/m² as determined by elemental carbon analysis

Solvents: S1-S7 = see Table LC 75

Detection: D1 = UV absorbance

REFERENCE

1. **Glajch, J. L. and Kirkland, J. J.,** *J. Chromatogr. Sci.*, 25, 4, 1987.

Table LC 77
PHENYLTHIOHYDANTOIN AMINO ACIDS ON A DIMETHYLPHENETHYL COLUMN

Packing	P1	P1	P1	P1	P1	P1	P1
Column							
Length (cm)	8.0	8.0	8.0	8.0	8.0	8.0	8.0
Diameter (mm, I.D.)	0.62	0.62	0.62	0.62	0.62	0.62	0.62
Solvent	S1	S2	S3	S4	S5	S6	S7
Flow rate (mℓ/min)	2.0	2.0	2.0	2.0	2.0	2.0	2.0
Temperature (°C)	35	35	35	35	35	35	35
Detection	D1	D1	D1	D1	D1	D1	D1
PTH amino acid				**t_r(min)**			
Asn	4.48	4.06	4.15	4.85	3.86	4.02	3.99
Ser	5.04	4.63	5.50	5.34	4.87	5.00	4.95
Asp	5.51	5.16	7.07	5.75	6.05	5.93	5.83
His	5.83	5.06	3.84	5.95	4.03	4.05	4.35
Thr	5.95	5.15	6.04	5.93	5.48	5.55	5.56
Gly	6.56	6.19	6.59	6.77	6.01	6.04	6.03
Gln	6.80	5.59	5.62	6.63	5.15	5.49	5.59
Glu	8.76	7.39	10.36	8.52	8.52	8.93	8.57
Arg	8.16	7.39	7.15	8.52	6.96	6.86	7.10
CMC	9.09	7.86	11.15	8.89	9.25	9.61	9.19
Ala	9.25	8.53	9.89	9.32	8.77	8.96	8.82
Tyr	12.54	11.21	16.14	12.51	13.23	14.09	13.21
Val	14.83	12.95	16.36	14.13	14.12	14.67	14.19
Met	15.82	13.57	17.54	14.71	14.74	15.77	14.95
Pro	16.26	13.93	13.59	15.26	13.13	13.39	13.59
Trp	17.37	15.99	21.33	17.48	17.92	19.30	18.21
Ile	17.73	15.47	19.82	16.73	16.94	17.89	17.05
Leu	18.29	16.08	20.89	17.36	17.59	18.56	17.72
Phe	18.30	15.98	19.72	17.60	17.12	18.53	17.71
Lys	20.84	17.43	24.91	21.24	19.42	21.40	19.97

Abbreviation:	CMS	=	carboxymethyl cysteine
Packing:	P1	=	the column was prepared using Zorbax® silica, mean particle size 3.04 μm, mean pore diameter 75 Å; a monofunctional silane was used to make the dimethylphenethyl (PE) packing, with a coverage of 2.88 μmol/m² as determined by elemental analysis
Solvent:	S1-S7	=	see Table LC 75
Detection:	D1	=	UV absorbance

REFERENCE

1. **Glajch, J. L. and Kirkland, J. J.,** *J. Chromatogr. Sci.*, 25, 4, 1987.

Table LC 78
o-PHTHALALDEHYDE DERIVATIVES OF AMINO ACIDS

Packing	P1	P2	P3	P4
Column				
Length (cm)	25	8	25	25
Diameter (mm, I.D.)	4.6	6.2	4	4
Solvent	S1	S2	S3	S4
Flow rate (mℓ/min)	2.0	3.6	2.0	1.2
Temperature (°C)	38 ± 0.1	rt	40	rt
Detection	D1	D1	D1	D1
Reference	1	2	3	4
Amino acid		t_r		
Alanine	13.58 ± 0.96	6.75	20.54	17.81
Arginine	13.21 ± 0.98	6.05	8.94	22.1
Asparagine	5.77 ± 1.39	2.10	3.55	
Aspartic acid	2.18 ± 0.92	0.63		2.18
Citrulline		5.37		
2,4-Diaminobutyric acid[a]		10.20		
Glutamic acid	3.31 ± 1.21	0.75	5.34	4.06
Glutamine	7.92 ± 1.23	3.58	11.92	
Glycine	9.69 ± 1.45	5.03	14.54	12.50
Histidine	8.49 ± 0.82	3.96		16.23
Isoleucine	25.93 ± 0.50	9.83		33.41
Leucine	26.76 ± 0.52	9.97		35.27
Lysine	33.36 ± 0.48	11.01		39.48
Methionine	21.65 ± 0.77	8.77		31.04
Ornithine		10.68		
Phenylalanine	24.15 ± 0.58	9.24	32.3	34.3
Putrescine		13.74		
Serine	6.78 ± 1.47	2.87	10.42	8.74
Taurine[a]		6.43		
Threonine	10.32 ± 1.26	5.37	13.67	13.34
Tryptophan	23.10 ± 0.74	8.89		
Tyrosine	16.11 ± 1.06	7.29		25.23
Valine	22.25 ± 1.12	902		29.43

[a] Internal standard.

Packing:
- P1 = Spheri® 5 RP-18 (Brownlee Laboratories)
- P2 = Zorbax® C_8 high-speed column (Du Pont Instruments)
- P3 = Spherisorb® C-8, 5 μm or Partisil® 5-ODS, 3.5 μm
- P4 = Lichrocart Superspher® CH-8, 4 μm (Merck)

Solvent:
- S1 = solution A, 0.05 *M* sodium acetate brought to pH 6.30 ± 0.05 with glacial acetic acid; solution B, methanol-tetra-hydrofuran, 90:10, solutions were degassed by continuous helium purging
  A linear gradient started at 15% B and 85% A and ended at 64% B and 36% A in 35 min; the gradient was initiated on injection of the sample
- S2 = solution A, potassium acetate buffer, pH 5.5, 20 mmol/ℓ containing 1% tetrahydrofuran; solution B, methanol containing 1% tetrahydrofuran; both solutions were vacuum filtered and degassed
  Isocratic elution with 20% solution B and 80% solution A was followed by a linear gradient from 20—90% solution B, the overall program time being 13.5 min

Table LC 78 (continued)
o-PHTHALALDEHYDE DERIVATIVES OF AMINO ACIDS

S3 = the initial eluent was 0.03 *M* Na_2HPO_4 brought to pH 6.8 with acetic and phosphoric acids, the solution being made 5% in tetrahydrofuran and 5% in methanol; a linear 35-min gradient to final conditions of methanol-water, 55:45, was used

S4 = solution A, 12.5 m*M* sodium phosphate, pH 7.2; solution B, 50% acetonitrile in 12.5 m*M* sodium phosphate, pH 7.2; both solutions were conditioned with helium; the gradient conditions are shown in Table LC 78a

Table LC 78a

Duration (min)	From	To (% buffer A)
0—4	100	94
4—6	94	92
6—9	92	92
9—17	92	84
17—23	84	80
23—34	80	67
34—42	67	50
42—45	50	0
45—47	0	0

Detection: D1 = fluorometric

REFERENCES

1. **Hodgin. J. E., Howard, P. Y., Ball, D. M., Cloete, C., and DeJager, L.,** *J. Chromatogr. Sci.,* 21, 503, 1983.
2. **Price, S. J., Palmer, T., and Griffin, M.,** *Chromatographia,* 18, 62, 1984.
3. **Halfpenny, A. P. and Brown, Ph. R.,** *J. High Res. Chromatogr. Chron. Commun.,* 8, 243, 1985.
4. **Godel, H., Graser, T., Foldi, P., Pfaender, P., and Furst, P.,** *J. Chromatogr.,* 297, 49, 1984.

Table LC 79
o-PHTHALALDEHYDE DERIVATIVES OF AMINO ACIDS ON NUCLEOSIL® RP-18

Packing	P1
Column	
Length (mm)	200
Diameter (mm, I.D.)	4.6
Solvent	S1
Flow rate (mℓ/min)	1.0
Detection	D1

Amino acid	**k′**
Cysteic acid	0.9
Aspartic acid	1.0
Glutamic acid	1.0
S-Carboxymethylcysteine	1.1
Asparagine	1.4
Serine	1.6
Glutamine	1.6
Histidine	1.7
Methionine sulfone	1.8
Threonine	1.9
Glycine	2.1
β-Alanine	2.6
Tyrosine	2.6
Arginine	2.7
Alanine	2.8
α-Aminobutyric acid	4.6
Tryptophan	5.5
Methionine	6.2
Valine	6.6
Phenylalanine	7.4
Ammonia	8.8
Isoleucine	10.3
Ornithine	25.8
Lysine	38.8

Packing: P1 = Nucleosil® RP-18, 5 μm
Solvent: S1 = methanol/0.1 *M* phosphate buffer, pH 7.3, 50:50
Detection: D1 = fluorescence, excitation wavelength 330 nm (Corning® 7—54 primary filter) and a 418-nm cutoff secondary filter

REFERENCE

1. **Lindroth, P. and Mopper, K.,** *Anal. Chem.*, 51, 1667, 1979.

Table LC 80
AMINO ACID -*o*-PHTHALALDEHYDE DERIVATIVES ON NUCLEOSIL® 5 C_{18}

Packing	P1
Column	
Length (mm)	250
Diameter (mm, I.D.)	4.6
Solvent	S1
Flow rate (mℓ/min)	1.0
Temperature	rt
Detection	D1

Amino acid	t_r (min)
Alanine	23.9 ± 0.2
Arginine	17.8 ± 0.1
Aspartic acid	10.0 ± 0.2
Cysteic acid	8.7 ± 0.1
Glutamic acid	13.7 ± 0.1
Glycine	19.7 ± 0.1
Histidine	16.7 ± 0.1
Isoleucine	34.9 ± 0.1
Leucine	35.8 ± 0.1
Lysine	40.1 ± 0.1
Methionine	30.9 ± 0.1
Phenylalanine	32.4 ± 0.1
Serine	16.1 ± 0.1
Threonine	19.3 ± 0.2
Tryptophan	30.2 ± 0.1
Tyrosine	23.2 ± 0.2
Valine	31.6 ± 0.1

Packing: P1 = Nucleosil® 5 C_{18}, 5 μm

Solvent: S1 = solvent A is 0.1 *M* sodium citrate, pH 6.5; solvent B is methanol; the initial concentration of 20% solvent B is increased linearly over 5 min to 45%, is held at this value for 5 min, and is then increased linearly over 20 min to 75%

Detection: D1 = fluorescence measurement

REFERENCE

1. **Burbach, J. P. H., Prins, A., Lebouille, J. L. M., Verhoef, J., and Witter, A.,** *J. Chromatogr.*, 237, 339, 1982.

Table LC 81
AMINO ACID *o*-PHTHALALDEHYDE DERIVATIVES USING AUTOMATED PRECOLUMN DERIVATIZATION

Packing	P1
Column	
Length (cm)	22
Diameter (mm, I.D.)	4.6
Solvent	S1
Flow rate (mℓ/min)	1.4
Temperature	rt
Detection	D1

Amino acid	t_r(min)	
	I[a]	II[b]
Alanine	15.39	16.54
Arginine	16.05	17.25
Asparagine	7.94	8.98
Aspartic acid	3.67	4.65
Glutamine	10.60	11.72
Glycine	11.64	12.77
Isoleucine	22.60	23.64
Leucine	22.93	24.03
Lysine	24.83	25.94
Phenylalanine	21.75	22.84
Serine	8.47	9.52
Threonine	12.18	13.29
Tyrosine	17.62	18.72
Valine	20.87	21.95

[a] Reaction time about 13 sec.
[b] Reaction time about 75 sec.

Packing: P1 = HPLC cartridge column, spheri 5, RP8, 5 μm

Solvent: S1 = solvent A, 0.05 *M* sodium phosphate, pH 7.0/tetrahydrofuran, 99:1; solvent B, methanol, gradient elution from 100% solvent A to 75% solvent B was performed over 26 min; derivatization was carried out in 10% solvent A with *o*-phthalaldehyde-mercaptoethanol

Detection: D1 = spectrofluorometric.

REFERENCE

1. **Buck, R. H. and Krummen, K.,** *J. Chromatogr.*, 303, 238, 1984.

Table LC 82
THE EFFECT OF SODIUM ION CONCENTRATION ON THE RETENTION OF *o*-PHTHALALDEHYDE DERIVATIVES OF TAURINE AND TYROSINE

	Ratio of retention times		
Sodium ion (*M*)	**Taurine/ arginine**	**Tyrosine/ arginine**	**Apparent resolution of taurine/tyrosine[a]**
0.010	0.797	0.774	0.85
0.015	0.862	0.830	1.20
0.020	0.908	0.870	1.45
0.025	0.946	0.902	1.66

[a] Apparent resolution, using gradient elution, is calculated as $t_{r(Tau)} - t_{r(Tyr)}$ $\frac{1}{2}$ $W_{(Tau)} + W_{(Tyr)}$ when t_r = retention time in minutes and W = the peak width in minutes.

Conditions: μBondapak® C_{18} column, 30 cm × 3.9 mm I.D., resin particle diameter 10 μm.
Solvent: Solvent a, pH 7.2 phosphate buffer; solvent b, acetonitrile; linear program from 9—49% (solvent b/solvent a) 40 min in duration with a 5-min isocratic hold at 15 min from the initiation of the program was established
Flow rate: 2.0 mℓ/min
Detection: Fluorescence, excitation wavelength, 229 nm, emission wavelength 470 nm cutoff filter

REFERENCE

1. **Hill, D. W., Walters, F. H., Wilson, T. D., and Stuart, J. D.,** *Anal. Chem.*, 51, 1338, 1979.

Table LC 83
RESOLUTION OF ENANTIOMERIC AMINO ACIDS AFTER DERIVATIZATION WITH *o*-PHTHALALDEHYDE IN THE PRESENCE OF *N*-ACETYL-L-CYSTEINE

Packing	P1		P1	
Column				
Length (cm)	15		15	
Diameter (cm)	0.42		0.42	
Solvent	S1		S2	
Flow rate (mℓ/min)	2.0		2.0	
Detection	D1		D1	
Amino acid	**k′$_L$**	**k′$_D$**	**k′$_L$**	**k′$_D$**
Alanine			11.5	15.6
Arginine			17.0	22.0
Asparagine	22.4	31.0	1.9	2.3
Aspartic acid	1.9	2.5		
Citrulline			8.6	13.8
Glutamic acid	5.8	9.6	0.8	0.8
Serine	17.8	24.3		
Threonine	48.6	58.7	3.8	4.0
Lysine			8.2	12.4

Table LC 83 (continued)
RESOLUTION OF ENANTIOMERIC AMINO ACIDS AFTER DERIVATIZATION WITH *o*-PHTHALALDEHYDE IN THE PRESENCE OF *N*-ACETYL-L-CYSTEINE

Packing:	P1	=	Nucleosil® 5 C_{18}
Solvent:	S1	=	5 m*M* L-proline, 2.5 m*M* $CuSO_4$; 5 H_2O and 2.0 g ammonium acetate, pH 7.0
	S2	=	as S1 but with 5% acetonitrile
Detection:	D1	=	fluorescence

REFERENCE

1. **Lam, S.,** *J. Chromatogr.*, 355, 157, 1986.

Table LC 84
RESOLUTION OF ENANTIOMERIC AMINO ACIDS AFTER DERIVATIZATION WITH *o*-PHTHALADEHYDE IN THE PRESENCE OF *N*-ACETYL-L-CYSTEINE AS A FUNCTION OF ELUENT COMPOSITION

Packing	P1		P1	
Column				
Length (cm)	15		15	
Diameter (cm)	0.42		0.42	
Solvent	S1		S2	
Flow rate (mℓ/min)	2.0		2.0	
Detection	D1		D1	
Amino acid	**k'_L**	**k'_D**	**k'_L**	**k'_D**
Ethionine	37.2	45.7	15.2	18.2
Leucine	48.4	65.3	21.0	26.7
Methionine	16.3	19.7	7.3	9.1
Norleucine	48.6	62.1	22.5	27.3
Norvaline	17.2	22.2	7.6	10.1
Phenylalanine	43.0	48.2	22.3	23.6
Tryptophan	39.9	60.1	17.2	23.8
Tyrosine	6.2	9.8	2.8	3.5
Valine	9.0	16.5	4.0	4.3

Packing:	P1	=	Nucleosil® 5 C_{18}
Solvent:	S1	=	10% acetonitrile in a buffer containing 5 m*M* L-proline, 2.5 m*M* $CuSO_4$; 5 H_2O and 2.0 g ammonium acetate, pH 7.0
	S2	=	as S1 but with 12.5% acetonitrile
Detection:	S1	=	fluorescence

REFERENCE

1. **Lam, S.,** *J. Chromatogr.*, 355, 157, 1986.

Table LC 85
PHOSPHOAMINO ACIDS AS *o*-PHTHALALDEHYDE DERIVATIVES

Packing	P1	P1
Column dimensions	na	na
Solvent	S1	S1
Flow rate (mℓ/min)	1.0	1.5
Detection	D1	D1

Phosphoamino acid	**t_r(min)**	
1-Phosphohistidine	2.4	1.5
Phosphoserine	5.7	3.2
3-Phosphohistidine	8.1	5.8
Phosphotyrosine	12.3	7.3
Phosphothreonine	16.2	9.7
Phosphoarginine	18.9	12.6

Packing: P1 = Hamilton® PRP-1 resin

Solvent: S1 = buffer A contains 100 mℓ of 0.3 *M* NaH_2PO_4 buffer, 25 mℓ of tetrahydrofuran, and 1875 mℓ of water, buffer B contains 45 mℓ of the phosphate buffer, 1100 mℓ of acetonitrile, and 855 mℓ of water; isocratic elution is carried out with 88% buffer A and 12% buffer B

Detection: D1 = fluorescence measurement of *o*-phthalaldehyde derivative

REFERENCE

1. **Carlomagno, L., Huebner, V. D., and Matthews, H. R.,** *Anal. Biochem.,* 149, 344, 1985.

Table LC 86
DANSYL AMINO ACIDS

Packing	P1
Column	
Length (mm)	250
Diameter (mm, I.D.)	4.6
Material	SS
Solvent	S1
Flow rate (mℓ/min)	1.0
Temperature (°C)	50
Detection	D1

Amino acid	**t_r (min)**
Alanine	24.27
Arginine	27.29
Asparagine	14.99
Aspartic acid	6.83
Cysteic acid	5.93
Cystine	54.76

Table LC 86 (continued)
DANSYL AMINO ACIDS

Amino acid	t_r (min)
Glutamic acid	7.46
Glutamine	16.93
Glycine	20.81
Histidine[a]	19.40
Hydroxyproline	16.34
Isoleucine	49.47
Leucine	51.09
Lysine[b]	36.94
Lysine[c]	59.26
Methionine	39.00
Norleucine	53.70
Phenylalanine	52.63
Proline	33.36
Serine	18.60
Threonine	22.06
Tryptophan	45.35
Tyrosine	60.68
Valine	38.28
Dansylamide	36.44
Dansyl sulfonic acid	11.46

[a] α-Dns-histidine.
[b] ε-Dns-lysine.
[c] di-Dns-lysine.

Packing: P1 = Develosil® C8-5, particle size 5 μm (Nomura Chemical Co., Seto, Japan)

Solvent: S1 = solvent A, 0.01 *M* Tris-HCl buffer, pH 7.75/methanol, 10:3; solvent B, methanol; both solvents are degassed for a few minutes by a water aspirator before use; the linear gradient is established by adding solvent B to solvent A at a rate of 0.5%/min until 25% of solvent B is reached — the rate is then increased to 5%/min; after 90% of solvent B is reached, the column is recycled with solvent A for 15 min

Detection: D1 = fluorometric excitation and emission wavelengths are 340 and 530 nm, respectively

REFERENCE

1. **Kaneda, M., Sato, M., and Yagi, K.,** *Anal. Biochem.*, 127, 49, 1982.

Table LC 87
THE RESOLUTIONS OF D- AND L-DANSYL-AMINO ACIDS BY MIXED CHELATE COMPLEXATION AS A FUNCTION OF ACETONITRILE CONCENTRATION

Packing	P1	P1	P1
Column			
Length (cm)	25	25	25
Diameter (cm)	0.26	0.26	0.26
Solvent	S1	S2	S3
Flow rate (mℓ/min)	1.0	1.0	1.0
Detection	D1	D1	D1
Dansyl amino acid		**k′**	
L-Leucine	2.36	6.57	31.97
D-Leucine	3.00	9.86	41.40
L-Methionine	1.36	3.43	18.03
D-Methionine	1.57	4.71	22.93
L-Norleucine	2.86	8.43	42.33
D-Norleucine	3.54	12.43	55.50
L-Norvaline	1.35	3.64	17.17
D-Norvaline	1.64	5.07	21.83
L-Phenylalanine	3.03	6.64	40.67
D-Phenylalanine	4.21	11.64	71.83
L-Serine	0.35	0.79	3.83
D-Serine	0.35	0.79	3.83
L-Threonine	0.34	0.71	4.17
D-Threonine	0.34	0.97	5.17
L-Tryptophan	2.57	9.80	46.07
D-Tryptophan	4.43	19.37	73.67
L-Valine	1.60	2.79	13.67
D-Valine	1.86	3.83	16.67

Packing: P1 = C_{18}, ODS-HC SIL-X-1 (Perkin-Elmer)

Solvent: S1, S2, and S3 = consisting of 25, 20, and 15%, respectively, of acetonitrile in an aqueous buffer containing 5×10^{-3} *M* L-proline and ammonium acetate and 2.5×10^{-3} *M* $CuSo_4$, 5 H_2O, pH 7

Detection: D1 = fluorescence at 480 nm was monitored with excitation at 340 nm.

REFERENCE

1. **Lam, S., Chow, F., and Karmen, A.,** *J. Chromatogr.*, 199, 295, 1980.

Table LC 88
THE RESOLUTION OF D- AND L-DANSYL-AMINO ACIDS BY MIXED CHELATE COMPLEXATION AS A FUNCTION OF L-PROLINE-Cu(II) CONCENTRATION

Packing	P1	P1	P1
Column			
Length (cm)	25	25	25
Diameter (cm)	026	0.26	0.26
Solvent	S1	S2	S3
Flow rate (mℓ/min)	1.0	1.0	1.0
Detection	D1	D1	D1
Dansyl amino acid		**k′**	
L-Leucine	3.00	6.57	7.20
D-Leucine	3.41	9.86	12.73
L-Methionine	1.76	3.43	4.53
D-Methionine	1.76	4.71	6.13
L-Norleucine	4.29	8.43	10.67
D-Norleucine	6.05	12.43	15.93
L-Norvaline	1.50	3.64	4.87
D-Norvaline	1.50	5.07	6.60
L-Phenylalanine	3.70	6.64	10.06
D-Phenylalanine	4.88	11.64	15.47
L-Threonine	0.18	0.71	1.00
D-Threonine	0.18	0.97	1.33
L-Tryptophan	3.88	9.80	10.40
D-Tryptophan	5.49	19.37	20.67
L-Valine	1.35	2.79	3.87
D-Valine	1.35	3.83	5.59

Packing: Pl = C_{18}, ODS-HC-SIL-X-1 (Perkin-Elmer)

Solvents: S1, S2, and S3 = consisting of 20% acetonitrile in aqueous buffer containing 5 × 10^{-3} *M* ammonium acetate, pH 7; S1 contains 1 × 10^{-3} *M* L-proline and 5 × 10^{-4} *M* $CuSo_4{\cdot}5H_2O$; S2 contains 5 × 10^{-3} *M* L-proline and 2.5 × 10^{-3} *M* $CuSo_4{\cdot}5H_2O$; S3 contains 2 × 10^{-1} *M* L-proline and 1 × 10^{-2} *M* $CuSo_4{\cdot}5\ H_2O$

Detection: D1 = fluorescence at 480 nm was monitored with excitation at 340 nm

REFERENCE

1. **Lam, S., Chow, F., and Karmen, A.,** *J. Chromatogr.*, 199, 295, 1980.

Table LC 89
RESOLUTION OF D- AND L-DANSYL AMINO ACIDS BY MIXED CHELATE COMPLEXATION

Packing	P1
Column	
Length (cm)	25
Diameter (cm)	0.26
Solvent	S1
Flow rate (mℓ/min)	1.0
Detection	D1

Dansyl amino acid	**k′**
L-Alanine	4.42
D-Alanine	4.17
L-α-Aminobutyric acid	6.16
D-α-Aminobutyric acid	5.67
L-Isoleucine	20.83
D-Isoleucine	17.16
L-Leucine	20.83
D-Leucine	18.50
L-Methionine	11.66
D-Methionine	10.33
L-Norleucine	28.16
D-Norleucine	22.33
L-Norvaline	13.00
D-Norvaline	11.00
L-Phenylalanine	23.00
D-Phenylalanine	20.33
L-Serine	3.00
D-Serine	3.00
L-Threonine	3.00
D-Threonine	3.00
L-Tryptophan	31.17
D-Tryptophan	24.83
L-Valine	9.50
D-Valine	8.33

Packing: P1 = Nucleosil® 5

Solvent: S1 = 1:4 ratio of acetonitrile/aqueous solution containing 5×10^{-3} *M* L-arginine, 2.5×10^{-3} *M* $CuSO_4 \cdot 5\ H_2O$, and $2 \cdot 5 \times 10^{-2}$ *M* ammonium acetate

Detection: D1 = fluorescence at 480 nm was monitored with excitation at 340 nm

REFERENCE

1. **Lam, S., Chow, F., and Karmen, A.,** *J. Chromatogr.*, 199, 295, 1980.

Table LC 90
THE SEPARATION OF D- AND L-DANSYL AMINO ACIDS AS MIXED CHELATE COPPER COMPLEXES

Packing	P1	P2	P2
Column			
Length (mm)	150	150	150
Diameter (mm)	4.2	4.2	4.2
Solvent	S1	S2	S3
Flow rate (mℓ/min)	2.0	2.0	2.0
Detection	D1	D1	D1
Dansyl amino acid		**k′**	
L-Alanine	4.1	4.4	6.6[a]
D-Alanine	4.3	4.1	8.2[a]
L-α-Aminobutyric acid	5.4	6.1	2.2
D-α-Aminobutyric acid	6.6	5.6	3.8
L-Isoleucine	14.7	20.8	7.2
D-Isoleucine	19.7	17.2	15.4
L-Leucine	17.0	20.8	7.2
D-Leucine	25.4	18.5	16.0
L-Methionine	9.7	11.7	4.4
D-Methionine	12.3	10.3	7.8
L-Norleucine	22.1	28.2	9.4
D-Norleucine	31.2	22.3	22.2
L-Norvaline	10.1	13.0	3.8
D-Norvaline	13.7	11.0	9.0
L-Phenylalanine	17.4	23.0	9.0
L-Serine	23.0[a]	3.0	5.6
D-Serine	20.0[a]	3.0	3.4
L-Threonine	19.0[a]	3.0	4.4[a]
D-Threonine	31.2[a]	3.0	3.4[a]
L-Tryptophan	23.0	31.2	7.4
D-Tryptophan	43.4	24.8	7.4
L-Valine	07.7	9.5	3.2
D-Valine	10.1	8.3	6.4

[a] 10% acetonitrile.

Packing:	P1	=	LiChrosorb® RP-8
	P2	=	Nucleosil® 5 C_{18}
Solvent:	S1	=	20% acetonitrile in a buffer containing 5 m*M* L-proline, 2.5 m*M* $CuSO_4 \cdot 5H_2O$, and 0.5 g ammonium acetate per liter, pH 7.0
	S2	=	1:4 acetonitrile/buffer containing 5 m*M* L-arginine, 2.5 m*M* $CuSO_4 \cdot 5H_2O$, and 25 m*M* ammonium acetate, pH 7.0
	S3	=	15% acetonitrile in buffer containing 5 m*M* L-histidine, 2.5 m*M* $CuSO_4 \cdot 5H_2O$, and 2.0 g ammonium acetate per liter of deionized water, pH 7.0
Detection:	D1	=	fluorometric

REFERENCE

1. **Lam, S.,** *J. Chromatogr. Sci.*, 22, 416, 1984.

Table LC 91
THE SEPARATION OF D- AND L-DANSYL AMINO ACIDS AS COPPER L-HISTIDINE METHYL ESTER COMPLEXES

Packing	P1
Column	
Length (mm)	150
Diameter (mm)	4.2
Solvent	S1
Flow rate (mℓ/min)	2.0
Detection	D1

Dansyl amino acid	k'_L	k'_D
Alanine	11.9	11.9
α-Aminobutyric acid	15.0	13.6
Asparagine	8.7	7.9
Aspartic acid	4.4	5.0
Glutamic acid	4.1	3.3
Isoleucine	44.7	40.4
Leucine	43.9	33.0
Methionine	30.7	27.2
Norleucine	58.4	40.4
Norvaline	27.6	23.3
Phenylalanine	77.9	60.4
Serine	10.7	9.3
Threonine	9.3	9.3
Tryptophan	121.8	93.3
Valine	20.1	17.3

Note: k'_L and k'_D refer to the L- and D-dansyl amino acids, respectively.

Packing:	P1	=	Nucleosil® 5 C_{18}
Solvent:	S1	=	20% acetonitrile in a buffer containing 5 m*M* L-histidine methyl ester, 2.5 m*M* $CuSo_4 \cdot 5H_2O$, and 2.0 g ammonium acetate per liter of deionized water, pH 7.0
Detection:	D1	=	fluorometric

REFERENCE

1. **Lam, S.,** *J. Chromatogr. Sci.,* 22, 416, 1984.

Table LC 92
CHROMATOGRAPHY OF DANSYL AMINO ACIDS USING DIFFERENT ELUTION SYSTEMS

Packing	P1	P1	P1	P2	P3	P4
Column						
Length (cm)	30	30	30	15	10	30
Diameter (cm, I.D.)	0.4	0.4	0.4	0.4	0.8	0.4
Material	SS	SS	SS	SS		SS
Solvent	S1	S2	S3	S4	S5	S6
Flow rate (mℓ/min)	2.0	2.0	1.0	1.0	7.0	2.0
Temperature	rt	rt	rt	rt	rt	rt
Detection	D1	D1	D1	D1	D1	D1

Dansyl amino acid	Order[a]	k′	Order	k′	Order	k′	Order	k′	Order	k′	Order
Cysteic acid	1	15	1	8.0	11	4.7	2	8.8	2	6.0	nd
Dns-NH_2	2	5.8	2	9.4	5	2.8	9	16.8	9	18.3	6
Asparagine	3	6.3	7	14.6	3	2.2	4	13.1	4	11.2	3
Serine	4	8.0	4	12.8	7	3.1	6	15.1	6	14.7	5
Arginine	5	8.1	5	12.9	1	0.6	14	22.7	17	37.3	11
Glutamine	5	8.1	9	16.2	4	2.5	5	14.5	5	13.0	4
Aspartic acid	7	9.8	3	10.0	10	3.9	1	7.6	1	5.0	1
ϵ-Lysine	7	9.8	8	14.7	2	1.3	12	20.5	12	27.0	17
Glutamic acid	9	10.0	5	12.9	8	3.3	3	9.k	2	6.0	2
Glycine	9	10.0	10	6.4	12	4.8	7	16.0	7	15.8	7
Threonine	11	11.0	11	17.0	9	3.7	8	16.3	8	16.9	6
Alanine	12	12.3	12	19.1	13	5.3	10	17.7	10	20.0	8
Valine	13	18.9	13	19.7	14	10.7	13	21.9	13	28.8	10
Methionine	14	20.4	14	20.6	14	10.7	15	23.3	14	31.7	12
Tryptophan	15	21.1	15	21.1	18	14.3	18	27.7	18	38.7	nd
Proline	16	21.2	16	21.4	6	2.8	11	20.1	11	24.7	9
Phenylalanine	17	23.2	17	23.0	19	14.5	19	28.7	19	39.5	15
Isoleucine	18	23.7	18	23.7	17	13.8	16	25.4	15	34.7	13
Leucine	19	24.4	19	24.1	16	13.7	17	26.1	16	35.7	14
Cysteine	20	28.3	20	26.8	20	16.6	20	32.6	20	46.3	nd
α, ϵ-Lysine	21	32.4	21	29.1	22	17.8	21	35.2	21	51.2	18
α, 3-Histidine	22	33.7	22	30.3	nd	nd	22	35.6	nd	nd	nd
α, *O*-Tyrosine	23	38.7	23	32.5	21	17.5	23	37.4	22	53.2	19

Table LC 92 (continued)
CHROMATOGRAPHY OF DANSYL AMINO ACIDS USING DIFFERENT ELUTION SYSTEMS

[a] Elution order.

Packing:	Pl	=	μBondapak®-alkylphenyl
	P2	=	Hypersil® ODS, 5 μm
	P3	=	Radial-Pak® A cartridge used in conjunction with a RCM module
	P4	=	μBondapak® C_{18}
Solvent:	S1	=	solvent A, water/15 m*M* H_3PO_4, pH 2.3; solvent B, 15 m*M* H_3PO_4, methanol/water, 80:20; a linear 60-min gradient from 10—100% B was established
	S2	=	solvent A, water/50 m*M* sodium acetate, pH 6.5; solvent B, 50 m*M* sodium acetate, methanol/water, 80:20; a linear 60-min gradient from 10—100% B was established
	S3	=	solvent A, 2 m*M* dodecylamine, 15 m*M* H_3PO_4, acetonitrile/water, 25:75; solvent B, 2 m*M* dodecylamine, 15 m*M* H_3PO_4, acetonitrile/water, 65:3; gradient, solvent A isocratic for 25 min followed by concave gradient to 100% B over 35 min
	S4	=	solvent A, 200 m*M* NH_4HCO_3, acetonitrile/water, 10:90; solvent B, 200 m*M* NH_4HCO_3, acetonitrile/water, 45:55; linear gradients from 0—30% B over 25 min, 30—45% B over 20 min, 45—100% B over 15 min, followed by 100% B for 15 min were established
	S5	=	solvent A, 200 m*M* NH_4HCO_3, acetonitrile/water, 10:90; solvent B, 200 m*M* NH_4HCO_3, acetonitrile/water, 45:55; linear gradients from 10—20% B over 8 min, 20—50% B over 12 min, 50—100% B over 3 min, followed by 100% B for 5 min were established
	S6	=	solvent A, 30 m*M* Na_3PO_4, pH 7.7; solvent B, acetonitrile; a inear gradient from 10—50% B over 23 min was established
Detection:	D1	=	UV at 254 nm

REFERENCE

1. **Grego, B. and Hearn, M. T. W.,** *J. Chromatogr.*, 255, 67, 1983.

Table LC 93
RETENTION OF DANSYL AMINO ACIDS AS A FUNCTION OF 1-PROPANOL CONTENT OF THE MOBILE PHASE

Packing	P1	P1	P1	P1	P1
Column					
Length (mm)	150	150	150	150	150
Diameter (mm, I.D.)	4.6	4.6	4.6	4.6	4.6
Material	SS	SS	SS	SS	SS
Solvent	S1	S2	S3	S4	S5
Flow rate (mℓ/min)	1.0	1.0	1.0	1.0	1.0
Detection	D1	D1	D1	D1	D1

Amino acid	**k'_1**	**k'_2**	**k'_1**	**k'_2**	**k'_1**	**k'_2**	**k'_1**	**k'_2**	**k'_1**	**k'_2**
Alanine			5.3	7.8	2.9	3.9	2.4	3.2	1.6	2.0
Aspartic acid			2.0	3.6	1.5	2.6	1.2	1.9	0.95	1.6
Glutamic acid			0.85	1.6	0.68	1.1	0.53	0.74	0.35	0.45
Methionine	11.2	11.2	7.4	7.4	4.1	4.1				
Phenylalanine			14.5	18.8	10.7	14.4	9.9	13.8	9.3	12.8
Serine			2.2	2.2	1.2	1.4	0.90	1.2	0.74	1.0
Threonine	2.2	2.6	1.6	1.6	1.3	1.3	1.0	1.0	0.55	0.55

Note: The D-enantiomer was eluted first except for dansyl alanine and dansyl glutamic acid.

Packing:	P1	=	BSA silica, 10-μm particle size (Resolvosil®)
Solvents:	S1, S2, S3, S4, and S5	=	50 m*M* phosphate buffer, pH 7.7, containing 0, 1, 2, 3, and 4%, respectively, of 1-propanol
Detection:	D1	=	UV at 250 nm

REFERENCE

1. **Allenmark, S. and Andersson, S.,** *J. Chromatogr.*, 351, 231, 1986.

Table LC 94
RETENTION OF DANSYL AMINO ACIDS AS A FUNCTION OF pH OF THE MOBILE PHASE

Packing	P1	P1	P1	P1
Column				
Length (mm)	150	150	150	150
Diameter (mm I.D.)	4.6	4.6	4.6	4.6
Material	SS	SS	SS	SS
Solvent	S1	S2	S3	S4
Flow rate (mℓ/min)	1.0	1.0	1.0	1.0
Detection	D1	D1	D1	D1

Amino acid	**k'_1**	**k'_2**	**k'_1**	**k'_2**	**k'_1**	**k'_2**	**k'_1**	**k'_2**
Alanine	21.5	27.3	8.6	11.8	7.5	10.8	6.4	8.9
Aspartic acid	9.4	14.9	4.0	6.4	3.2	5.4	3.4	6.9
Glutamic acid	4.5	5.1	1.6	1.9	1.7	2.3	1.6	3.4
Methionine	17.0	17.0	9.8	9.8	8.4	8.4	7.2	7.2
Phenylalanine	38.2	62.8	28.6	45.4	22.2	34.8	26.8	41.1
Serine	7.1	9.4	3.7	4.6	3.0	3.7	2.5	2.5
Threonine	4.9	4.9	3.2	3.2	2.2	2.2	1.9	1.9

Note: The D-enantiomer was eluted first, except for dansyl alanine and dansyl glutamic acid.

Table LC 94 (continued)
RETENTION OF DANSYL AMINO ACIDS AS A FUNCTION OF pH OF THE MOBILE PHASE

Packing: Pl = BSA silica, 10-μm particle size (Resolvosil®)
Solvent: S1, S2, S3, and S4 = 50 m*M* phosphate buffer/2%-1-propanol, the pHs of the buffers being 5.7, 6.8, 7.3, and 7.9, respectively
Detection: D1 = UV at 250 nm

REFERENCE

1. **Allenmark, S. and Andersson, S.,** *J. Chromatogr.*, 351, 231, 1986.

Table LC 95
RETENTION OF DANSYL AMINO ACIDS AS A FUNCTION OF BUFFER STRENGTH OF THE MOBILE PHASE

Packing	Pl	Pl	Pl	Pl
Column				
Length (mm)	150	150	150	150
Diameter (mm, I.D.)	4.6	4.6	4.6	4.6
Material	SS	SS	SS	SS
Solvent	S1	S2	S3	S4
Flow rate (mℓ/min)	1.0	1.0	1.0	1.0
Detection	D1	D1	D1	D1

Amino acid	k'_1	k'_2	k'_1	k'_2	k'_1	k'_2	k'_1	k'_2
Alanine	4.0	5.8	4.0	5.4	3.6	5.0	3.5	4.7
Aspartic acid	2.8	6.0	1.8	3.1	1.3	2.2	1.3	2.0
Glutamic acid	1.3	3.0	0.79	1.1	0.68	0.95	0.67	0.89
Methionine	5.1	5.1	4.8	4.8	4.5	4.5	4.2	4.2
Phenylalanine	15.5	25.2	14.0	20.4	12.3	16.9	15.7	21.5
Serine	1.8	2.1	1.8	2.0	1.6	2.0	1.6	1.9
Threonine	1.4	1.4	2.2	2.2	1.7	1.7	1.4	1.4

Note: The D-enantiomer was eluted first, except for dansyl alanine and dansyl glutamic acid.

Packing: Pl = BSA silica, 10-μm particle size (Resolvosil®)
Solvents: S1, S2, S3, and S4 = phosphate buffers, pH 7.7/2% 1-propanol, the buffer strengths being 25, 50, 75, and 100 m*M* respectively
Detection: D1 = UV at 250 nm

REFERENCE

1. **Allenmark, S. and Andersson, S.,** *J. Chromatogr.*, 351, 231, 1986.

Table LC 96
ENATIOMERIC RESOLUTION OF DANSYL AMINO ACIDS WITH β-CYCLODEXTRIN INCLUSION COMPLEXES

Amino acid	Solvent	k'_D	k'_L
Alanine	S1	28.9	30.4
α-Aminobutyric acid	S3	19.5	20.9
Arginine	S2	21.2	22.6
Asparagine	S2	12.5	13.2
Aspartic acid	S1	6.4	7.1
Glutamic acid	S1	7.1	7.7
Glutamine	S2	17.5	18.5
Isoleucine	S4	18.7	20.0
Leucine	S4	21.7	23.2
Lysine	S1	17.6	18.7
Methionine	S3	83.9	86.9
Norleucine	S4	30.3	31.7
Norvaline	S4	13.5	14.1
Phenylalanine	S4	30.7	32.4
Proline	S3	21.0	22.1
Serine	S1	18.1	19.5
Threonine	S1	21.7	24.3
Tryptophan	S4	29.6	29.6
Valine	S4	10.1	10.8

Conditions: Packing: ODS-Hypersil®, 3 μm
Column: 108 × 0.26 mm I.D.
Solvent: S1 = 10% acetonitrile, pH 6.1
S2 = 10% acetonitrile, pH 6.5
S3 = 15% acetonitrile, pH 6.5
S4 = 20% acetonitrile, pH 6.1
The pH of the solvent was adjusted with potassium phosphate and phosphoric acid (0.01 *M*); all the solvents contained β-cyclodextrins, 12.5 *M*
Detection: UV at 220 nm

REFERENCE

1. **Takeuchi, T., Asai, H., and Ishii, D.,** *J. Chromatogr.*, 357, 409, 1986.

Table LC 97
DANSYL-AMINO ACID AMIDES AND AMINO ACIDS

Packing	P1
Column	
Length (mm)	250
Diameter (mm, I.D.)	4.6
Solvent	S1
Flow rate (mℓ/min)	1.0
Detection	D1

Amino acid or amide	t_r (min)
Alanine	5.1
Alanine amide	24.0
Asparagine	3.3
Asparagine amide	18.0
Aspartic acid	1.9
Aspartic acid amide	3.7
Glutamic acid	1.9
Glutamic acid amide	3.5
Glycine	4.4
Glycine amide	22.8
Histidine amide	25.5
Histidine amide[a]	41.8
Isoleucine	14.5
Isoleucine amide	33.9
Leucine	15.2
Leucine amide	34.7
Lysine[a]	25.2
Lysine amide[a]	40.7
Methionine	11.3
Methionine amide	30.2
Phenylalanine	17.4
Phenylalanine amide	36.2
Proline	6.6
Proline amide	28.9
Serine	3.9
Serine amide	19.9
Threonine	4.5
Threonine amide	22.0
Tryptophan	17.1
Tryptophan amide	35.3
Tyrosine[a]	30.7
Tyrosine amide[a]	43.3
Valine	8.9
Valine amide	29.5
Dansyl sulfonic acid	3.0
Dansyl amide	25.0

[a] Didansyl derivative.

Packing	P1	=	Altex Ultrasphere® ODS
Solvent:	S1	=	solvent A, 5 m*M* sodium phosphate, pH 7.0; solvent B, acetonitrile; gradient program, 0 min, 20% B; 8 min, 20% B; 20 min, 35% B; 30 min, 40% B; 40 min, 80% B; 45 min, 80% B
Detection:	D1	=	UV at 254 nm

Table LC 97 (continued)
DANSYL-AMINO ACID AMIDES AND AMINO ACIDS

REFERENCE

1. **Simmons, W. G. and Meisenberg, G.,** *J. Chromatogr.*, 266, 483, 1983.

Table LC 98
DANSYLATED LYSINES

Packing	P1
Column	
Length (mm)	250
Diameter (mm)	4.6
Solvent	S1
Flow rate (mℓ/min)	1.0
Detection	D1

Compound	k′
ε-Dns-Lysine	2.0
bis-Dns-Lysine	2.4
α-Dns-Lysine	3.25
Dns-Alanine	4.9

Packing: P1 = Partisil® PAC, 10 μm
Solvent: S1 = acetonitrile/water/acetic acid, 35:65:1
Detection: D1 = UV at 254 nm

REFERENCE

1. **Szokan, G.,** *J. Liquid Chromatogr.*, 5, 1493, 1982.

Table LC 99
RETENTION OF DNP AMINO ACIDS AS A FUNCTION OF 1-PROPANOL CONTENT OF THE MOBILE PHASE

Packing	P1	P1	P1	P1	P1
Column					
Length (mm)	150	150	150	150	150
Diameter (mm, I.D.)	4.6	4.6	4.6	4.6	4.6
Material	SS	SS	SS	SS	SS
Solvent	S1	S2	S3	S4	S4
Flow rate (mℓ/min)	1.0	1.0	1.0	1.0	1.0
Detection	D1	D1	D1	D1	D1

Amino acid	k'_1	k'_2	k'_1	k'_2	k'_1	k'_2	k'_1	k'_2	k'_1	k'_2
Alanine			1.4	2.3	1.0	1.3	1.0	1.1	0.68	0.68
Aspartic acid			2.0	5.4	1.0	2.4	0.72	1.7	0.42	0.84
Glutamic acid			2.2	9.5	1.1	4.2	0.68	1.7	0.42	1.0
Methionine			2.1	8.6	1.8	5.6	1.3	4.1	1.0	2.5
Phenylalanine			7.5	7.5	4.7	5.3	4.2	4.8	2.8	3.5
Phenylglycine			6.6	33.4	5.1	18.4	4.6	11.4	3.7	7.8
Serine	0.48	0.70	0.37	0.47	0.21	0.25	0.16	0.16	0.10	0.10
Tyrosine	4.3	5.4	2.9	3.3	2.2	2.2	1.7	1.7	1.2	1.2

Table LC 99 (continued)
RETENTION OF DNP AMINO ACIDS AS A FUNCTION OF 1-PROPANOL CONTENT OF THE MOBILE PHASE

Note: When separation occurred, the L-enantiomer eluted first.

Packing:	P1	=	BSA silica, 10 μm particle size (Resolvosil®)
Solvents:	S1, S2, S3, S4, and S5	=	50 m*M* phosphate buffer, pH 7.7, containing 0, 1, 2, 3, and 4%, respectively, of 1-propanol
Detection:	D1	=	UV at 340 nm

REFERENCE

1. **Allenmark, S. and Andersson, S.,** *J. Chromatogr.*, 351, 231, 1986.

Table LC 100
RETENTION OF DNP AMINO ACIDS AS A FUNCTION OF pH OF THE MOBILE PHASE

Packing	P1		P1		P1		P1	
Length (mm)	150		150		150		150	
Diameter (mm, I.D.)	4.6		4.6		4.6		4.6	
Material	SS		SS		SS		SS	
Solvent	S1		S2		S3		S4	
Flow rate (m/min)	1.0		1.0		1.0		1.0	
Detection	D1		D1		D1		D1	
Amino acid	k'_1	k'_2	k'_1	k'_2	k'_1	k'_2	k'_1	k'_2
Alanine	2.6	2.9	1.8	2.1	1.3	1.6	1.0	1.2
Aspartic acid	2.4	19.2	1.2	8.1	0.95	5.1	0.63	2.4
Glutamic acid	2.5	2.5	1.2	1.2	0.80	1.1	0.58	1.1
Methionine	3.2	12.2	1.8	7.1	1.6	6.8	1.4	4.7
Phenylalanine	10.1	12.8	5.9	7.7	4.6	5.8	3.8	4.1
Phenylglycine	7.2	14.9	6.8	14.5	5.8	13.2	5.2	11.3
Serine	0.47	0.47	0.32	0.32	0.25	0.25	0.21	0.21
Tyrosine	3.9	5.2	2.5	3.3	2.0	2.4	1.7	1.9

Note: When separation occurred, the L-enantiomer eluted first.

Packing:	P1	=	BSA silica, 10 μm particle size (Resolvosil®)
Solvents:	S1, S2, S3, and S4	=	50 m*M* phosphate buffer/2% 1-propanol, the pHs of the buffers being 5.8, 6.4, 6.9, and 7.4, respectively
Detection:	D1	=	UV at 340 nm

REFERENCE

1. **Allenmark, S. and Andersson, S.,** *J. Chromatogr.*, 351, 231, 1986.

Table LC 101
RETENTION OF DNP AMINO ACIDS AS A FUNCTION OF BUFFER STRENGTH OF THE MOBILE PHASE

Packing	P1		P1		P1		P1	
Column								
Length (mm)	150		150		150		150	
Diameter (mm, I.D.)	4.6		4.6		4.6		4.6	
Material	SS		SS		SS		SS	
Solvent	S1		S2		S3		S4	
Flow rate (mℓ/min)	1.0		1.0		1.0		1.0	
Detection	D1		D1		D1		D1	
Amino acid	k'_1	k'_2	k'_1	k'_2	k'_1	k'_2	k'_1	k'_2
Alanine	1.2	1.4	1.2	1.4	1.0	1.4	0.90	1.2
Aspartic acid	1.2	5.5	0.84	2.9	0.68	2.1	0.50	1.6
Glutamic acid	2.0	5.2	0.79	2.0	0.64	1.7	0.50	1.3
Methionine	1.6	6.5	1.6	5.8	1.5	5.4	1.6	5.3
Phenylalanine	6.5	7.3	5.6	5.6	5.3	5.3	4.9	4.9
Phenylglycine	5.2	15.6	4.3	11.3	5.3	13.0	6.8	16.1
Serine	0.21	0.21	0.21	0.21	0.15	0.15	0.15	0.15
Tyrosine	2.0	2.4	1.9	1.9	1.9	1.9	2.0	2.2

Note: The L-enantiomer eluted first.

Packing:	P1	=	BSA silica, 10-μm particle size (Resolvosil®)
Solvents:	S1, S2, S3, and S4	=	phosphate buffers, pH 7.7/2% 1-propanol, the buffer strengths being 25, 50, 75, and 100 m*M*, respectively
Detection:	D1	=	UV at 340 nm

REFERENCE

1. **Allenmark, S. and Andersson, S.,** *J. Chromatogr.*, 351, 231, 1986.

Table LC 102
DABSYL (DIMETHYLAMINOAZOBENZENESULFONYL) AMINO ACIDS

Packing	P1	P1
Column		
Length (cm)	15	15
Diameter (cm, I.D.)	0.45	0.45
Solvent	S1[a]	S2[b]
Flow rate (mℓ/min)	1.0	1.0
Temperature (°C)	50	50
Detection	D1	D1
Amino acid	**t_r (min)**	
Alanine	15.48	44.80
β-Alanine		45.60
Ammonia	21.86	63.40
Arginine	17.31	49.65
Asparagine		37.53
Aspartic acid	9.17	26.05
Carboxymethyl cysteine		28.82
Carboxymethyl histidine		56.72
Cysteic acid		25.66

Table LC 102 (continued)
DABSYL (DIMETHYLAMINOAZOBENZENESULFONYL) AMINO ACIDS

Amino acid	t_r (min)	
Cysteine		49.01
Cystine	20.39	59.65
2,4-Diaminobutyric acid		66.20
2,3-Diaminopropionic acid		65.67
Glutamic acid	9.50	27.82
Glutamine		38.54
Glycine	15.07	43.67
Histidine	23.61	69.12
Homoserine		42.63
Homoserine lactone		48.01
δ-Hydroxylysine		66.65
4-Hydroxyproline		30.42
Isoleucine	18.34	52.67
Kynurenine		54.22
Leucine	18.63	53.57
Lysine	23.15	67.97
Methionine	17.79	51.62
Methionine sulfone		42.68
Methionine sulfoxide		42.15
3-Methylhistidine		45.47
N-ε-Methyllysine		70.87
Norleucine		54.87
3-Nitrotyrosine		74.75
Phenylalanine	19.47	55.82
Phosphoserine		24.90
Phosphothreonine		26.80
Phosphotyrosine		28.10
Proline	16.47	47.77
Serine	14.18	41.45
Threonine	14.69	42.77
Tyrosine	24.82	73.65
Tryptophan	19.03	54.82
Valine	16.87	48.90

[a] High speed system.
[b] High resolution system.

Packing: P1 = Novapak® C_{18} (4-μm spherical support, Waters Assoc.)

Solvent: S1 = solvent A is 9.0 m*M* sodium phosphate, pH 6.5 containing 4% dimethylformamide; solvent B is acetonitrile containing 4% dimethylformamide; a gradient in linear steps unless indicated otherwise is established as follows: 9% B at 0 min to 23% B at 8 min with a hyperbolic step, 37% B at 16 min, 46% B at 17 min, 49% B at 20 min, and 60% B at 21 min

S2 = solvents A and B are as for S1; the gradient is established as follows: 7% B from 0—5 min, 19% B at 32 min, 24% B at 42 min, 34% B at 55 min, 38% B at 56 min, 53% B at 78 min, and 60% B at 80 min

Table LC 102 (continued)
DABSYL (DIMETHYLAMINOAZOBENZENESULFONYL) AMINO ACIDS

Detection: D1 = absorption at 436 nm

REFERENCE

1. **Vendrell, J. and Aviles, F. X.,** *J. Chromatogr.*, 358, 401, 1986.

Table LC 103
AMINO ACIDS DERIVATIZED WITH 9-FLUORENYLMETHYL CHLOROFORMATE (FMOC-Cl)

Packing	P1
Column	
Length (mm)	500
Diameter (mm)	0.226
Material	Fused silica
Solvent	S1
Flow rate (μℓ/min)	2.0
Detection	D1

FMOC amino acid	t_r
Alanine	0.28
β-Alanine	0.26
β-Aminoisobutyric acid	0.28
Anserine	0.31
Arginine	0.22
Asparagine	0.47
Ethanolamine	0.22
Glutamic acid	0.28
Glutamine	0.46
Glycine	0.19
Histidine	0.26
Leucine	0.32
Lysine	0.20
3-Methylhistidine	0.27
Phenylalanine	0.39
Serine	0.44
Taurine	0.42
Threonine	0.23
Tyrosine	0.29
Valine	0.34

Packing: P1 = Spherisorb® octyl, 5 μm

Solvent: S1 = gradient, 0—15 min, 25% acetonitrile, 60% buffer, pH 4.08, 15% water (25:60:15), and 15—90 min, 25:60:15—65:28:7; the buffer is made of acetic acid, 50 m*M* in double-distilled water, the pH being adjusted with NaOH

Detection: D1 = fluorescence measurement, excitation at 265 nm, emission at 315 nm

REFERENCE

1. **Einarsson, S., Folestad, S., Josefsson, B., and Lagerkvist, S.,** *Anal. Chem.*, 58, 1638, 1986.

Table LC 104
COBALT(III) BIS(ETHYLENEDIAMINE) AMINO ACID COMPLEXES

Packing	P1
Column	
Length (cm)	30
Diameter (mm, I.D.)	3.9
Solvent	S1
Flow rate (mℓ/min)	2.5
Temperature (°C)	18
Detection	D1
Amino acid	**t_r(min)**
Glycine	5.6
Proline	6.8
Valine	8.8
Leucine	10.8
Phenylalanine	13.8

Packing: P1 = μBondapak® C_{18}, 10 μm
Solvent: S1 = 15-min linear gradient from 0—100% methanol containing 5 m*M* *p*-toluenesulfonate, pH 3.5
Detection: D1 = absorption at 480 nm

REFERENCE

1. **Buckingham, D. A., Clark, C. R., Tasker, R. F., and Hearn, M. T. W.,** *J. Liquid Chromatogr.*, 4, 689, 1981.

Table LC 105
THE RETENTION ENHANCEMENT EFFECT OF 18-CROWN-6

Compound	k'_o	k′
Phenylalanine	1.50	3.19
Tryptophan	2.68	6.28
Tyrosine	0.62	1.19
Phenylethylamine	1.40	13.9
Tryptamine	2.12	27.3
Tyramine	0.42	3.73

Note: k'_o and k′ are the capacity factors in the absence and presence, respectively, of 10 m*M* 18-crown-6.

Conditions:
- Packing: Chemcosorb® 5 ODSH (Chemco Scientific, Osaka, Japan)
- Column: 15 cm × 4.6 mm I.D.
- Solvent: Water/methanol, 80:20 brought to pH 2.5 with HCl
- Flow rate: 1.0 mℓ/min
- Temperature: 40°C
- Detection: UV at 200 or 210 nm

REFERENCE

1. **Nakagawa, T., Shibukawa, A., Kaihara, A., Itamochi, T., and Tanaka, H.,** *J. Chromatogr.*, 353, 399, 1986.

Table LC 106
AMINO ACID ENANTIOMERS

Packing	P1	P1	P1
Column			
Length (mm)	125	125	125
Diameter (mm)	4	4	4
Solvent	S1	S2	S3
Flow rate (mℓ/min)	1.0	1.0	1.0
Detection	D1	D1	D1
Amino acid		**k′**	
D-Alanine	0.40	0.24	0.36
L-Alanine	0.48	0.34	0.63
D-Alloisoleucine	6.64	4.73	5.44
L-Alloisoleucine	12.08	7.98	18.25
D-Aminobutyric acid	0.83	0.62	0.73
L-Aminobutyric acid	1.24	0.98	2.14
D-Arginine	0.15	0.46	0.22
L-Arginine	0.52	0.74	0.66
D-Asparagine	1.20	0.33	0.54
L-Asparagine	1.00	0.33	0.54
D-Glutamic acid	1.23	0.60	1.99

Table LC 106 (continued)
AMINO ACID ENANTIOMERS

Amino acid	k′		
L-Glutamic acid	1.33	0.69	3.24
Glycine	0.49	0.23	0.36
D-Histidine	6.78	4.54	1.64
L-Histidine	5.19	2.11	0.85
D-5-Hydroxytryptophan[a]	8.97	8.43	14.78
L-5-Hydroxytryptophan[a]	14.42	13.52	20.31
D-Isoleucine	7.96	5.43	5.66
L-Isoleucine	[b]	[b]	20.75
D-Leucine	10.64	6.84	7.05
L-Leucine	16.62	11.36	18.32
D-Methionine	7.53	5.44	6.78
L-Methionine	10.25	6.73	11.05
d-Methyl-DOPA[c]	8.00	6.85	9.21
L-Methyl-DOPA[c]	11.40	6.85	15.42
D-Norvaline	3.37	1.74	2.32
L-Norvaline	5.58	3.75	7.25
D-Norleucine	10.84	7.33	8.32
L-Norleucine	19.26	13.61	27.40
D-Phenylalanine[a]	16.87	15.35	25.58
L-Phenylalanine[a]	d	29.62	57.41
D-Phenylglycine	6.78	4.02	6.29
L-Phenylglycine	12.53	8.47	17.81
D-Proline	0.82	0.58	0.80
L-Proline	4.25	1.71	3.06
D-Serine	0.79	0.39	0.46
L-Serine	0.71	0.30	0.67
D-Threonine	1.42	0.66	0.68
L-Threonine	1.10	0.39	1.02
D-*tert*-leucine	6.36	5.88	7.26
L-*tert*-leucine	12.20	8.11	34.70
D-Trifluroalanine	3.22	1.34	1.68
L-Trifluoroalanine	5.53	3.68	5.20
D-Tryptophan[a]	42.87	41.57	92.12
L-Tryptophan[a]	60.07	60.31	100.29
D-Tyrosine	8.44	7.65	8.54
L-Tyrosine	17.40	14.26	24.28
D-Valine	2.77	1.52	1.91
L-Valine	4.73	2.74	7.12

[a] 10% methanol was added to the mobile phase.
[b] The peaks of L-alloisoleucine and L-isoleucine were not resolved; it was only possible to detect one broad peak.
[c] α-Methyl-1-3,4-dihydroxyphenylalanine.
[d] Only one peak.

Packing:	P1	=	LiChrosorb® RP-18, particle diameter 7 μm
Solvent:	S1	=	1 m*M* cupric acetate + 2 m*M* L-phenylalanine, adjusted to pH 4.5 with acetic acid
	S2	=	1 m*M* cupric acetate + 2 m*M* -diamethyl-L-phenylalanine adjusted to pH 4.5 with acetic acid
	S3	=	1 m*M* cupric acetate + 2 m*M* *N,N*-diamethyl-L-phenylalanine adjusted to pH 4.5 with acctic acid
Detection:	D1	=	UV at 254 nm

Table LC 106 (continued)
AMINO ACID ENANTIOMERS

REFERENCE

1. **Wernicke, R.,** *J. Chromatogr. Sci.*, 23, 39, 1985.

Table LC 107
THE EFFECT OF pH ON THE RETENTION OF AMINO ACID ENANTIOMERS

Packing	P1	P1	P1
Column			
Length (mm)	125	125	125
Diameter (mm)	4	4	4
Solvent	S1	S2	S3
Flow rate (mℓ/min)	1.0	1.0	1.0
Detection	D1	D1	D1
Amino acid		**k′**	
D-Alanine	0.06	0.15	0.40
L-Alanine	0.06	0.15	0.48
D-Histidine	0.59	3.79	6.78
D-Histidine	0.46	2.91	5.19
D-Leucine	3.41	7.12	10.64
L-Leucine	4.59	10.36	16.62
D-Methionine	1.86	4.92	7.53
L-Methionine	3.11	6.84	10.25
D-Phenylglycine	1.58	3.74	6.78
L-Phenylglycine	2.94	6.62	12.53
D-Proline	0.30	0.46	0.82
L-Proline	0.65	2.62	4.25
D-Threonine	0.20	0.70	1.42
L-Threonine	0.20	0.49	1.10
D-Trifluoroalanine	0.82	1.88	3.22
L-Trifluoroalanine	1.48	3.54	5.53
D-Tyrosine	3.41	5.95	8.44
L-Tyrosine	5.26	11.55	17.40
D-Valine	0.72	1.23	2.77
L-Valine	1.03	2.64	4.73

Packing: P1 = LiChrosorb® RP-18, particle diameter 7 μm

Solvent: S1 = 1 m*M* cupric acetate, 2 m*M* L-phenylalanine, adjusted to pH 3.5 with acetic acid

S2 = 1 m*M* cupric acetate, 2 m*M* L-phenylalanine adjusted to pH 4.0 with acetic acid

S3 = 1 m*M* cupric acetate, 2 m*M* L-phenylalanine adjusted to pH 4.5 with acetic acid

Detection: D1 = UV at 254 nm

REFERENCE

1. **Wernicke, R.,** *J. Chromatogr. Sci.*, 23, 39, 1985.

Table LC 108
THE EFFECT OF METHANOL CONCENTRATION ON THE RETENTION OF AMINO ACID ENANTIOMERS

Packing	P1	P1
Column		
Length (mm)	125	125
Diameter (mm)	4	4
Solvent	S1	S2
Flow rate (mℓ/min)	1.0	1.0
Detection	D1	D1
Amino acid	**k′**	
D-Leucine	7.05	3.87
L-Leucine	18.32	8.51
D-Norleucine	8.32	4.67
L-Norleucine	27.40	12.73
D-Norvaline	2.32	1.32
L-Norvaline	7.25	3.32
D-Phenylglycine	6.29	2.89
L-Phenylglycine	17.81	7.59
D-Tyrosine	8.54	3.55
L-Tyrosine	24.28	7.52
D-Valine	1.91	1.16
L-Valine	7.12	3.34

Packing: P1 = LiChrosorb® RP-18, particle diameter 7 μm

Solvent: S1 = 1 m*M* cupric acetate, 2 m*M*, *N*, *N*-dimethyl-L-phenylalanine, adjusted to pH 4.5 with acetic acid

S2 = 1 m*M* cupric acetate, 2 m*M*, *N*,*N*-dimethyl-L-phenylalanine, adjusted to pH 4.5 with acetic acid + 10% methanol.

Detection: D1 = UV at 254 nm

REFERENCE

1. **Wernicke, R.,** *J. Chromatogr. Sci.*, 23, 39, 1985.

Table LC 109
THE EFFECT OF AMMONIUM ACETATE CONCENTRATION ON THE RETENTION OF AMINO ACID ENANTIOMERS

Packing	P1	P1	P1
Column			
Length (mm)	125	125	125
Diameter (mm)	4	4	4
Solvent	S1	S2	S3
Flow rate (mℓ/min)	1.0	1.0	1.0
Detection	D1	D1	D1
Amino acid		**k′**	
D-Alanine	0.40	0.30	0.26
L-Alanine	0.48	0.38	0.33
D-Proline	0.82	0.70	0.66
L-Proline	4.25	2.95	2.65
D-Threonine	1.42	1.16	1.12
L-Threonine	1.10	0.89	0.80
D-Tyrosine	8.44	7.97	7.93
L-Tyrosine	17.40	15.50	15.36
D-Valine	2.77	2.09	2.00
L-Valine	4.73	3.50	3.29

Packing: P1 = LiChrosorb® RP-18, particle diameter 7 μm

Solvent: S1 = 1 m*M* cupric acetate + 2 m*M* L-phenylalanine, adjusted to pH 4.5 with acetic acid

S2 = 1 m*M* cupric acetate + 2 m*M* L-phenylalanine + 5 m*M* ammonium acetate, adjusted to pH 4.5 with acetic acid

S3 = 1 m*M* cupric acetate + 2 m*M* L-phenylalanine + 10 m*M* ammonium acetate adjusted to pH 4.5 with acetic acid

Detection: D1 = UV at 254 nm

REFERENCE

1. **Wernicke, R.,** *J. Chromatogr. Sci.*, 23, 39, 1985.

Table LC 110
RESOLUTION OF AMINO ACID ENANTIOMERS USING A CHIRAL MOBILE PHASE

Packing	P1	P1	P1	P2
Column				
Length (cm)	15	15	15	15
Diameter (mm)	4.6	4.6	4.6	4.6
Material	SS	SS	SS	SS
Solvent	S1	S2	S3	S4
Flow rate (mℓ/min)	0.2	0.25	0.5	0.25
Temperature	rt	rt	rt	rt
Detection	D1	D1	D1	D2
Amino acid		**t_r(min)[a]**		
D-Alanine	7.0	4.2		2.5
L-Alanine	10.0	6.2		3.1
D-Arginine	5.2			
L-Arginine	8.4			
D-Asparagine[b]	6.1			
L-Asparagine[b]	7.0			
D-Aspartic acid	21.0			
L-Aspartic acid	24.6			
D-Cysteic acid	12.3			
L-Cysteic acid	13.6			
D-Glutamic acid	40.0			
L-Glutamic acid	70.2			
D-Glutamine	8.7			
L-Glutamine	12.4			
Glycine	6.4	3.8		
D-Histidine	10.4			
L-Histidine	13.8			
D-Isoleucine		45.4		
L-Isoleucine		121.4		
D-Leucine		49.0		
L-Leucine		112.6		
D-Lysine	3.4			
L-Lysine	4.4			
D-Methionine	49.9	34.6		
L-Methionine	86.6	62.6		
D-Phenylalanine			14.7	
L-Phenylalanine			20.3	
D-Proline				4.5
L-Proline				10.5
D-Serine	6.8			
L-Serine	8.4			
D-Threonine	7.8			
L-Threonine	9.6			
D-Tryptophan			29.1	
L-Tryptophan			32.7	
D-Tyrosine			2.7	
L-Tyrosine			3.5	
D-Valine	22.6	15.4		
L-Valine	54.2	38.6		

[a] Retention times from dead volume, determined for each amino acid separately.

[b] The column was immersed in an ice bath.

Packing: P1 = Spherisorb®/LC-18, 5 μm
P2 = Supelco® LC-18

Table LC 110 (continued)
RESOLUTION OF AMINO ACID ENANTIOMERS USING A CHIRAL MOBILE PHASE

Solvent: S1 = water
S2 = 1% acetonitrile in water
S3 = 14% acetonitrile in water
S4 = 5% acetonitrile in water
The mobile phases contained 8.0×10^{-3} *M*, *N*,*N*-di-*n*-propyl-L-alanine and 4.0×10^{-3} *M* cupric acetate, pH 5.5

Detection: D1 = *o*-phthalaldehyde and fluorometric detection
D2 = ninhydrin derivatization and detection, alanine at 570 nm and proline at 440 nm

REFERENCE

1. **Weinstein, S., Engel, M. H., and Hare, P. E.,** *Anal. Biochem.*, 121, 370, 1982.

Table LC 111
ENANTIOMERS OF α-METHYL-α-AMINO ACIDS USING ELUENT CONTAINING THE COPPER COMPLEX OF *N*, *N*-DIPROPYL-L-ALANINE

Methylamino acid	**Acetonitrile conc (%)**	**t_r (min) I**	**t_r (min) II**
Methylaspartic acid	0	8.7	9.9
Methylglutamic acid	0	18.0	18.0
Methylhistidine	0	9.0	9.0
Methylnorvaline	2.5	6.0	6.0
Methylserine	0	4.7	5.7

Note: Peak assignment of all the methylamino acids had not been made.

Conditions:
Packing: Nucleosil® C_{18}
Column: 24 × 4.6 mm I.D., the column was equilibrated with an aqueous solution prepared from 2.5 mℓ of 0.32 *M*, *N*,*N*-dipropyl-L-alanine (DPA) plus 2.5 mℓ of 0.16 *M* Cu(II) acetate solution diluted to 100 mℓ.
Solvent: 8 m*M* DPA, 16 m*M* Cu(II) acetate with acetonitrile added as indicated in the table
Flow rate: 0.3 mℓ/min
Detection: *o*-Phthalaldehyde

REFERENCE

1. **Weinstein, S. and Grinberg, N.,** *J. Chromatogr.*, 318, 117, 1985.

Table LC 112
ENANTIOMERS OF α-METHYL-α-AMINO ACIDS USING ELUENT pH 5.5 CONTAINING THE COPPER COMPLEX OF *N, N*-DIMETHYL-L-VALINE

Methylamino acid	Acetonitrile conc (%)	t_r (min) I	t_r (min) II
Methylaspartic acid	0	7.52	9.10
Methylglutamic acid	0	11.5	13.7
Methylleucine	10	15.9	21.5
Methylmethionine	5	17.5	20.1
Methylnorleucine	10	19.1	25.5
Methylphenylalanine	15	18.1	21.9
Methylserine	0	7.7	10.9
Methyltryptophan	5	31.5	32.5

Note: Peak assignment of all the methylamino acids had not been made.

Conditions:
- Packing: Nucleosil® C_{18}
- Column: 24 × 4.6 mm I.D., the column was equilibrated with an aqueous solution prepared from 2.5 mℓ of 0.32 *M N,N*-dimethyl-L-valine (DMV) plus 2.5 mℓ of 0.16 *M* Cu(II) acetate solution diluted to 100 mℓ
- Solvent: Copper complex of DMV, (Cu)/(DMV) = 8.16 (m*M*/m*M*); pH = 5.5; acetonitrile was added as indicated in the table
- Flow rate 0.3 mℓ/min
- Detection: *o*-Phthalaldehyde

REFERENCE

1. **Weinstein, S. and Grinberg, N.,** *J. Chromatogr.*, 318, 117, 1985.

Table LC 113
ENANTIOMERS OF α-METHYL-α-AMINO ACIDS USING ELUENT pH 7 CONTAINING THE COPPER COMPLEX OF *N, N*-DIMETHYL-L-VALINE

Methylamino acid	Acetonitrile conc (%)	t_r(min) I	t_r(min) II
Methylaspartic acid	0	10.5	13.1
Methylglutamic acid	0	12.9	16.1
Methylleucine	10	16.3	22.3
Methylmethionine	5	23.9	28.9
Methylnorleucine	10	20.7	30.5
Methylphenylalanine	10	25.5	31.7
Methylserine	0	11.3	15.9
Methyltryptophan	10	41.3	50.7

Note: Peak assignment of all the methylamino acids had not been made.

Table LC 113 (continued)
ENANTIOMERS OF α-METHYL-α-AMINO ACIDS USING ELUENT pH 7 CONTAINING THE COPPER COMPLEX OF *N, N*-DIMETHYL-L-VALINE

Conditions: Packing: Nucleosil® C_{18}

Column: 25 × 4.6 mm I.D., the column was equilibrated with an aqueous solution prepared from 2.5 mℓ of 0.32 *M N,N*-dimethyl-L-valine (DMV) plus 2.5 mℓ of 0.16 *M* Cu(II) acetate solution diluted to 100 mℓ

Solvent: Copper complex of DMV dissolved in 0.1 *M* sodium acetate; (Cu)/(DMV) = 8.16 (m*M*/m*M*), pH = 7; acetonitrile was added as indicated in the table

Flow rate: 0.3 mℓ/min

Detection: *o*-Phthalaldehyde

REFERENCE

1. **Weinstein, S. and Grinberg, N.,** *J. Chromatogr.*, 318, 117, 1985.

Table LC 114
AMINO ACID ENANTIOMERS ON CHEMICALLY BONDED CHIRAL PHASES

Packing	P1	P2
Column		
Length (mm)	250	250
Diameter (mm)	4.6	4.6
Solvent	S1	S1
Flow rate (mℓ/min)	2	2
Temperature (°C)	50	50
Detection	D1	D1

Amino acid	k'_D	k'_L	k'_D	k'_L
Alanine	1.3	1.3	1.4	1.4
Arginine	2.4	3.0	2.4	2.4
Asparagine	3.0	4.1	2.6	3.6
Aspartic acid	3.2	4.1	5.9	7.2
Citrulline	1.8	2.1	6.87	6.87
Cystine	5.4	5.4		
DOPA	3.4	11.2	4.4	19.2
DOPS			5.4	14.4
Ethionine	4.0	4.6	5.2	5.2
Glutamic acid	2.0	2.0	4.0	4.0
Glutamine	1.0	1.0		
Histidine	6.7	12.1	7.6	17.6
Hydroxyproline			2.8	2.4
5-Hydroxytryptophan	7.4	25.2	7.33	37.4
Isoleucine	3.0	3.6	3.4	3.4
Kynurenine			12.4	14.8
Leucine	3.1	3.1	3.4	2.9
Lysine	2.0	2.2	1.6	1.6
Methionine	3.0	3.4	3.9	3.9
α-Methyl-DOPA			3.4	29.6

Table LC 114 (continued)
AMINO ACID ENANTIOMERS ON CHEMICALLY BONDED CHIRAL PHASES

Amino acid	k'_D	k'_L	k'_D	k'_L
Norleucine	2.9	3.2	2.8	2.8
Norvaline	2.0	2.2	2.2	2.2
Ornithine	0.6	0.75	1.4	1.4
Phenylalanine	3.2	9.4	5.2	12.2
Phenylglycine	6.7	12.1	3.4	3.8
Phenylserine	6.2	10.2	7.4	14.6
Pipecolic acid			0.2	0.2
Proline	2.4	1.4	3.1	1.3
Serine	2.0	3.2	2.2	3.4
Threonine	2.0	3.2	3.2	4.6
Tryptophan	7.8	27.4	9.2	39.8
o-Tyrosine	2.8	9.0	4.8	17.0
m-Tyrosine	2.2	6.0	4.4	14.0
p-Tyrosine	3.3	10.2	3.6	15.0
Valine	2.5	3.8	2.2	2.6

Abbreviation: DOPS = β,3-dihydroxytyrosine

Packing: P1 = The phase was synthesized by treating silica gel with 3-glycidoxyproplytrimethoxysilane and bonding L-proline to the product, which was then treated with aqueous Cu(II) sulfate solution

P2 = similar to P1, with L-hydroxyproline replacing L-proline

Solvent: S1 = 0.5 *M* KH_2PO_4, pH 4.5, containing 0.1 mmol Cu(II)

Detection: D1 = UV at 223 nm

REFERENCE

1. **Gübitz, G., Juffmann, F., and Jellenz, W.,** *Chromatographia,* 16, 103, 1982.

Table LC 115
AMINO ACID ENANTIOMERS ON CHEMICALLY BONDED CHIRAL PHASES

Packing	P1	P2
Column		
Length (mm)	250	250
Diameter (mm)	4.6	4.6
Solvent	S1	S1
Flow rate (mℓ/min)	2	2
Temperature (°C)	50	50
Detection	D1	D1

Amino acid	k'_D	k'_L	k'_D	k'_L
Alanine	2.08	2.08	1.40	1.25
Arginine	1.85	1.85	1.55	2.30
Asparagine	3.05	3.85	2.60	3.65
Aspartic acid	4.05	4.05	2.52	3.13
Citrulline			1.33	2.08

Table LC 115 (continued)
AMINO ACID ENANTIOMERS ON CHEMICALLY BONDED CHIRAL PHASES

Amino acid	k′D	k′L	k′D	k′L
Cystine			6.43	6.43
DOPA	2.55	7.40	4.48	16.63
DOPS	5.0	9.2	4.48	11.45
Ethionine	3.95	3.95	3.05	4.18
Glutamic acid	2.45	2.45	1.85	2.30
Histidine			5.08	7.93
Hydroxyproline	2.30	1.78	2.60	2.60
5-Hydroxytryptophan	8.23	20.45	11.6	31.4
Isoleucine	2.60	2.60	1.70	2.75
Kynurenine			6.28	11.15
Leucine	3.50	31.05	2.08	2.60
Lysine	1.33	1.33	1.10	1.55
Methionine	3.20	3.20	2.52	3.35
α-Methyl-DOPA	3.35	12.65	5.08	20.45
Norleucine	2.45	2.45	1.85	2.83
Norvaline	2.75	2.75	1.55	2.15
Ornithine	1.70	1.70	1.63	2.15
Phenylalanine	4.70	7.85	4.03	8.99
Phenylglycine			2.30	3.13
Phenylserine	6.70	10.55	4.63	10.25
Pipecolic acid	0.35	0.35	0.28	0.35
Proline	1.63	0.65	1.55	1.18
Serine	2.08	2.68	1.63	2.83
Threonine	3.13	3.65	2.38	3.88
Tryptophan	9.50	23.6	12.2	31.1
o-Tyrosine	4.1	10.1	4.85	19.25
m-Tyrosine	3.65	7.10	3.88	12.58
p-Tyrosine	3.50	8.30	4.33	14.45
Valine	3.05	3.05	1.40	2.68

Abbreviation: DOPS = β,3-dihydroxy tyrosine

Packing: P1 = the phase was synthesized by treating silica gel with 3-glycidoxypropyltrimethoxysilane and bonding L-azetidine carboxylic acid to the product, which was then treated with aqueous Cu(II) sulfate solution

P2 = similar to Pl, with L-pipecolic acid replacing L-azetidine carboxylic acid

Solvent: S1 = 0.5 *M* KH_2PO_4, pH 4.5, containing 0.1 mmol Cu(II)

Detection: D1 = UV at 223 nm

REFERENCE

1. **Gübitz, G., Juffmann, F., and Jellenz, W.,** *Chromatographia,* 16, 103, 1982.

Table LC 116
AMINO ACID ENANTIOMERS ON A CHEMICALLY BONDED CHIRAL PHASE

Packing	P1
Column	
Length (mm)	250
Diameter (mm)	4.6
Solvent	S1
Flow rate (mℓ/min)	2
Temperature	rt
Detection	D1

Amino acid	k′$_D$	k′$_L$
Arginine	0.94	0.94
Asparagine	4.17	3.67
Aspartic acid	10.0	8.0
DOPA	4.9	4.9
Ethionine	5.57	6.14
Glutamic acid	6.86	5.57
Lysine	0.62	0.62
Methionine	4.58	4.08
Phenylalanine	6.00	6.00
Proline	2.50	1.60
Serine	2.71	2.36
Threonine	4.00	3.50
Tryptophan	22.7	22.7
Tyrosine	7.00	7.00
Valine	1.50	1.50

Packing: P1 = the phase was synthesized by treating silica gel with 3-glycidoxypropyltrimethoxysilane and bonding L-phenylalanine to the product which was then treated with aqueous Cu(II) sulfate solution

Solvent: S1 = 0.05 *M* KH_2PO_4, pH 4.5, containing 0.1 mmol Cu(II)

Detection: UV at 223 nm

REFERENCE

1. **Gübitz, G., Juffmann, F., and Jellenz, W.,** *Chromatographia,* 16, 103, 1982.

Table LC 117
THE INFLUENCE OF pH ON THE RESOLUTION OF AMINO ACID ENANTIOMERS BY MIXED CHELATE COMPLEXATION

Packing	P1
Column	
Length (cm)	10
Diameter (mm, I.D.)	4.0
Material	SS
Solvent	S1
Flow rate (mℓ/min)	1.0
Temperature (°C)	30
Detection	D1

	k′		
Amino acid	**pH 5.0**	**pH 6.0**	**pH 7.0**
L-Alanine	2.00	2.40	3.60
D-Alanine	2.00	3.24	5.00
L-Glutamic acid	1.60	2.63	3.40
D-Glutamic acid	1.60	2.20	2.70
L-Tyrosine	6.44	10.51	16.25
D-Tyrosine	8.38	21.86	36.40
L-Valine	3.00	8.25	11.20
D-Valine	3.80	16.90	23.30

Packing: P1 = chemically bonded *n*-octylsilyl silica gel Develosil® C8, particle size 5 μm (Nomura Chemical, Seto-shi, Japan); unreacted accessible silanol groups were capped with trimethylsilane

Solvent: S1 = aqueous solution containing 1m*M* *N*-(*p*-toluenesulfonyl)-L-phenylalanine and 0.5 m*M* $CuSO_4{\cdot}5H_2O$; the pH is adjusted with 5% aqueous Na_2CO_3

Detection: D1 = fluorescence measurement after reaction of the eluent with *o*-phthalaldehyde

REFERENCE

1. **Nimura, N., Suzuki, T., Kasahara, Y., and Kinoshita, T.,** *Anal. Chem.,* 53, 1380, 1981.

Table LC 118
RESOLUTION OF RACEMIC AMINO ACIDS USING CHIRAL ELUENTS

Amino acid	k′ TosPhe[a]	k′ TosPhG[b]	Acetonitrile (%)
L-Alanine	5.50	7.60	0
D-Alanine	7.20	6.00	0
L-Arginine	14.00	6.34	15
D-Arginine	12.60	8.38	15
L-Asparagine	6.40	5.50	0
D-Asparagine	5.90	6.14	0
L-Aspartic acid	6.20	2.30	0
D-Aspartic acid	5.00	2.60	0
L-Glutamic acid	8.60	5.60	0
D-Glutamic acid	6.60	6.80	0
L-Glutamine	7.40	6.20	0
D-Glutamine	7.40	8.00	0
L-Histidine	6.00	1.64	15
D-Histidine	3.40	2.05	15
L-Isoleucine	4.80	8.12	15
D-Isoleucine	8.20	3.90	15
L-Leucine	7.20	9.40	15
D-Leucine	10.40	5.60	15
L-Lysine	11.60	6.05	15
D-Lysine	10.60	7.90	15
L-Norleucine	7.60	11.40	15
D-Norleucine	12.30	6.00	15
L-Norvaline	2.70	4.40	15
D-Norvaline	4.00	2.55	15
L-Phenylalanine	16.00	24.71	15
D-Phenylalanine	26.00	12.19	15
L-Phenylglycine	4.90	10.08	15
D-Phenylglycine	10.00	3.90	15
L-Tryptophan	22.00	34.20	15
D-Tryptophan	36.10	17.70	15
L-Valine	7.80	14.10	10
D-Valine	13.80	10.55	10

[a] TosPhe = *N*-(*p*-toluenesulfonyl)-L-phenylalanine.
[b] TosPhG = *N*-(*p*-toluenesulfonyl)-L-phenylglycine.

Conditions:
- Column: Develosil® ODS, particle size 5 μm (Nomura Chemical, Seto-shi, Japan) in a stainless steel column 10 cm × 4.0 mm I.D.
- Solvent: An aqueous solution containing 1 m*M* TosPhe or TosPhG, 0.5 m*M* $CuSO_4 \cdot 5H_2O$, and acetonitrile (percentage indicated in the table), pH 6.0; the pH is adjusted with 5% aqueous Na_2CO_3
- Flow rate: 1.0 mℓ/min
- Temperature: 30°C
- Detection: Fluorescence measurement after reaction of the eluent with *o*-phthalaldehyde

REFERENCE

1. **Nimura, N., Toyama, A., Kasahara, Y., and Kinoshita, T.,** *J. Chromatogr.*, 239, 671, 1982.

Table LC 119
THE RESOLUTION OF AMINO ACID ENANTIOMERS WITH COPPER-ASPARTAME ELUANTS

Packing	P1	P1	P1	P1
Column				
Length (cm)	25	25	25	25
Solvent	S1	S2	S3	S4
Flow rate (mℓ/min)	2.0	2.0	2.0	2.0
Temperature (°C)	32	32	32	32
Detection	D1	D1	D1	D1
Amino acid		**k′**		
L-DOPA	3.2			
D-DOPA	4.8			
L-Phenylalanine	[a]	4.8	4.3	
D-Phenylalanine	[a]	7.8	5.4	
L-Tryptophan	[a]	[a]	13.2	5.6
D-Tryptophan	[a]	[a]	16.7	6.8
L-Tyrosine	5.4	1.2	1.0	
D-Tyrosine	8.9	1.8	1.7	

[a] Retention times too long for accurate measurement.

Packing:	P1	=	ODS
Solvent:	S1	=	10^{-3} *M* copper/aspartame (L-aspartyl-L-phenylalanine methyl ester) complex in water
	S2	=	7% acetonitrile in 10^{-3} *M* copper/aspartame complex solution
	S3	=	8% acetonitrile in 10^{-3} *M* copper/aspartame complex solution
	S4	=	10% acetonitrile in 10^{-3} *M* copper/aspartame complex solution
Detection:	D1	=	UV at 275 or 254 nm.

REFERENCE

1. **Gilon, C., Leshem, R., Tapuhi, Y., and Grushka, E.,** *J. Am. Chem. Soc.*, 101, 7612, 1979.

Table LC 120
THE RESOLUTION OF AMINO ACID ENANTIOMERS WITH ZINC-ASPARTAME ELUANTS

Packing	P1	P1	P1
Column			
Length (cm)	25	25	25
Solvent	S1	S2	S3
Flow rate (mℓ/min)	2.0	2.0	2.0
Temperature (°C)	32	32	32
Detection	D1	D1	D1
Amino acid		**k′**	
L-Phenylalanine	5.5	9.3	16.0
D-Phenylalanine	9.0	16.0	30.0
L-Tryptophan	19.4	34.0	a
D-Tryptophan	23.0	42.0	a
L-Tyrosine	1.9	2.9	5.3
D-Tyrosine	2.5	4.2	7.9

[a] Retention times too long for accurate measurement.

Packing: P1 = ODS

Solvent: S1 = 10^{-3} *M* zinc-aspartame (L-aspartyl-L-phenylalanine methyl ester) complex in water

S2 = 5×10^{-4} *M* zinc/aspartame complex in water

S3 = 2.5×10^{-4} *M* zinc/aspartame complex in water

Detection: D1 = UV at 275 or 254 nm.

REFERENCE

1. **Gilon, C., Leshem, R., Tapuhi, Y., and Grushka, E.,** *J. Am. Chem. Soc.*, 101, 7612, 1979.

Table LC 121
THE RESOLUTION OF AMINO ACID ENANTIOMERS USING L-ASPARTYLCYCLOHEXAMIDE-Cu(II) (AspcHex-Cu[II])

Packing	P1	P1	P1
Column			
Length (cm)	25	25	25
Diameter (mm, I.D.)	4.1	4.1	4.1
Solvent	S1	S2	S3
Flow rate (mℓ/min)	2.0	2.0	2.0
Temperature (°C)	34	34	34
Detection	D1	D1	D1
Amino acid		**k′**	
L-Cystine	0.76	1.81	
D-Cystine	1.15	2.43	
L-DOPA	1.78	3.21	5.67
D-DOPA	2.33	3.90	6.86
L-Ethionine	4.65	7.14	10.7
D-Ethionine	6.64	9.43	12.8
L-Isoleucine	1.76	2.86	4.5
D-Isoleucine	3.76	5.0	7.0
L-Leucine	2.35	3.23	5.19
D-Leucine	3.88	4.63	6.81
L-Methionine	1.41	2.52	3.7
D-Methionine	2.06	3.29	4.4
L-Norleucine	2.47	3.67	5.13
D-Norleucine	4.23	5.40	6.76
L-Norvaline	0.56	1.14	1.74
D-Norvaline	0.89	1.43	2.0
L-Proline	0.24	0.55	2.17
D-Proline	1.12	1.42	3.0
L-Tyrosine	3.03	5.02	7.03
D-Tyrosine	4.5	6.98	8.86
L-Valine	0.41	1.07	1.83
D-Valine	0.82	1.52	2.17

Packing: P1 = ODS bonded to Partisil® 10, it is prepared by refluxing chlorodimethyloctadecyl silane with the silica gel in dry toluene

Solvent: S1 = 1×10^{-3} *M* AspcHex-Cu(II)
S2 = 2.5×10^{-4} *M* AspcHex-Cu(II)
S3 = 1×10^{-4} *M* AspcHex-Cu(II)
These mobile phases are prepared by dissolving the chiral reagent and Cu(II) in triple-distilled water.

Detection: D1 = UV at 230 nm

REFERENCE

1. **Gilson, C., Leshem, R., and Grushka, E.,** *Anal. Chem.*, 52, 1206, 1980.

Table LC 122
THE CHROMATOGRAPHY OF AMINO ACIDS USING AQUEOUS MOBILE PHASES CONTAINING COPPER IONS AND ALKYSULFONATE

Packing	P1	P1	P1
Column			
Length (mm)	250	250	250
Diameter (mm)	4	4	4
Solvent	S1	S2	S3
Temperature (°C)	30	30	30
Detection	D1	D1	D1

Amino acid		**k′**	
Arginine	6.78	0.55	33.5
Asparagine	0.11	0.05	0.52
Glycine	0.09	0.03	0.35
Histidine	1.93	0.35	1.65
Hydroxyproline	0.28	0.21	0.61
Lysine	1.31	0.17	5.19
Methionine	0.87	1.95	6.0
Tyrosine	1.42	3.8	9.87
Valine	0.44	1.16	3.85

Packing: P1 = Merck LiChrosorb® RP-18 cartridge packed with 7-μm particles; a Merck 4 × 4-mm guard cartridge was connected to the column

Solvent: S1 = 0.1 *M* sodium acetate buffer, pH 5.6 + 5 m*M* sodium heptanesulfonate

S2 = 0.1 *M* sodium acetate buffer, pH 5.6 + 0.5 m*M* Cu(II)

S3 = 0.1 *M* sodium acetate buffer, pH 5.6 + 5 m*M* sodium heptanesulfonate + 0.5 m*M* Cu(II)

Detection: D1 = UV at 235 nm

REFERENCE

1. **Levin, S. and Grushka, E.,** *Anal. Chem.*, 57, 1830, 1985.

Table LC 123
AMINO ACIDS USING DIFFERENT CHIRAL REAGENTS

Packing	P1	P1	P1
Column			
Length (cm)	25	25	25
Diameter (mm, I.D.)	4.1	4.1	4.1
Solvent	S1	S2	S3
Flow rate (mℓ/min)	2	2	2
Temperature (°C)	34	34	34
Detection	D1	D1	D1

Amino acid		**k′**	
L-DOPA	2.08[a]	2.39	3.25
D-DOPA	2.41[a]	3.26	4.42
L-Leucine	2.53[a]	2.63	3.37
D-Leucine	2.94[a]	3.95	5.89
L-Methionine		1.93	2.43
D-Methionine		2.51	3.39
L-Proline	0.31	0.50	0.55
D-Proline	0.34	1.89	2.4
L-Tyrosine	1.58	3.81	5.11
D-Tyrosine	1.71	5.69	7.39

[a] Results obtained with phosphate buffer at pH 6.

Packing: S1 = ODS

Solvent: S1 = L-aspartyl ethylamide (AspnEth) 6×10^{-4} *M*, Cu(II) 3×10^{-4} *M*, 0.05 *M* phosphate buffer, pH 5.0

S2 = L-aspartyl-*n*-butylamide (AspnBut), 6×10^{-4} *M*, Cu(II) 3×10^{-4} *M*, 0.05 *M* phosphate buffer, pH 5.0

S3 = L-aspartyl-*n*-hexylamide (AspnHex), Cu(II) 3×10^{-4} *M*, 0.05 *M* phosphate buffer, pH 5.0

The complexes were prepared by dissolving the appropriate amount of $CuCl_2$ and the aspartates in phosphate buffer

Detection: D1 = UV at 230 nm

REFERENCE

1. **Gilon, C., Leshem, R., and Grushka, E.,** *J. Chromatogr.*, 203, 365, 1981.

Table LC 124
AMINO ACIDS USING DIFFERENT CHIRAL REAGENTS

Packing	P1	P1
Column		
Length (cm)	25	25
Diameter (mm, I.D.)	4.1	4.1
Solvent	S1	S2
Flow rate (mℓ/min)	2	2
Temperature (°C)	34	34
Detection	D1	D1

Amino acid	**k′**	
L-DOPA	1.53	1.68
D-DOPA	1.89	1.99
L-Leucine	1.60	1.70
D-Leucine	2.40	2.70
L-Methionine	1.14	1.37
D-Methionine	1.55	1.81
L-Tyrosine	2.53	2.43
D-Tyrosine	3.33	3.16
L-Valine	0.52	0.53
D-Valine	0.80	0.81

Packing: P1 = ODS

Solvent: S1 = L-aspartyl ethylamide (AspnEth) 6×10^{-4} *M*, Cu(II) 3×10^{-4} *M*, 0.05 *M* phosphate buffer, pH 4.0

S2 = L-aspartyl-*n*-octylamide (Aspn-Oct), 6×10^{-4} *M*, Cu(II) 3×10^{-4} *M*, 0.05 *M* phosphate buffer, pH 4.0, containing 10% methanol

The complexes were prepared by dissolving the appropriate amount of $CuCl_2$ and the aspartates in phosphate buffer

Detection: D1 = UV at 230 nm

REFERENCE

1. **Gilon, C., Leshem, R., and Grushka, E.,** *J. Chromatogr.,* 203, 365, 1981.

Table LC 125
THE EFFECT OF pH ON THE SEPARATION OF AMINO ACIDS USING CHIRAL ELUENTS

Packing	P1	P1	P1	P1
Column				
Length (cm)	25	25	25	25
Diameter (mm, I.D.)	4.1	4.1	4.1	4.1
Solvent	S1	S2	S3	S4
Flow rate (mℓ/min)	2	2	2	2
Temperature (°C)	34	34	34	34
Detection	D1	D1	D1	D1

Amino acid		**k′**		
L-Alanine	0	0.22		0.38
D-Alanine	0	0.22		0.47
L-DOPA	1.53	3.25		
D-DOPA	1.89	4.42		
L-Histidine	0.31	1.42		5.68
D-Histidine	0.31	1.42		4.68
L-Leucine	1.60	3.37		
D-Leucine	2.40	5.89		
L-Proline	0.32	0.55	0.57	1.06
D-Proline	0.79	2.43	3.20	5.38
L-Threonine	0.13	0.49		1.02
D-Threonine	0.13	0.49		0.94
L-Tyrosine	2.53	5.11	5.54	10.2
D-Tyrosine	3.33	7.39	9.49	18.1
L-Valine	0.52	0.91	0.99	1.79
D-Valine	0.80	1.55	1.73	4.21

Packing: P1 = ODS

Solvent: S1 = L-aspartyl-*n*-hexylamide (AspnHex), 6 × 10^{-4} *M*, Cu(II) 3 × 10^{-4} *M*, 0.05 *M* phosphate buffer, pH 4

S2 = L-aspartyl-*n* hexylamide (AspnHex), 6 × 10^{-4} *M*, Cu(II) 3 × 10^{-4} *M*, 0.05 *M* phosphate buffer, pH 5

S3 = L-aspartyl cyclohexamide (AspcHex), 6 × 10^{-4} *M*, Cu(II) 3 × 10^{-4} *M*, 0.05 *M* phosphate buffer, pH 5

S4 = L-aspartyl cyclohexamide (AspcHex), 6 × 10^{-4} *M*, Cu(II) 3 × 10^{-4} *M*, 0.05 *M* phosphate buffer, pH 6

The complexes were prepared by dissolving the appropriate amount of $CuCl_2$ and the aspartates in phosphate buffer.

Detection: D1 = UV at 230 nm

REFERENCE

1. **Gilon, C., Leshem, R., and Grushka, E.,** *J. Chromatogr.*, 203, 365, 1981.

Table LC 126
THE EFFECT OF THE METAL ION ON THE SEPARATION OF AMINO ACIDS USING CHIRAL ELUENTS

Packing	P1	P1
Column		
Length (cm)	25	25
Diameter (mm, I.D.)	4.1	4.1
Solvent	S1	S2
Flow rate (mℓ/min)	2	2
Temperature (°C)	34	34
Detection	D1	D1
Amino acid	**k′**	
L-Proline	0.57	0.26
D-Proline	3.20	0.62
L-Tyrosine	5.54	1.92
D-Tyrosine	9.5	3.04
L-Valine	0.99	0.36
D-Valine	1.73	0.56

Packing: P1 = ODS

Solvent: S1 = L-aspartyl cyclohexamide (AspcHex), 6×10^{-4} *M*, Cu(II) 3×10^{-4} *M*, 0.05 *M* phosphate buffer, pH5

S2 = L-aspartyl cyclohexamide, 6×10^{-4} *M*, Ni(II) 3×10^{-4} *M*, 0.05 *M* phosphate buffer, pH 5

The complexes were made by dissolving appropriate amounts of $CuCl_2$ or $NiCl_2$ and the aspartate in phosphate buffer

Detection: D1 = UV at 230 nm

REFERENCE

1. **Gilon, C., Leshem, R., and Grushka, E.,** *J. Chromatogr.*, 203, 365, 1981.

Table LC 127
THE EFFECT OF pH ON THE SEPARATION OF AMINO ACIDS USING A CHIRAL ELUENT

Packing	P1	P1
Column		
Length (cm)	25	25
Diameter (mm, I.D.)	4.1	4.1
Solvent	S1	S2
Flow rate (mℓ/min)	2	2
Temperature (°C)	34	34
Detection	D1	D1
Amino acid	**k′**	
L-Proline	0.26	0.74
D-Proline	0.62	2.86
L-Valine	0.36	1.45
D-Valine	0.56	3.42

Table LC 127 (continued)
THE EFFECT OF pH ON THE SEPARATION OF AMINO ACIDS USING A CHIRAL ELUENT

Packing:	P1 =	ODS
Solvent:	S1 =	L-aspartyl cyclohexamide (AspcHex), 6×10^{-4} *M*, Ni(II) 3×10^{-4} *M*, 0.05 *M* phosphate buffer, pH 5
	S2 =	L-aspartyl cyclohexamide (AspcHex), 6×10^{-4} *M*, Ni(II) 3×10^{-4} *M*, 0.05 *M* phosphate buffer, pH 7
	The complex is prepared by dissolving the appropriate amount of $NiCl_2$ and the aspartate in phosphate buffer	
Detection:	D1 =	UV at 230 nm

REFERENCE

1. **Gilon, C., Leshem, R., and Grushka, E.,** *J. Chromatogr.*, 203, 365, 1981.

Table LC 128
AMINO ACID ENANTIOMERS AT LOW CHIRAL REAGENT CONCENTRATIONS

Packing	P1
Column	
Length (cm)	25
Diameter (mm, I.D.)	4.1
Solvent	S1
Flow rate (mℓ/min)	1.0
Temperature (°C)	35
Detection	D1

Amino acid	**k′**
L-Proline	0.32
D-Proline	0.43
L-Tyrosine	2.36
D-Tyrosine	2.71
L-Valine	0.62
D-Valine	0.72

Packing:	P1 =	ODS, 10 μm
Solvent:	S1 =	0.05 m*M* aspartyl-c-hexylamide, 0.025 m*M* $CuCl_2$ in phosphate buffer, pH 4.8
Detection:	D1 =	UV at 230 nm

REFERENCE

1. **Grushka, E., Leshem, R., and Gilon, C.,** *J. Chromatogr.*, 255, 41, 1983.

Table LC 129
LIGAND EXCHANGE CHROMATOGRAPHY OF AMINO ACID ENANTIOMERS

Packing	P1
Column	
Length (cm)	15
Diameter (cm, I.D.)	0.46
Solvent	S1
Temperature	rt
Detection	UV

Amino acid	**k'_2**	**α**
Alanine	13.9	1.12
Isoleucine	6.58	1.16
Leucine	6.12	1.29
Phenylalanine	2.61	1.39
Pipecolic acid	4.11	1.27
Proline	25.5	2.4
Tryptophan	7.11	1.49
Valine	6.66	1.30

Note: k'_2 refers to the most retained isomer.

Packing: P1 = a linear polymer based on a copolymer of acrylamide and a vinylpyridine derivative, whose structure is shown below
L-Proline units formed 48% of the polymer which was treated with a copper salt solution before chromatography; number of copper atoms/number of L-proline residues = 0.8

Solvent: S1 = water/acetonitrile, 30:70, ammonia 0.4 *N*, NH_4Cl 0.002 *N*

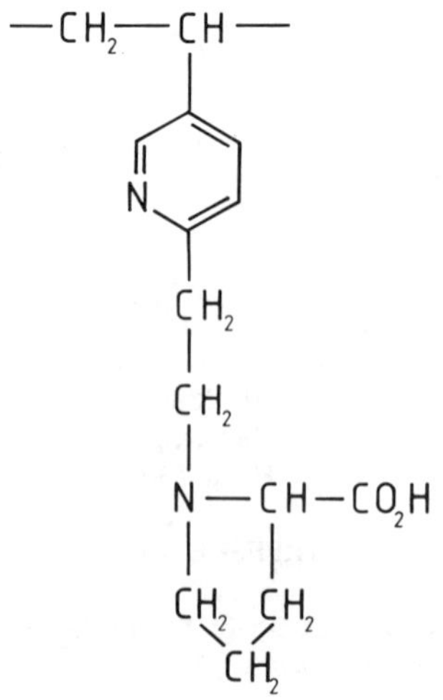

REFERENCE

1. **Charmot, D., Audebert, R., and Quivoron, C.,** *J. Liquid Chromatogr.*, 8, 1769, 1985.

Table LC 130
RESOLUTION OF DL-AMINO ACIDS ON NATIVE CELLULOSE

Amino acid	t_r(min) D	t_r(min) L	Solvent
Alanine	1126	1167	S1
α-Aminobutyric acid	759	827	S1
Cysteine	990	1040	S3
β-3,4-Dihydroxyphenylalanine	503	592	S2
Histidine	925	820	S5
Isoleucine	512	547	S1
Leucine	523	540	S1
Methionine	634	695	S1
Norleucine	558	625	S1
Norvaline	648	718	S1
Phenylalanine	641	700	S2
Proline	710	770	S4
Serine	1263	1337	S2
Threonine	1006	927	S2
Tryptophan	380	345	S5
Tyrosine	596	674	S2
Valine	625	663	S1

Solvent: S1 = pyridine/ethanol/water, 5:1:1
S2 = pyridine/ethanol/water, 4:1:1
S3 = ethanol/water, 6:1
S4 = pyridine/ethanol/water, 3:1:1
S5 = pyridine/ethanol/water, 1:1:1

Conditions: A column, 0.85 × 250 cm, was packed with washed cellulose and equilibrated with elution buffer; separations were made at room temperature, except for cysteine (0—5°C), with a flow rate of 30 mℓ/hr; detection was performed using ninhydrin, peaks being identified by circular dichroism spectra or cochromatography

REFERENCE

1. **Yuasa, S., Itoh, M., and Shimada, A.,** *J. Chromatogr. Sci.,* 22, 288, 1984.

Table LC 131
RESOLUTION OF AMINO ACID ENANTIOMERS USING A STATIONARY PHASE CONTAINING L-PROLINE RESIDUES

Packing	P1
Column	
Length (cm)	25
Diameter (cm, I.D.)	0.48
Solvent	S1
Flow rate (mℓ/min)	0.4
Detection	D1

Amino acid	**k′$_L$**	**k′$_D$**
Alanine	1.80	1.80
Aminobutyric acid	1.94	1.49
Asparagine	4.16	3.29
Isoleucine	4.37	2.84
Norleucine	4.81	3.94
Norvaline	2.34	1.80
Phenylalanine	12.35	6.10
Proline	2.06	3.20
Serine	4.20	2.27
Threonine	5.31	2.80
Tryptophan	38.28	15.73
Tryosine	36.27	16.31
Valine	3.09	1.76

Packing: P1 = a chiral polymer obtained by treating a linear polyacrylamide with L.-proline was adsorbed on a silica support and complexed with Cu(II)

Solvent: S1 = 0.1 *M* KNO_3 with 0.2 × 10^{-4} *M* Cu(II)

Detection: D1 = absorption in UV or refractive index measurements

REFERENCE

1. **Boué, J., Audubert, R., and Quivoron, C.,** *J. Chromatogr.*, 204, 185, 1981.

Table LC 132
ENANTIOMERS OF *N*-ACETYL AMINO ACID *TERT*-BUTYL ESTERS

Amino acid	Solvent	k'_D	k'_L
Leu	S1	3.12	4.33
Val	S1	3.41	4.69
Nle	S1	3.13	4.28
Nva	S1	3.63	4.89
Abu	S1	4.12	5.37
Ala	S1	4.83	5.95
Ile	S1	3.16	4.32
O-*t*-BuSer	S1	2.13	2.82
O-AcTyr	S1	7.67	9.33
O-*t*-BuAsp	S1	2.57	3.10
O-*t*-BuGlu	S1	3.57	4.33
S-BzlCys[a]	S2	1.77	2.31
N-*t*-BuTrp[a]	S2	1.88	2.58
PheGly[a]	S3	2.28	3.02
Phe[a]	S3	1.96	2.71
N-AcLys[b]	S4	6.83	7.17
Gln[b]	S5	17.63	
Pro[b]	S6	2.77	2.77

[a] Detection at 254 nm.
[b] Column temperature 40°C.

Conditions:	Column:	Stainless steel, 20 × 0.4 cm, I.D., containing approximately 2 g of (*N*-formyl-L-valylamino) propyl (FVA) silica gel
	Solvent:	The binary solvents contained the following percentages (v/v) of second solvent in the first solvent, *n*-hexane; S1, Et_2O, 80; S2, CH_2Cl_2, 30; S3, $CHCl_3$, 30; S4, 2-PrOH, 12; S5, 2-PrOH, 8; S6, 2-PrOH, 4
	Flow rate:	1 mℓ/min
	Temperature:	20°C
	Detection:	UV at 230 nm

REFERENCE

1. **Dobashi, A., Oka, K., and Hara, S.,** *J. Am. Chem. Soc.*, 102, 7122, 1980.

Table LC 133
SEPARATION OF ENANTIOMERIC AMINO ACID DERIVATIVES USING A TARTARIC ACID DERIVATIVE

R_1	R_2	X	Y	k'_1	k'_2
Ph	*tert*-Bu	NH	AcNH	2.14	2.53
Bzl	*tert*-Bu	NH	AcNH	2.26	2.90

Note: R_1, R_2, X, and Y refer to groups in the compound

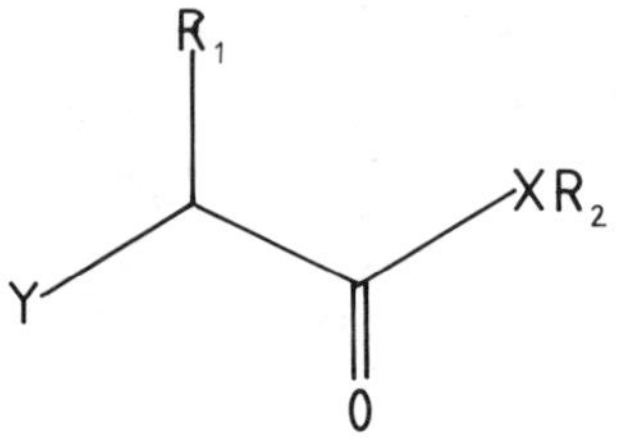

k'_1 is the capacity ratio of the lesser retained enantiomer, k'_2 that of the enantiomer retaining more.

Conditions:

Conditions:	Packing:	Nucleosil® 100-5, 5-μm spherical silica gel
	Column:	stainless steel, 50 × 0.1 cm I.D.
	Solvent:	chloroform-*n*-hexane, 8:7 containing 0.165 mol % of L (+)-*N,N'*-diisopropyl tartramide
	Flow rate:	60 μℓ/min
	Temperature:	30°C
	Detection:	UV at 254 nm

REFERENCE

1. **Dobashi, Y., Dobashi, A., and Hara, S.,** *Tetrahedron Lett.*, 25, 329, 1984.

Table LC 134
THE SEPARATION OF RACEMIC ALKYLAMINO ACIDS BY HPLC USING A CHIRAL MOBILE PHASE

Packing	P1
Column	
Length (mm)	300
Diameter (mm, I.D.)	4
Material	SS
Solvent	S1
Flow rate (mℓ/min)	2.0
Temperature	r t
Detection	RI

Alkylamino acid	k'
L-2-Aminobutyric acid	1.21
D-2-Aminobutyric acid	1.58
L-Norvaline	2.63
D-Norvaline	4.13
L-Norleucine	9.25
D-Norleucine	16.33

Table LC 134 (continued)
THE SEPARATION OF RACEMIC ALKYLAMINO ACIDS BY HPLC USING A CHIRAL MOBILE PHASE

Alkylamino acid	k′
L-Valine	2.67
D-Valine	4.25
L-Leucine	9.42
D-Leucine	16.25
L-Isoleucine	6.58
D-Isoleucine	13.00

Packing: P1 = Nucleosil® 10 C-18 (Macherey-Nagel)

Solvent: S1 = a solution containing copper sulfate and L-aspartyl-L-phenylalanine methyl ester, both at a concentration of 1 mmol/ℓ, in distilled water is filtered, and treated with ultrasound for 10 min before use

REFERENCE

1. **Gundlach, G., Sattler, E. L., and Wagenbach, U.,** *Fresenius Z. Anal. Chem.*, 311, 684, 1982.

Table LC 135
RESOLUTION OF *N*-2,4-DINITROPHENYL AMINO ACID ESTERS (DNP-AMINO ACID ESTERS)

Amino acid	Ester group	k'_1	k'_2	Solvent, *n*-PrOH in *n*-C_6H_{14} (%)
Alanine	Methyl	10.07	11.07	2
Alanine	*n*-Propyl	2.23	2.53	2
Leucine	Methyl	2.47	2.47	1
Isoleucine	Methyl	2.73	3.20	1
Valine	Methyl	2.83	3.23	2
Valine	*n*-Propyl	1.33	1.73	1
Phenylalanine	Methyl	6.87	7.27	2
Phenylalanine	*n*-Propyl	3.27	3.80	1
Phenylglycine	Methyl	3.80	4.33	2
Phenylglycine	*n*-Propyl	1.65	1.95	1

Note: Except for the DNP-phenylglycine esters the D-isomer emerged first.

Conditions: The chiral support is prepared by neutralizing 100 mg of (+)-hexahelicene-7,7′-dicarboxylic acid dissolved in tetrahydrofuran/methanol/water, 1:1:1, with 0.3 *M* NaOH and coating on 3 g of Lichrosorb® Si 100 (Merck); the silica gel, after evaporation of the solvent and drying, is thoroughly washed with chloroform

Table LC 135 (continued)
RESOLUTION OF *N*-2,4-DINITROPHENYL AMINO ACID ESTERS (DNP-AMINO ACID ESTERS)

Column:	21 × 0.46 cm, is run at room temperature with a flow rate of 1 mℓ/min
Detection:	Performed at 254 nm

REFERENCE

1. **Kim, Y. H., Balan, A., Tishbee, A., and Gil-Av, E.,** *J. Chem. Soc. Chem. Commun.*, p. 1336, 1982.

Table LC 136
ENANTIOMERS OF 3,5-DINITROBENZOYL AMINO ACID METHYL ESTERS

Packing	P1	P2	P3
Column			
Length (cm)	15	15	15
Diameter (cm)	0.4	0.4	0.4
Material	SS	SS	SS
Solvent	S1	S1	S1
Flow rate (mℓ/min)	1.0	1.0	1.0
Detection	D1	D1	D1

Amino acid	**k′**	**α**	**k′**	**α**	**k′**	**α**
Ala	1.57	1.32	1.69	1.0	1.57	1.20
Asp	1.04	1.27	1.12	1.0	1.15	1.14
Met	1.22	1.60	1.39	1.0	1.23	1.43
Phe	1.11	1.38	1.05	1.0	0.95	1.20
Val	0.67	1.85	0.71	1.0	0.59	1.35

Note: The first eluted enantiomer has the R configuration, k′ of the first eluted enantiomer.

Packings:	P1, P2, and P3	=	chiral phases based on (α)-1-naphthylethylamine; P1 and P2 are diastereomeric phases, P3 is a single enantiomer phase; the structure of the phases is shown below (P1 = I, P2 = II, P3 = III)
Solvent:	S1	=	dichloromethane-hexane, 8:20
Detection:	D1	=	UV at 254 nm

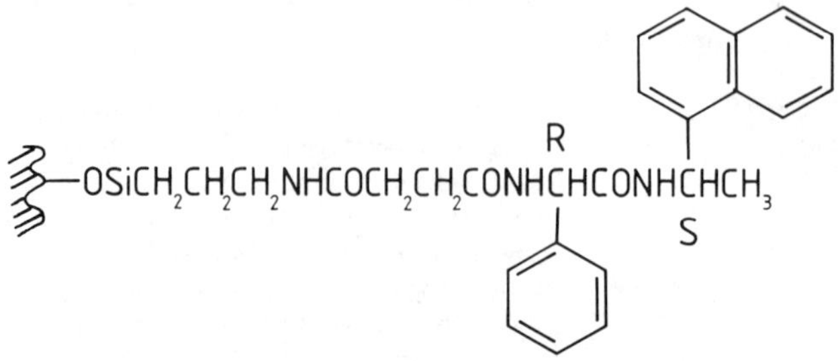

I

Table LC 136 (continued)
ENANTIOMERS OF 3,5-DINITROBENZOYL AMINO ACID METHYL ESTERS

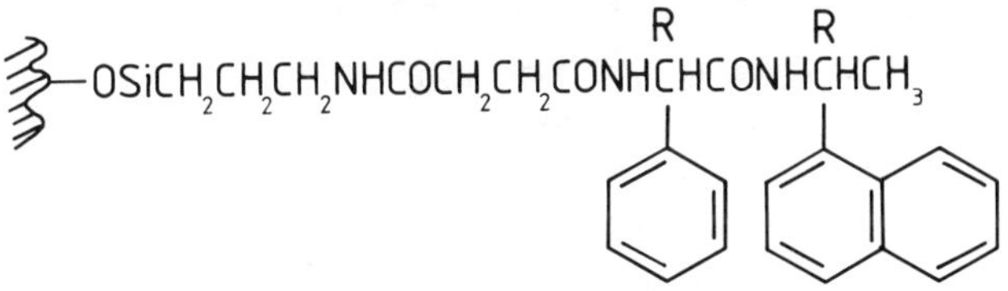

II

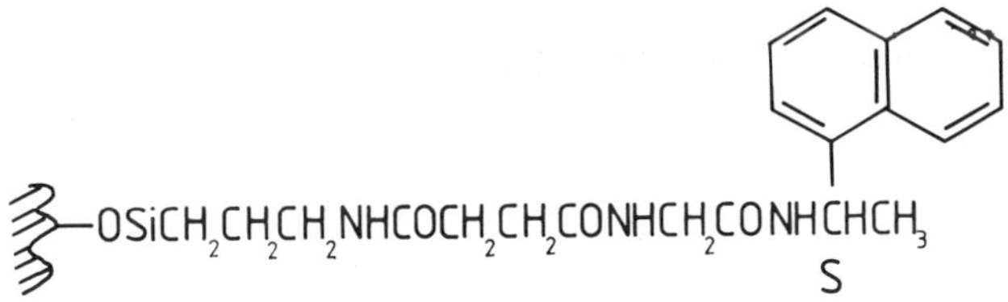

III

REFERENCE

1. **Lloyd, M. J. B.,** *J. Chromatogr.*, 351, 219, 1986.

Table LC 137
ENANTIOMERS OF 3,5-DINITROBENZOYL AMINO ACID METHYL ESTERS

Amino acid	Solvent	k′	α
	Phase I		
Alanine	S1	8.35	1.54
Aspartic acid	S1	14.77	1.20
Isohistidine	S1	9.21	1.33
Methionine	S2	4.55	1.58
Phenylalanine	S1	13.4	1.26
Proline	S1	4.01	1.00
Serin e	S2	11.41	1.31
Threonine	S2	6.07	1.40
Tryptophan	S3	10.46	1.19
Tyrosine	S3	12.11	1.47
Valine	S1	5.49	2.03
	Phase II		
Alanine	S1	7.23	1.54
Aspartic acid	S1	13.61	1.19
Isohistidine	S1	9.72	1.42
Methionine	S2	3.75	1.59
Phenylalanine	S1	9.5	1.54
Proline	S1	4.08	1.00
Serine	S2	11.46	1.28
Threonine	S2	6.15	1.24
Tryptophan	S3	8.6	1.17
Tyrosine	S3	11.77	1.25
Valine	S1	4.26	1.79

Table LC 137 (continued) ENANTIOMERS OF 3,5-DINITROBENZOYL AMINO ACID METHYL ESTERS

Amino acid	Solvent	k′	α
	Phase III		
Alanine	S1	6.73	1.52
Aspartic acid	S1	13.09	1.22
Isohistidine	S1	8.31	1.32
Methionine	S2	3.55	1.70
Phenylalanine	S1	9.24	1.40
Proline	S1	3.60	1.00
Serine	S2	9.89	1.24
Threonine	S2	5.58	1.33
Tryptophan	S3	7.26	1. 33
Tyrosine	S3	9.39	1.39
Valine	S1	4.05	1.89

Note: The first eluted enantiomer has the R configuration on phases I and III and the S configuration on phase II, k′ of the first eluted enantiomer.

Conditions: Phases I, II, and III are chiral phases based on (α)-I-naphthylethylamine; I and II are diastereomeric phases, III is a single enantiomer phase; their structures are shown in Table LC 136

Column: Stainless steel, 15 × 0.4 cm

Solvent: S1 = isopropanol/hexane, 10:90

S2 = isopropanol/hexane, 20:80

S3 = isopropanol/hexane, 30:70

Flow rate: 2 mℓ/min

Detection: UV at 254 nm

REFERENCE

1. **Lloyd, M. J. B.,** *J. Chromatogr.*, 351, 219, 1986.

Table LC 138
THE SEPARATION OF D AND L LEUCINE DERIVATIVES

R^1	R^2	Solvent, % (v/v) of 2-PrOH in *n*-hexane	k'_D	k'_L
Me	H	6	3.25	3.25
Me	Me	6	3.35	3.66
Me	Et	4	2.36	2.63
Me	i-Pr	2	3.00	3.39
Me	t-Bu	0.5	2.65	2.93
Me	Me	4	5.50	6.05
Et	Me	3	4.98	5.74
i-Pr	Me	2.5	4.71	5.65
t-Bu	Me	4	2.03	2.54
t-Bu	Me	2	4.84	6.70

Note: R^1 and R^2 refer to groups in the compound $(CH_3)_2CHCH_2CH(COOR^1)NHCOR^2$.

Conditions:
- Column: Stainless steel, 20 × 0.4 cm I.D., containing approximately 2 g of (*N*-formyl-L-valylamino) propyl (FVA) silica gel
- Flow rate: 1 mℓ/min
- Temperature: 40°C
- Detection: UV at 230 nm

REFERENCE

1. **Dobashi, A., Oka, K., and Hara, S.,** *J. Am. Chem. Soc.*, 102, 7122, 1980.

Table LC 139
THE SEPARATION OF ENANTIOMERS OF *N*-AROYL-DL AMINO ACID DERIVATIVES Ar-CONH-CHR-COOR′

Compound					
Ar	R	R′	Solvent	k'_L	k'_D
Ph	CH_3	H	S1	0.5	1.8
Ph	$PhCH_2$	H	S1	3.05	2.3
Ph	$HOCH_2$	H	S1	0.36	0.56
Ph	$(CH_3)_2CH$	H	S1	2.5	2.5
Ph	$(CH_3)_2CHCH_2$	H	S1	1.7	1.5
2-C_4H_3S	CH_3	H	S1	0.56	0.78
2-$C_{10}H_7$	CH_3	H	S2	7.1	3.5
2-$C_{10}H_7$	$PhCH_2$	H	S2	2.95	9.32
Ph	CH_3	CH_3	S1	0.3	0.3

Conditions:
- Packing: Silica, 10 μm, modified to contain a surface layer of bovine serum albumin (Resolvosil®)
- Solvent: S1 = 0.05 *M* phosphate, pH 7.0 S2 = 0.05 *M* phosphate + 5% 1-propanol, pH 8.1
- Detection: UV at 235 nm

Table LC 139 (continued)
THE SEPARATION OF ENANTIOMERS OF *N*-AROYL-DL AMINO ACID DERIVATIVES Ar-CONH-CHR-COOR′

REFERENCE

1. **Allenmark, S., Bomgren, B., and Boren, H.,** *J. Chromatogr.*, 264, 63, 1983.

Table LC 140
ENANTIOMERIC AMINO ACID DERIVATIVES ON A β-CYCLODEXTRIN BONDED PHASE COLUMN

Packing	P1
Column	
Length (mm)	100
Diameter (mm)	4.6
Solvent	S1
Flow rate (mℓ/min)	1.0
Temperature (°C)	20
Detection	D1

Compound	**k′**
L-Alanine β-naphthylamide	5.1
D-Alanine β-naphthylamide	6.1
L-Alanine β-naphthyl ester	1.0
D-Alanine β-naphthyl ester	1.8
N-Benzoyl-L-arginine β-naphthylamide	28.5
N-Benzoyl-D-arginine β-naphthylamide	30.4
Dansyl-L-leucine	3.0
Dansyl-D-leucine	4.2
Dansyl-L-phenylalanine	3.1[a]
Dansyl-D-phenylalanine	3.8[a]
L-Methionine β-naphthylamide	2.7
D-Methionine β-naphthylamide	3.6

[a] Solvent methanol/water, 55:45

Packing: P1 = cyclodextrin-bonded column, the support being 5 μm diameter silica gel (Advanced Separation Technologies)
Solvent: S1 = methanol/water, 50:50
Detection: D1 = UV at 254 nm

REFERENCE

1. **Armstrong, D. W. and Demond, W.,** *J. Chromatogr. Sci.*, 22, 411, 1984.

Table LC 141
SEPARATION OF THE ENANTIOMERS OF AMINO ACID DERIVATIVES WITH QUININE AS CHIRAL COUNTER ION

Packing	P1
Column	
Length (mm)	100
Diameter (mm, I.D.)	3.0
Material	SS
Solvent	S1
Detection	D1

Compound	$\alpha_{D/L}$
N-Benzoxycarbonylphenylalanine	1.26
N-Benzoxycarbonylvaline	1.08
N-tert-Butoxycarbonylphenylalanine	1.27
Dansyl-phenylalanine	1.34
Densyl-valine	1.28

Packing: P1 = LiChrosorb® DIOL, 5 μm
Solvent: S1 = 0.35 n*M* quinine and acetic acid in dichloromethane(dry)-1-pentanol, 99:1
Detection: D1 = UV at 337 nm

REFERENCE

1. **Petterson, C.,** *J. Chromatogr.*, 316, 553, 1984.

Table LC 142
THE CHIRAL SEPARATION OF *N*-PROTECTED AMINO ACIDS AFTER DERIVATIZATION WITH L-*O*-(4-NITROBENZYL)-TYROSINE METHYL ESTER

Packing	P1
Column	
Length (cm)	25
Diameter (mm, I.D.)	4.6
Solvent	S1
Flow rate (mℓ/min)	2
Detection	D1

Amino acid derivative	k'_L	k'_D
BOC-Cys(Bz)OPFP	4.0	4.9
BOC-Phe-OSu	4.0	5.8
Z-Asp(OBu^t)-OH	6.6	7.4
Z-Glu(OBu^t)-OH	7.2	8.3
Z-Ser(OBu^t)-OH	6.6	7.4
Z-Thr(OBu^t)-OH	4.3	4.8
Z-Phe-OPFP	5.4	7.0

Abbreviations: Su = succinimide, PFP = pentafluorophenyl

Packing: P1 = LiChrosorb® Si 60-10
Solvent: S1 = the gradient elution program was as follows: the starting solution was a 98:2 mixture of hexane and 2-propanol; the proportion of the latter was increased to 5% in the first 20 min and then to 40% within 10 min, the cycle was terminated with a 15-min isocratic elution with a 60:40 solvent ratio
Detection: D1 = UV at 270 nm

REFERENCE

1. **Gorog, S., Herenyi, B., and Low, M.,** *J. Chromatogr.*, 353, 417, 1986.

Table LC 143
AMINES ON A BONDED ION EXCHANGE COLUMN

Packing	P1	P1
Column		
Length (cm)	25	25
Diameter (mm, I.D.)	4.5	4.5
Material	SS	SS
Solvent	S1	S2
Flow rate (mℓ/min)	1.0	1.0
Detection	D1	D1

Amine	**k′**	
Benzylamine	0.74	
2-Phenylethylamine	1.17	
3-Phenylpropylamine	2.48	
4-Phenylbutylamine	4.65	
Aniline		1.17
N-Methylaniline		2.26
N,*N*-Dimethylaniline		4.22

Packing: P1 = Nucleosil® SA($-SO_3H$, 10 μm) spherical porous silica gel particles coated *in situ* with octadecyltrichlorosilane reagent

Solvent: S1 = 0.05 *M* KH_2PO_4, 0.10 *M* KCl, pH 6.80 with 20% methanol by volume

S2 = 0.25 *M* potassium acetate, pH 4.5 with 40% methanol by volume

Detection: D1 = UV at 254 nm

REFERENCE

1. **Liao, J. C. and Vogt, C. R.,** *J. Chromatogr. Sci.,* 17, 237, 1979.

Table LC 144
ALIPHATIC AMINES ON A DIAMINE-BONDED PHASE

Packing	P1	P1	P1	P1
Column				
Length (mm)	200	200	250	200
Diameter (mm, I.D.)	4	4	4.6	4
Material	SS	SS	SS	SS
Solvent	S1	S2	S3	S4
Flow rate (mℓ/min)	1.0	1.0	0.7	1.0
Temperature	r t	r t	r t	r t
Detection	D1	D1	D1	D1
Amine			**k′**	
n-Butylamine		7.54	7.54	21.68
tert-Butylamine		3.55	1.28	19.61
Di-*n*-butylamine	9.98	1.18	2.25	2.08
Diisobutylamine	9.55	0.84	1.00	0.82
n-Propylamine		9.71	8.26	28.39
Isopropylamine		5.34	3.05	26.82
Di-*n*-propylamine	13.00	1.36	2.73	3.66
Diisopropylamine	14.22	1.43	1.42	4.68
n-Octylamine		3.46	6.94	9.39
Di-*n*-octylamine	2.66	0.45	0.89	0.26

Packing: P1 = silica gel, dried at 140°C in vacuo, was treated with a 5% boiling solution of *N*-(3-trimethoxysilylpropyl) ethylenediamine in toluene giving diamine-bonded (N2-)silica; the modified silica gel was packed in the column by the balanced viscosity slurry technique, the slurry solvent being methanol/2-propanol/carbon tetrachloride/toluene/cyclohexanol, 5:5:18:1.2:10.8, and the pumping solvent being methanol. Cu(II) ion was loaded on the column by pumping a 0.01 *M* methanolic solution of Cu(II) nitrate, the remaining metal solution in the column being replaced with hexane before conditioning; the column was then conditioned with mobile phase solvent for 5 hr

Solvent: S1 = hexane containing 10% methanol and 4% 1-butanol

S2 = hexane containing 4% methanol and 2 m*M* ammonia

S3 = acetonitrile containing 5% methanol and 3 m*M* ammonia

S4 = hexane/methanol, 97:3, and 3 m*M* ammonia

Detection: D1 = UV at 225 nm

REFERENCE

1. **Takayanagi, H., Tokuda, H., Uehira, H., Fujimura, K., and Ando, T.,** *J. Chromatogr.*, 356, 15, 1986.

Table LC 145
ALIPHATIC AMINES ON A TRIAMINE-BONDED PHASE

Packing	P1	P1	P1	P1	P1
Column					
Length (mm)	100	100	100	100	100
Diameter (mm, I.D.)	4	4	4	4	4
Material	SS	SS	SS	SS	SS
Solvent	S1	S2	S3	S4	S5
Flow rate (mℓ/min)	1.0	1.0	1.0	1.0	1.0
Temperature	r t	r t	r t	r t	r t
Detection	D1	D1	D1	D1	D1

Amine			**k′**		
n-Butylamine	15.68	15.04	13.80	12.79	25.86
tert-Butylamine	13.66	13.56	12.86	12.14	23.91
Di-*n*-butylamine	1.73	1.71	1.70	1.70	3.82
Diisobutylamine	0.84	0.82	0.82	0.82	1.43
n-Propylamine	20.75	19.68	18.45	16.93	
Isopropylamine	18.84	18.09	17.16	15.50	28.70
Di-*n*-propylamine	2.52	2.50	2.48	2.48	4.88
Diisopropylamine	2.75	2.73	2.71	2.70	7.36
n-Octylamine	6.95	6.55	5.52	5.23	13.48
Di-*n*-octylamine	0.75	0.68	0.64	0.57	0.61

Packing: P1 = silica gel, dried at 140°C in vacuo, was treated with a 5% boiling solution of *N*-(3-trimethoxysilylpropyl)-diethylenetriamine in toluene giving triamine-bonded (N3 −) silica; the modified silica gel was packed in the column by the balanced viscosity slurry technique, the slurry solvent being methanol/2-propanol-carbon tetrachloride/toluene/cyclohexanol, 5:5:18:1.2:10.8, and the pumping solvent methanol; Cu(II) ion was loaded on the column by pumping a 0.01 *M* methanolic solution of Cu(II) nitrate, the remaining metal solution in the column being replaced with hexane before conditioning; the column was then conditioned with mobile phase solvent for 5 hr

Solvent: S1, S2, S3, and S4 = hexane containing 2.3 m*M* ammonia with 1.5, 2.0, 2.5, and 3.0%, respectively, of methanol

S5 = hexane/methanol, 97:3, containing 4.4 m*M* ammonia

Detection: D1 = UV at 225 nm

REFERENCE

1. **Takayanagi, H., Tokuda, H., Uehira, H., Fujimura, K., and Ando, T.,** *J. Chromatogr.*, 356, 15, 1986.

Table LC 146
AROMATIC AMINES

Packing	P1
Column	
Length (cm)	25
Diameter (mm, I.D.)	4.6
Solvent	S1
Flow rate (mℓ/min)	1.0
Detection	D1

Amine	t_r (min)	Minimum detection limit (pg)
4-Methyl-3-nitroaniline	15.11	260
2-Nitroaniline	8.46	151
4-Methoxy-3-nitroaniline	9.46	168
3-Nitroaniline	12.8	136
4-Chloro-2-nitroaniline	7.58	117
2-Methoxy-4-nitroaniline	14.56	260
2-Chloro-4-nitroaniline	10.81	368
2-Methyl-4-nitroaniline	17.76	647
4-Nitroaniline	20.40	865

Packing: P1 = cyanopropyl-bonded phase, Zorbax® CN
Solvent: S1 = 14% isopropanol in hexane
Detection: D1 = UV at 254 nm

REFERENCE

1. **Narang, A. S., Choudhury, D. R., and Richards, A.,** *J. Chromatogr. Sci.,* 20, 235, 1982.

Table LC 147
PRIMARY AROMATIC AMINES AND DERIVATIVES

Packing	P1
Column	
Length (cm)	25
Diameter (mm, I.D.)	4.0
Material	SS
Solvent	S1
Flow rate (mℓ/min)	1.0
Detection	D1

Amine	V_r	k′
o-Nitroaniline	6.58	1.86
m-Nitroaniline	9.58	3.17
p-Nitroaniline	6.75	1.93
2,4-Dinitroaniline	18.00	6.83
p-Aminobenzoic acid	9.33	3.06
4,4′-Methylenedianiline	NE[a]	
o-Anisidine	3.67	0.66
p-Anisidine	13.00	4.65

Table LC 147 (continued)
PRIMARY AROMATIC AMINES AND DERIVATIVES

Amine	V_r	k′
p-Aminophenol	8.00	2.48
o-Chloroaniline	3.30	0.44
p-Chloroaniline	4.38	0.90
2,3-Dichloroaniline	4.08	0.77
2,6-Dichloroaniline	2.75	0.20
3,4-Dichloroaniline	6.00	1.61
3,5-Dichloroaniline	5.50	1.39
2-Amino-1,4-naphthoquinone	4.83	1.10

[a] NE = not eluted after 40 min.

Packing: P1 = 15.78% by weight of 3-(acetylphenoxy) propylsilane reagent on 10 μm RSil® silica
Solvent: S1 = hexane/isopropanol, 95:5
Detection: D1 = UV at 254 nm

REFERENCE

1. **Pharr, D. Y., Uden, P. C., and Siggia, S.,** *J. Chromatogr. Sci.*, 23, 391, 1985.

Table LC 148
SECONDARY AMINES AND AMIDES

Packing	P1
Column	
Length (cm)	25
Diameter (mm, I.D.)	4.0
Material	SS
Solvent	S1
Flow rate (mℓ/min)	1.0
Detection	D1

Compound	V_r	k′
Acetanilide	6.17	1.68
o-Nitroacetanilide	4.00	0.74
m-Nitroacetanilide	13.80	5.00
p-Nitroacetanilide	17.00	6.39
2,4-Dinitroacetanilide	6.55	1.85
N-Methylbenzylamine	2.92	0.27
N-Methylaniline	3.17	0.38
Carbazole	6.08	1.64
1-*N*-Methyl-1-naphthylamine	3.42	0.49
o-Toluamide	6.00	1.61
p-Toluamide	7.42	2.23
Pyrazinamide	14.33	5.23
3,5-Dinitrobenzamide	11.33	3.93

Table LC 148 (continued)
SECONDARY AMINES AND AMIDES

Packing: P1 = 15.78% by weight of 3-(*p*-acetylphenoxy) propylsilane reagent on 10 μm RSil® silica
Solvent: S1 = hexane/isopropanol, 95:5
Detection: D1 = UV at 254 nm

REFERENCE

1. **Pharr, Y. P., Uden, P. C., and Siggia, S.,** *J. Chromatogr. Sci.*, 23, 391, 1985.

Table LC 149
NITROANILINES

Packing	P1	P2	P3	P4	P5
Column					
Length (cm)	25	25	25	25	25
Diameter (mm, I.D.)	4.6	4.6	4.6	4.6	4.6
Solvent	S1	S1	S1	S1	S1
Temperature	r t	r t	r t	r t	r t
Detection	D1	D1	D1	D1	D1
Aniline			**k′**		
Aniline	1.10	1.19	0.96	1.91	2.30
o-Nitroaniline	0.26	1.07	1.48	1.72	2.44
m-Nitroaniline	1.03	2.15	2.95	4.06	5.52
p-Nitroaniline	1.11	4.35	6.21	7.75	12.86

Packing: P1 = silica
P2 = amino (Alltech Associates)
P3 = ethylenediamine-bonded phase prepared using Partisil®-10 and 3-(2-[aminoethyl]amino)propyltrimethoxysilane
P4 = (Co[en]edda)Cl, (ethylenediamine) (ethylenediamine-*N*,*N′*-diacetate) Co(III)-bonded phase
P5 = $(Co[en]_3)Cl_3$, *tris*(ethylenediamine)Co(III)chloride-bonded phase
Solvent: S1 = 2-propanol/*n*-hexane, 25:75
Detection: D1 = UV at 254 nm

REFERENCE

1. **Chang, C. A., Tu, C.-F., and Huang, C.-S.,** *J. Chromatogr. Sci.*, 22, 321, 1984.

Table LC 150
CHLOROANILINES

Packing	P1	P2	P3	P4	P5	P6
Column						
Length (cm)	25	25	25	25	25	25
Diameter (mm, I.D.)	4.6	4.6	4.6	4.6	4.6	4.6
Sovlent	S1	S1	S1	S1	S1	S1
Temperature	r t	r t	r t	r t	r t	r t
Detection	D1	D1	D1	D1	D1	D1
Aniline			k′			
Aniline	1.33	1.27	1.08	1.08	2.25	2.77
o-Chloroaniline	0.18	0.26	0.42	0.65	0.76	0.97
p-Chloroaniline	1.33	1.63	1.40	1.58	2.67	3.16

Packing: P1 = C-18 (PX5, ODS-2)(Whatman)
P2 = silica
P3 = amino (Alltech Associates)
P4 = ethylenediamine-bonded phase prepared using Partisil®-10 and 3-(2-[aminoethyl]amino) propyltrimethoxysilane
P5 = (Co[en][edda])Cl, (ethylenediamine)(ethylenediamine-*N,N′*-diacetato)Co(III)-bonded phase
P6 = $(Co[en]_3)Cl_3$, *tris*(ethylenediamine)Co(III) chloride-bonded phase
Solvent: S1 = 2-propanol/*n*-heptane, 20:80
Detection: D1 = UV at 254 nm

REFERENCE

1. **Chang, C. A., Tu, C.-F., and Huang, C.-S.,** *J. Chromatogr. Sci.*, 22, 321, 1984.

Table LC 151
THE SEPARATION OF METHOXYANILINES BY COMPETING-ION REVERSED-PHASE HPLC

Packing	P1
Column	
Length (cm)	30
Diameter (mm)	3.9
Solvent	S1
Flow rate (mℓ/min)	1.0
Detection	D1
Compound	t_r
Aniline (internal standard)	6.98
2,4-Dimethoxyaniline	8.78
2,5-Dimethoxyaniline	19.10
3,4-Dimethoxyaniline	5.50
3,5-Dimethoxyaniline	15.82
2-Methoxyaniline	12.50
3-Methoxyaniline	10.42
4-Methoxyaniline	4.96

Table LC 151 (continued)
THE SEPARATION OF METHOXYANILINES BY COMPETING-ION REVERSED-PHASE HPLC

Packing: P1 = μBondapak® C_{18}
Solvent: S1 = acetonitrile/0.01 *M* dibutylamine, 10:90, with 0.1% acetic acid, pH 4.5
Detection: D1 = UV at 254 nm

REFERENCE

1. **Hill, R. A.,** *J. Chromatogr.*, 316, 227, 1984.

Table LC 152
PRIMARY, SECONDARY, AND TERTIARY ANILINES

Packing	P1	P2	P2	P3
Column				
Length (mm)	200	200	200	200
Diameter (mm, I.D.)	4.4	4.4	4.4	4.4
Material	SS	SS	SS	SS
Solvent	S1	S2	S3	S4
Temperature	r t	r t	r t	r t
Detection	D1	D1	D1	D1

Aniline substituent		**k′**		
Primary Anilines				
None	7.53	0.41	0.55	0.60
2-Methyl	4.84	0.55	0.85	0.92
2-Ethyl	3.65	0.71	1.20	1.40
2-Isopropyl	3.14	0.86	1.63	2.07
3-Methyl	8.07	0.51	0.83	0.94
3-Ethyl	7.02	0.67	1.20	1.49
4-Methyl	9.62	0.54	0.85	0.97
4-Ethyl	8.72	0.71	1.26	1.59
4-Isopropyl	8.87	0.90	1.75	2.46
4-*n*-Butyl	8.72	1.38	3.11	4.84
2,3-Dimethyl	5.48	0.74	1.18	1.31
2,4-Dimethyl	6.04	0.78	1.25	1.49
2-Methyl-4-*n*-butyl	5.07	1.91	4.71	7.40
2,5-Dimethyl	4.45	0.75	1.26	1.43
2,6-Dimethyl	2.83	0.78	1.31	1.43
3,4-Dimethyl	9.51	0.69	1.16	1.40
2,4,6-Trimethyl	2.96	1.08	1.99	2.34
Secondary Anilines				
N-Methyl	2.32	0.68	1.12	1.22
N-Methyl-2-methyl	1.43	0.97	1.68	1.86
N-Methyl-3-methyl	2.16	0.89	1.62	1.86
N-Methyl-4-methyl	2.73	0.90	1.63	1.93

Table LC 152 (continued) PRIMARY, SECONDARY, AND TERTIARY ANILINES

Aniline substituent	k′			
N-Ethyl	1.49	0.87	1.54	1.77
N-Ethyl-2-methyl	0.82	1.34	2.57	2.91
N-Ethyl-3-methyl	1.40	1.14	2.26	2.71
N-Ethyl-4-methyl	1.72	1.15	2.26	2.73
N-Ethyl-2-ethyl	0.67	1.86	3.80	4.61
N-Acetyl		0.41	0.62	0.71
N-Acetyl-4-methyl		0.54	0.87	1.09
N-Propyl	1.06	1.23	2.35	3.00
N-*n*-Butyl	0.90	1.69	3.70	5.04
N-Phenyl	2.50	1.52	3.45	4.86
N-Benzyl	1.65	1.50	3.42	4.79
Tertiary Anilines				
N, *N*-Dimethyl	0.65	1.35	2.42	2.68
N, *N*-Dimethyl-2-methyl	0.45	1.76	3.48	3.99
N, *N*-Dimethyl-3-methyl	0.68	1.82	3.61	4.17
N, *N*-Dimethyl-4-methyl	0.79	1.80	3.64	4.17
N, *N*-Dimethyl-2,6-dimethyl	0.14	4.20	10.1	
N, *N*-Diethyl	0.50	2.32	5.21	6.44
N, *N*-Diethyl-2-methyl	0.29	3.45	7.71	8.89
N, *N*-Diethyl-3-methyl	0.54	3.08	7.31	9.18
N, *N*-Diethyl-4-methyl	0.90	2.95	6.81	8.06
N, *N*-Diethyl-2-ethyl	0.21	5.21	12.8	16.8
N, *N*-Diphenyl	0.39	7.40	25.9	

Packing: P1 = Nucleosil® CN, 5 μm
P2 = Nucleosil® C_{18}, 5 μm
P3 = LiChrosorb® RP-8, 10 μm
Solvent: S1 = isooctane containing 0.25 (v/v) of 2-propanol
S2 = methanol/aqueous buffer, 80:20, pH 7.0; 15 mℓ of 0.05 *M* Na_2HPO_4 + 250 mℓ of 0.05 *M* NaH_2PO_4 + 1060 mℓ of methanol; the pH was adjusted to 7.0 with small amounts of H_3PO_4 and 1 *M* NaOH
S3 = methanol/aqueous buffer, 70:30, pH 7.0; 15 mℓ of 0.025 *M* Na_2HPO_4 + 250 mℓ of 0.025 *M*NaH_2PO_4 + 618.33 mℓ of methanol; the pH was adjusted as above
S4 = methanol/aqueous buffer, 60:40, pH 7.0; 15 mℓ of 0.025 *M* Na_2HPO_4 + 250 mℓ of 0.025 *M*NaH_2PO_4 + 397.5 mℓ of methanol; the pH was adjusted as above
Detection: D1 = UV at 280 nm

REFERENCE

1. **Truedsson, L.-A. and Smith, B. E. F.,** *J. Chromatogr.*, 214, 291, 1981.

Table LC 153
THE RETENTION OF *N*-ALKYLANILINES UNDER DIFFERENT CHROMATOGRAPHIC CONDITIONS

	k′			
Conditions	***N*-Butyl aniline**	***N*-Propyl aniline**	***N*-Ethyl aniline**	***N*-Methyl aniline**
Methanol (%)				
30		46.32	15.99	7.31
35		30.55	11.17	5.32
40		20.75	7.96	3.95
45		13.08	5.42	2.81
50		8.38	3.72	2.02
Temperature (°C)				
10	66.92	25.05	9.41	4.73
20	56.20	21.72	8.34	4.16
25	50.15	19.72	7.67	3.82
30	45.67	18.09	7.12	3.54
40	36.24	15.12	6.19	3.08
Eluent pH				
7.58	44.38	17.87	7.07	3.53
8.52	44.52	18.09	7.12	3.54
9.73	45.22	18.17	7.20	3.58

Conditions: Packing: Hypersil® ODS, 5 μm
Column: Size not apparent
Solvent: Methanol/phosphate buffer, pH 8.5, 40:60; the buffer is prepared by dissolving 1.56 g $NaH_2PO_4 \cdot 2H_2O$ and 12.78 g Na_2HPO_4 in 1000 mℓ of water
Flow rate: 2 mℓ/min
Temperature: 30°C
Detection: UV at 254 nm
Variations on these conditions are recorded in the table

REFERENCE

1. **Smith, R. M., Hurdley, T. G., Gill, R., and Moffat, A. C.,** *J. Chromatogr.*, 351, 259, 1986.

Table LC 154
THE RETENTION OF *N*-ALKYLANILINES IN ELUENT CONTAINING *N*-HEXYLAMINE

	k′			
Conditions	***N*-Butyl aniline**	***N*-Propyl aniline**	***N*-Ethyl aniline**	***N*-Methyl aniline**
Hexylamine (%)				
0	20.44	8.63	3.82	2.37
0.35	3.97	1.28	0.36	0.18
0.7	3.90	1.27	0.34	0.22
1.40	2.57	0.83	0.21	0.13
Methanol (%)				
10	5.93	1.86	0.46	0.27
15	3.90	1.27	0.34	0.22
20	2.73	0.94	0.27	0.19

Table LC 154 (continued)
THE RETENTION OF *N*-ALKYLANILINES IN ELUENT CONTAINING *N*-HEXYLAMINE

	k′			
Conditions	*N*-Butyl aniline	*N*-Propyl aniline	*N*-Ethyl aniline	*N*-Methyl aniline
Temperature (°C)				
10	3.61	1.14	0.34	0.19
30	3.90	1.27	0.34	0.22
40	4.26	1.42	0.37	0.26
Eluent pH				
3.0	6.86	2.28	0.59	0.47
2.5	3.90	1.27	0.34	0.22
2.0	2.64	1.83	0.24	0.12

Conditions: Packing: Hypersil® ODS, 5 μm
Column: size not apparent
Solvent: methanol/water/1% phosphoric acid/*n*-hexyl amine (pH 2.5), 30:70:100:1.4
Flow rate: 2 mℓ/min
Temperature: 30°C
Detection: UV at 254 nm
Variations on these conditions are recorded in the table

REFERENCE

1. **Smith, R. M., Hurdley, T. G., Gill, R., and Moffat, A. C.,** *J. Chromatogr.*, 351, 259, 1986.

Table LC 155
CHLOROANILINES

Packing	P1
Column	
Length (cm)	10
Diameter (mm, I.D.)	4.6
Material	SS
Solvent	S1
Flow rate (mℓ/min)	1.0
Detection	D1

Compound	k′
3-Chloroaniline	1.7
4-Chloroaniline	1.4
3,4-Dichloroaniline	4.4

Packing: P1 = Polygosil® 60-10 CN (Macherey, Nagel & Co., Düren)
Solvent: S1 = 0.05 *M*, pH 3, acetate buffer/acetonitrile, 95:5
Detection: D1 = fluorescamine

REFERENCE

1. **Scholten, A. H. M. T., Brinkman, U. A. Th., and Frei, R. W.,** *J. Chromatogr.*, 218, 3, 1981.

Table LC 156
HPLC SEPARATION OF PRIMARY α,ω-DIAMINES AFTER CONVERSION TO BIS(DIPHENYLBORON) CHELATES OF α,ω-BIS(SALICYLIDENEAMINO) ALKANES

Packing	P1
Column	
Length (mm)	250
Diameter (mm, I.D.)	4.6
Solvent	S1
Flow rate (mℓ/min)	1.0
Detection	D1

Chelate of	t_r(min)
1,2-Diaminoethane	34.3
1,3-Diaminopropane	90.0
1,4-Diaminobutane	144.5
1,5-Diaminopentane	157.8
1,6-Diaminohexane	179.1

Packing: P1 = LiChrosorb® Si 60, particle size 7 μm
Solvent: S1 = dichloromethane-*n*-heptane, 55:45
Detection: D1 = spectrofluorometrically

REFERENCE

1. **Claas, K. E., Hohaus, E., and Monien, H.,** *Fresenius Z. Anal. Chem.*, 316, 781, 1983.

Table LC 157
BIOGENIC AMINES AS THEIR DANSYL DERIVATIVES

Packing	P1
Column	
Length (cm)	25
Diameter (mm, I.D.)	5
Solvent	S1
Flow rate (mℓ/min)	1.5
Temperature (°C)	35
Detection	D1

Amine	t_r(min)
Diethanolamine	6.75
Ethanolamine	7.00
Ammonium chloride	7.50
5-Aminopentan-1-ol	9.25
Methylamine	10.75
6-Aminohexan-1-ol	11.50
Ethylamine	14.00
Isopropylamine	18.00
Dimethylamine	19.00
n-Propylamine	19.75
Norephedrine	20.00
Isobutylamine	26.50
Agmatine	26.50
Lysine	27.25
n-Butylamine	27.50
Benzylamine	27.50
6-Methoxytryptamine	28.50
Ephedrine	29.00
Tryptamine	30.00
2-Phenylethylamine	34.50
Isoamylamine	35.50
N-Methylmescaline	36.00
1,3-Diaminopropane	41.00
1,2-Diaminopropane	41.50
N-Methyltryptamine	43.00
Putrescine	44.50
n-Hexylamine	47.00
Cadaverine	49.50
Normetanephrine	54.00
Cystamine	54.50
1,6-Diaminohexane	55.00
Octopamine	55.75
5-Hydroxytryptamine	59.00
Metanephrine	59.25
Synephrine	60.00
n-Octylamine	60.50
3-Methoxy-*p*-tyramine	61.25
p-Tyramine	62.25
n-Tyramine	62.25
o-Tyramine	62.75
Spermidine	65.00
Norepinephrine	69.50
Epinephrine	73.00
Dopamine	74.00
Spermine	75.50

Table LC 157 (continued)
BIOGENIC AMINES AS THEIR DANSYL DERIVATIVES

Packing: P1 = Spherisorb® C_{18} (5 μm ODS), protected by a guard column, 5 cm × 5 mm I.D. with Vydac® reversed-phase packing
Solvent: S1 = the pH 7.17, 17.5 m*M* phosphate buffer used as one component of the mobile phase is prepared by titrating aqueous Na_2HPO_4 to pH with glacial acetic acid; this buffer is mixed with acetonitrile to give a series of isocratic and linear steps which have the following acetonitrile concentrations at the time stated: 0 min, 30%; 2 min, 42%; 24 min, 42%; 25 min, 45%; 32 min, 45%; 52 min, 55%; 53 min, 65%; 64 min, 65%; 70 min, 80%; 78 min, 80%; 84 min, 30%; the shape of the gradient is critical
Detection: D1 = fluorescence measurement

REFERENCE

1. **Hayman, A. R., Gray, D. A., and Evans, S. V.,** *J. Chromatogr.*, 325, 462, 1985.

Table LC 158
POLYAMINES AS DANSYL DERIVATIVES

Packing	P1
Column	
Diameter (mm, I.D.)	5
Solvent	S1
Flow rate (mℓ/min)	1.5
Detection	D1

Amine	**t_r(min)**
Cadaverine	9.56 ± 0.15
1,6-Hexanediamine[a]	9.95 ± 0.18
Putrescine	9.25 ± 0.06
Spermidine	12.08 ± 0.17
Spermine	14.26 ± 0.19

[a] Internal standard.

Packing: P1 = cartridge Radialpak® C8, 10 μm
Solvent: S1 = triethylammonium phosphate (TEAP), 0.25 *N*, pH 3.5 and methanol were filtered through Millipore membranes; solution A was TEAP/methanol, 50:50 and solution B was TEAP/methanol, 20:80; the mixtures were de-

Table LC 158 (continued)
POLYAMINES AS DANSYL DERIVATIVES

gassed with helium or in an ultrasound bath immediately prior to chromatography; a linear gradient was realized between solution A and B going from (0—100%) of B in 10 min

Detection: D1 = spectrofluorometer equipped with a continuous flow cell and recorder; excitation wavelength, 350 nm; emission wavelength, 500 nm

REFERENCE

1. **Brossat, B., Straczek, J., Belleville, F., Nabet, P., and Metz, R.,** *J. Chromatogr.*, 277, 87, 1983.

Table LC 159
ANALYSIS OF POLYAMINES AND ACETYL DERIVATIVES BY AN AMINO ACID ANALYZER TECHNIQUE

Packing	P1
Column	
Length (cm)	11.5
Diameter (mm, I.D.)	1.7
Solvent	S1
Flow rate (mℓ/hr)	19.2
Temperature (°C)	54 ± 1
Detection	D1

Amine	t_r(min)	RFR[a]
Acetylputrescine	9.48	0.63
N[8]-Acetylspermidine	27.70	0.74
N[1]-Acetylspermidine	30.77	0.87
Cadaverine	41.08	1.48
3,3′-Iminobispropylamine	44.43	1.0
Putrescine	35.25	1.13
Spermidine	47.75	1.53
Spermine	53.90	0.79

[a] RFR = relative fluorescence response. The RFR of each polyamine (PA) and acetylpolyamine was calculated as follows.

$$RFR_{PA/IS} = \frac{\text{area PA}}{\text{nmol/m}\ell\text{ PA}} \times \frac{\text{nmol/m}\ell\text{ IS}}{\text{area IS}}$$

where IS is the internal standard 3,3′-iminobispropylamine.

Packing: P1 = Aminex® A-9 cation exchange resin (>8% cross-linked styrene-divinylbenzene copolymer, bead diameter 11.5 ± 0.5 μm); the column was

Table LC 159 (continued)
ANALYSIS OF POLYAMINES AND ACETYL DERIVATIVES BY AN AMINO ACID ANALYZER TECHNIQUE

packed from a slurry of resin in 0.2 *N* sodium citrate, 2.1 *N* NaCl, 2.5% ethanol, and 0.5% thiodiglycol, pH 5.6, at a pressure of 126 bars at ambient temperature; it was used in conjunction with a model D-500 amino acid analyzer (Dionex) equipped with a fluorescence detector

Solvent: S1 = a five-buffer system described in Table 159a was used

Table 159a
BUFFER COMPOSITION

	Buffer				
	1	2	3	4	5
Sodium citrate (N)	0.125	0.2	0.2	0.2	0.2
NaCl (N)	0	0.4	1.2	2.1	1.8
NaOH (N)	0.5	0	0	0	0.05
Ethanol (%)	0	2.5	2.5	2.5	2.5

Note: The pH values of buffers 1, 2, 3, 4, and 5 were 10.2, 5.7, 5.65, 5.6, and 10.2, respectively. Buffers 2, 3, and 4 were adjusted to pH 5.6—5.7 using concentrated HCl, while buffers 1 and 5 were adjusted using granular boric acid. All buffers were prepared and brought to volume prior to pH adjustment. They all contained 0.5% thiodiglycol with four drops of pentachlorophenol per liter. Before use buffers were filtered and degassed through 0.45-μm filters. Buffers were used according to the following time schedule: buffer 1, 0.38 min; buffer 2, 38—43 min; buffer 3, 43—59 min; buffer 4, 59—73.5 min; buffer 5, 73.5—83 min. Column injection occurred at 25 min.

Detection: D1 = *o*-phthalaldehyde and fluorescence measurement

REFERENCE

1. **Russell, D. H., Ellingson, J. D., and Davis, T. P.,** *J. Chromatogr.*, 273, 263, 1983.

Table LC 160
ALIPHATIC AMINES DERIVATIZED WITH NBD-Cl (7-CHLORO-4-NITRO-2,1,3-BENZOXADIAZOLE)

Packing	P1
Column	
Length (cm)	20
Diameter (mm, I.D.)	4.0
Material	SS
Solvent	S1
Flow rate (mℓ/min)	1.0
Detection	D1

Amine	t_r(min)
Ammonia	3.21 ± 0.007
Methylamine	3.90 ± 0.006
Ethylamine	4.88 ± 0.005
Dimethylamine	5.43 ± 0.005
Isopropylamine	6.15 ± 0.007
n-Propylamine	6.91 ± 0.005
tert-Butylamine	8.82 ± 0.004
sec-Butylamine	8.82 ± 0.004
Diethylamine	9.65 ± 0.009
Isobutylamine	10.09 ± 0.007
n-Butylamine	11.23 ± 0.006

Packing: P1 = LiChrosorb® RP-18, 5 μm
Solvent: S1 = methanol/acetonitrile/water, 50:5:45
Detection: D1 = fluorescence, excitation at 470 nm and detection at 530 nm

REFERENCE

1. **Nishikawa, Y. and Kuwata, K.,** *Anal. Chem.,* 56, 1790, 1984.

Table LC 161
CHIRAL AMINES AFTER DERIVATIZATION WITH (*S*)-(+)-BENOXAPROFEN CHLORIDE

Column packing	P1
Column	
Length (mm)	250
Diameter (mm, I.D.)	4.6
Solvent	S1
Flow rate (mℓ/min)	1.0
Temperature	r t
Detection	D1

Derivative of	t_r(min)
(S)-(+)-Amphetamine	9.5
(R)-(−)-Amphetamine	8.0
(S)-(+)-Methamphetamine	11.5
(R)-(−)-Methamphetamine	10.5
(R)-(+)-α-Methylbenzylamine	6.7
(S)-(−)-α-Methylbenzylamine	10.8
(R)-(+)-Tranylcypromine	10.7
(S)-(−)-Tranylcypromine	9.0

Column packing: P1 = Zorbax-Sil®, 7 μm
Solvent: S1 = cyclohexane/dichloromethane/tetrahydrofuran, 5:1:1
Detection: D1 = fluorometric, excitation wavelength 312 nm, emission wavelength 365 nm

REFERENCE

1. **Weber, H., Spahn, H., Mutschler, E., and Möhrke, W.,** *J. Chromatogr.,* 307, 145, 1984.

Table LC 162
AMINES DERIVATIZED WITH BENOXAPROFEN CHLORIDE

Packing	P1
Column	
Length (mm)	250
Diameter (mm)	4.6
Solvent	S1
Flow rate (mℓ/min)	1.0
Temperature	rt
Detection	D1

Compound	**t_r(min)**
Amphetamine	8.0
	9.5
Benoxaprofen	4.3
Benoxaprofen chloride	4.3
Benzylamine	10.3
Maprotiline	8.6
Methamphetamine	10.5
	11.5
α-Methylbenzylamine	6.7
	10.8
Phenylbutylamine	11.3
β-Phenylethylamine	11.8
Procaine	8.5
Tolylethylamine	10.9
Tranylcypromine	9.0
	10.7

Note: Two peaks with different retention times represent diastereomers obtained by reaction of the optically active compound with racemic benoxaprofen chloride.

Packing: P1 = Zorbax-sil®, 7 μm
Solvent: S1 = cyclohexane/chloromethane/tetrahydrofuran, 5:1:1
Detection: D1 = fluorescence measurement, excitation wavelength 312 nm, emission wavelength 365 nm

REFERENCE

1. **Spahn, H., Weber, H., Mutschler, E., and Möhrke, W.,** *J. Chromatogr.*, 310, 167, 1984.

Table LC 163
ENANTIOMERS OF R,S-*N*(1-PHENYLETHYL)-3,5-DINITROBENZYLAMIDE (PEDA)

	Column I	Column II
A — Solvent S1		
k'_1	8.07	11.0
α	1.31	1.25
First eluted peak	S(+)	R(−)
B — Solvent S2		
k'_1	0.91	9.06
α	1.21	1.08
First eluted peak	S(+)	R(−)

Note: K'_1 = capacity factor of the first eluted peak; $\alpha = k'_2 k'_1$ = separation factor; k'_2 = capacity factor of the second eluted peak.

Conditions: Columns I and II are stainless steel tubes, 250 × 3.2 mm I.D., they are packed with phase I in the R configuration and phase II in the S configuration, respectively; the structures of the phases are shown below; a slurry prepared from 2 g of the phase and 30 mℓ of methanol/triethylene glycol, 1:9, is filled into the column under pressure; the columns are conditioned with methanol, ethyl acetate, and *n*-hexane

Solvent: S1 = isopropanol/*n*-hexane, 1:4
S2 = water/methanol, 1:3

Flow rate: 1 mℓ/min

Detection: UV at 254 nm

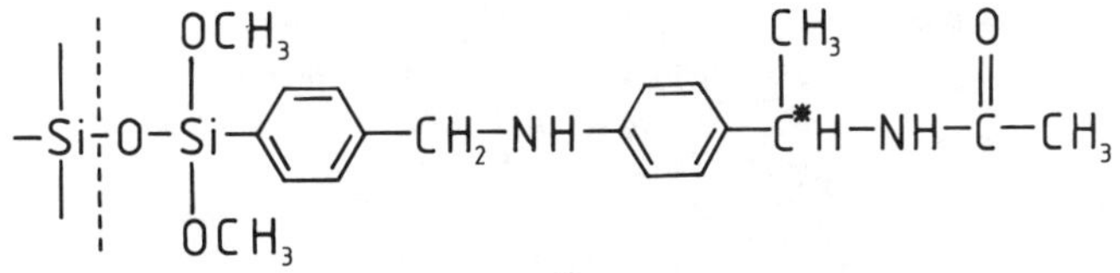

I

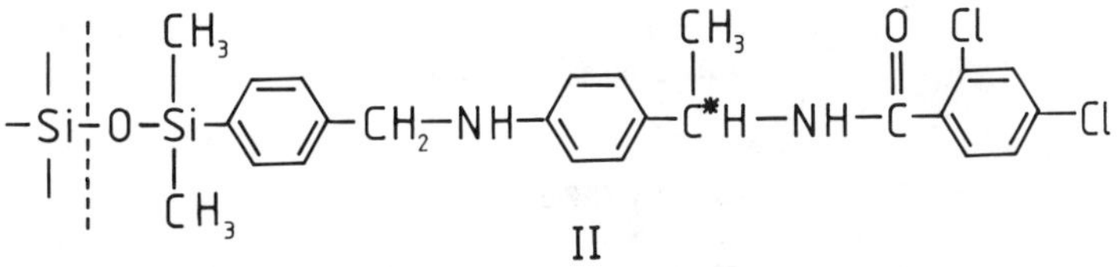

II

REFERENCE

1. **Däppen, R., Meyer, V. R., and Arm, H.,** *J. Chromatogr.*, 295, 367, 1984.

Table LC 164
ENANTIOMERIC AMINES

Packing	P1	P2
Column		
Length (mm)	250	250
Diameter (mm, I.D.)	3.2	3.2
Material	SS	SS
Solvent	S1	S1
Flow rate (mℓ/min)	1	1
Detection	D1	D1

Compound	k'_1	α	k'_1	α
1	6.87	1.39	13.0	1.25
2.	6.25	1.41	13.5	1.22
3	6.18	1.40	12.7	1.23
4	5.71	1.45	12. 5	1.22
5	4.94	1.48	10.8	1.23
6	4.71	1.51	10.8	1.26
7	4.32	1.55	9.51	1.26
8	4.09	1.55	8.85	1.25
9	3.48	1.60	7.75	1.28
10	3.32	1.63	7.38	1.28
11	3.01	1.64	6.94	1.30
12	1.72	1.07	2.41	1.05
13	4.06	1.05	6.94	1.07
14	9.96	1.40	20.3	1.27
15	8.95	1.49	19.3	1.26
16	7.33	1.80	15.7	1.65

Note: The formulas of the compounds are shown below in Table 164a.

k'_1 = capacity factor of the first eluted peak; $\alpha = k'_2\ k'_1$ = separation factor; k'_2 = capacity factor of the second eluted peak. On column I the first eluted peak had the S(+) configuration, the second peak the R(−) configuration.

Packing: P1 = phase I in the R configuration
P2 = phase II in the S configuration; the structures of the phases are shown in Table LC 163; a slurry prepared from 2 g of the phase and 30 mℓ of methanol/triethylene glycol, 1:9, is filled into the column under pressure; the columns are conditioned with methanol, ethyl acetate, and *n*-hexane
Solvent: S1 = isopropanol/*n*-hexane, 1:4
Detection: D1 = UV at 254 nm

REFERENCE

1. **Däppen, R., Meyer, V. R., and Arm, H.,** *J. Chromatogr.*, 295, 367, 1984.

Table LC 164a

Compound	R	R′	R″
1	C_6H_5	C_2H_5	$(3,5\text{-}[NO_2]_2)C_6H_3$
2		$n\text{-}C_3H_7$	
3		$n\text{-}C_4H_9$	
4		$n\text{-}C_5H_{11}$	
5		$n\text{-}C_7H_{15}$	
6		$n\text{-}C_8H_{17}$	
7		$n\text{-}C_9H_{19}$	
8		$n\text{-}C_{10}H_{21}$	
9		$n\text{-}C_{13}H_{27}$	
10		$n\text{-}C_{15}H_{31}$	
11		$n\text{-}C_{16}H_{35}$	
12	C_6H_5	CH_3	$(2,4\text{-}Cl_2)C_6H_3$
13	$(4\text{-}NO_2)O_6H_4$	CH_3	$(2,4\text{-}Cl_2)C_6H_3$
14	$(4\text{-}CH_3O)C_6H_4$	CH_3	$(3,5\text{-}[NO_2]_2)C_6H_3$
15	$(4\text{-}CH_3O)C_6H_4$	$n\text{-}C_4H_9$	$(3,5\text{-}[NO_2]_2)C_6H_3$
16	1-Naphthyl	CH_3	$(3,5\text{-}[NO_2]_2)C_6H_3$

Note: R, R′, and R″ are groups in the compound of structure:

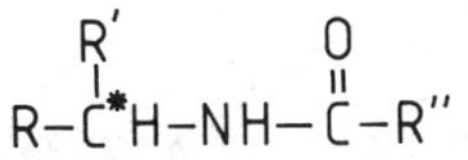

Table LC 165
CATION EXCHANGE CHROMATOGRAPHY OF HISTAMINE IN THE PRESENCE OF ETHYLAMMONIUM CHLORIDE

Packing	P1	P1	P1	P1
Column				
Length (cm)	25	25	25	25
Diameter (mm, I.D.)	4.6	4.6	4.6	4.6
Solvent	S1	S2	S3	S4
Temperature	r t	r t	r t	r t
Detection	D1	D1	D1	D1
Compound		**k′**		
Histamine	6.85	5.45	5.14	3.48
Histidine	2.92	2.58	2.33	2.16
Tryptamine[a]	4.80	4.25	4.15	3.09

[a] Internal standard.

Packing: P1 = Partisil® 10 SCX

Solvent: S1, S2, S3, and S4 = 0.03 $M K_2HPO_4$ in distilled water, pH 4.5, containing 0.02, 0.25, and 0.5%, respectively, of ethylammonium chloride

Detection: D1 = UV at 210 nm

REFERENCE

1. **Mett, C. L. and Sturgeon, R. J.,** *J. Chromatogr.*, 235, 536, 1982.

Table LC 166
THE EFFECT OF POTASSIUM DIHYDROGEN PHOSPHATE CONCENTRATION ON THE RETENTION OF HISTAMINES

Histamine	KH_2PO_4 conc (mol/ℓ) 0.5	0.4	0.3	0.25
Histamine	0.55	0.68	0.83	1.04
N^α-Methylhistamine	1.32	1.55	1.96	2.96
3-Methylhistamine	1.96	2.24	3.35	5.98
1-Methylhistamine	3.60	4.43	6.17	

Conditions:
- Packing: Partisil® 10 SCX, particle size 10 μm
- Column: 25 × 0.4 cm I.D.
- Solvent: 0.4 *M* KH_2PO_4, pH 4.5
- Flow rate 2 mℓ/min
- Detection: UV at 210 nm

REFERENCE

1. **Robert, J. C., Vatier, J., Nguyen Phuoc, B. K., and Bonfils, S.,** *J. Chromatogr.*, 273, 275, 1983.

Table LC 167
HISTAMINE AND OTHER ENDOGENOUS SUBSTANCES

Packing	P1
Column	
Length (mm)	50
Diameter (mm, I.D.)	4
Solvent	S1
Flow rate (mℓ/min)	0.56
Temperature (°C)	70
Detection	D1

Compound	t_r(min)
Arginine	2.55
Cadaverine	6.36
Cystine	1.47
Dopamine	9.00
Histamine	7.30
Histidine	2.48
Lysine	1.90
Putrescine	5.04
Serine, threonine	1.51
Serotonin	21.80
Spermidine	9.66
Tryptophan	3.84

Packing: P1 = Hitachi®-gel 2619F, with a μBondapak® C18 Guard-PAK® pre-column

Solvent: S1 = trisodium citrate dihydrate, 27.5 g; NaCl, 117 g; 35% HCl, 5.26

Table LC 167 (continued)
HISTAMINE AND OTHER ENDOGENOUS SUBSTANCES

mℓ; Brij 35, 0.2 g; 10% EDTA disodium salt dihydrate, 3.7 mℓ; water, 800 mℓ; methanol, 250 mℓ

Detection: D1 = *o*-phthalaldehyde, excitation filter, 370 nm, emission filter, 418 nm UV-cut filter

REFERENCE

1. **Arakawa, Y. and Tachibana, S.,** *Anal. Biochem.,* 158, 20, 1986.

Table LC 168
HISTAMINE AND DERIVATIVES

Packing	P1
Column	
Length (mm)	250
Diameter (mm, I.D.)	4.6
Solvent	S1
Flow rate (mℓ/min)	1.0
Temperature	r t
Detection	ECD

Amine	**t_r(min)**
Histamine	8.4
N^{τ}-Methylhistamine	10.0
N^{α}-Methylhistamine	11.6
Histidine	4.2
Spermidine	32.6

Packing: P1 = Beckman Ultrasphere® ODS; a Whatman® CSKI guard column packed with Waters μBondapak® C_{18}/Corasil® was employed

Solvent: S1 = 0.14 *M* sodium acetate/methanol, 17:73, containing 3.89 *M* 1-octanesulfonic acid and 56 mg of EDTA; the pH was adjusted to 3.48 with glacial acetic acid and the solution degassed under vacuum

REFERENCE

1. **Mine, K., Jacobson, K. A., Kirk, K. L., Kitajima, Y., and Linnoila, M.,** *Anal. Biochem.,* 152, 127, 1986.

Table LC 169
HISTAMINE AND COMPOUNDS INTERFERING WITH ITS FLUOROMETRIC DETERMINATION

Packing	P1
Column	
Length (mm)	100
Diameter (mm, I.D.)	6
Material	SS
Solvent	S1
Flow rate (mℓ/min)	0.6
Detection	D1

Compound	**t_r(min)**	**Relative fluorescence intensity[a]**
Ammonia	6.8	0.01
Glutathione	4.5	2
Histamine	12.0	100
Histidine	6.0	4
Putrescine	11.0	0.005
Spermidine	17.0	0.5
Spermine	17.0	0.02
Serotonin	17.0	0.2

[a] Peak heights obtained by analysis of appropriate amounts of each substance were compared with the calculated peak heights of the same amount of histamine. Values are expressed as percentages of that of the histamine peak.

Packing: P1 = cation exchanger, TSK gel SP-2SW, 5 μm Toyo Soda

Solvent: S1 = 0.025 *M* KH_2PO_4; after the elution of histamine this was altered to 0.5 *M* KH_2PO_4

Detection: D1 = reaction with *o*-phthalaldehyde followed by fluorometric measurement, excitation at 360 nm, emission at 450 nm

REFERENCE

1. **Yamatodani, A., Fukuda, H., Wada, H., and Watanabe, T.,** *J. Chromatogr.*, 344, 115, 1985.

Table LC 170
EFFECT OF THE ION-PAIRING AGENT ON THE RETENTION OF HISTAMINE-*O*-PHTHALALDEHYDE COMPLEXES

Solvent	l-Heptane sulfonic acid (× 10^{-3} mol/ℓ)	k′	1-Octane sulfonic acid (× 10^{-3} mol/ℓ)	k′
S1	0	0	0	0
	0.025	0.5	0.050	0.9[a]
	0.050	0.65	0.100	1.4,1.6[b]
	0.075	0.75[a]	0.150	1.7,1.9[b]
	0.100	0.85[a]		
	0.150	0.91[a]		
S2	0	0	0	0
	0.025	0.2	0.050	0.5
	0.050	0.35	0.100	0.8
	0.075	0.45	0.150	1.0
	0.100	0.5	0.200	1.15
	0.150	0.55		
	0.200	0.6		

Note: 1-Pentanesulfonic acid, sodium salt, had no effect on the k′ values of histamine-*o*-phthalaldehyde complexes.

[a] A shoulder appeared.
[b] Two peaks appeared in the ratio 6:1.

Conditions:
- Packing: μBondapak® C_{18}, particle size 10 μm
- Column: 30 × 0.4 cm I.D.
- Solvent: S1 = methanol/0.020 *M* sodium acetate/acetic acid, 50:48:2
  S2 = methanol/0.020 *M* sodium acetate/acetic acid, 55:43:2
- Flow rate: 1.5 mℓ/min
- Detection: Fluorometric, emission wavelength 450 nm, excitation wavelength 360 nm

REFERENCE

1. **Robert, J. C., Vatier, J., Nguyen Phuoc, B. K., and Bonfils, S.,** *J. Chromatogr.*, 273, 275, 1983.

Table LC 171
AMINES AND AMINO ACIDS AS *o*-PHTHALALDEHYDE DERIVATIVES

Column packing	P1
Column	
Length (mm)	250
Diameter (mm, I.D.)	5
Solvent	S1
Flow rate (mℓ/min)	1.0
Temperature	r t
Detection	D1

Compound	**k′**	**Fluorescence yield**[a]
Agmatine	4.4	3.5×10^3
Histamine	6.0	10
Spermidine	5.2	525
Cysteine	6.6	107
Histidine	6.6	320
Ornithine	6.6	76
Phenylalanine	6.6	62
His-Leu	12.6	1.7×10^3
His-Lys	5.4	9.7×10^3
His-Ser	7.2	701
Carnosine	8.0	8.4×10^3
Homocarnosine	8.0	14.4×10^3

[a] Values expressed as the amount (ng) that gives the same peak height as 10 ng of histamine. Derivatization at −20°C in a total volume of 2.5 mℓ.

Column packing: P1 = Nucleosil® C_{18}, 5 μm
Solvent: S1 = 1.0 g concentrated H_2SO_4/ℓ, 25% methanol (pH 2.25)
Detection: D1 = fluorescence with excitation at 353 nm and emission read at 451 nm

REFERENCE

1. **Rönnberg, A. L., Hansson, C., and Håkanson, R.,** *Anal. Biochem.*, 139, 338, 1984.

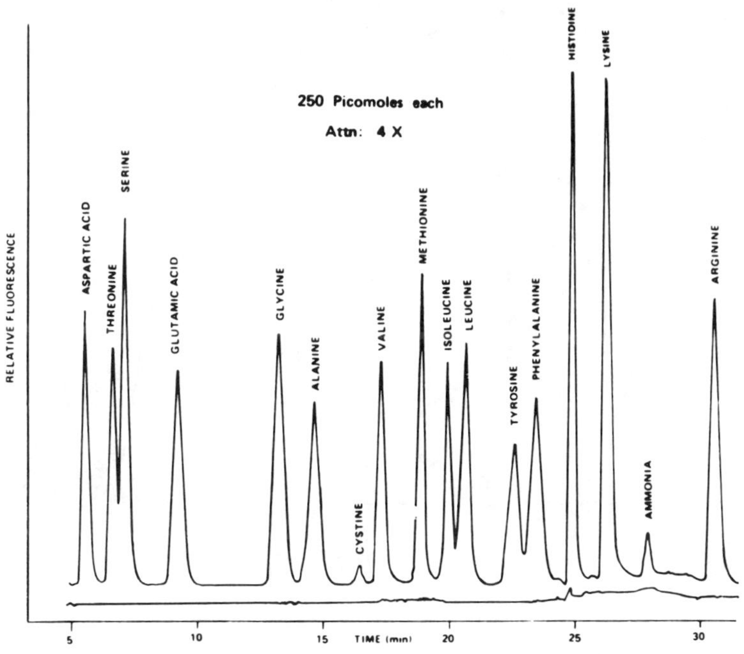

FIGURE LC 1. Separation of amino acids on a Pierce high-speed AA511 column. Column: 0.46 × 12 cm packed with a special sulfonated polystyrene-divinylbenzene copolymer. Column temperature: 60°C. Buffer program: Buffelute hydrolyzate buffer A, 9 min, Buffelute hydrolyzate buffer B, 5 min, Buffelute hydrolyzate buffer C, 17 min. Flow rate: 0.5 mℓ/min. Detection: *o*-phthalaldehyde, excitation filter 360 nm, emission filter 455 nm. Reference: Pierce Chemical Company, 1985/1986 Handbook and General Catalog, p. 259. (Reproduced with permission.)

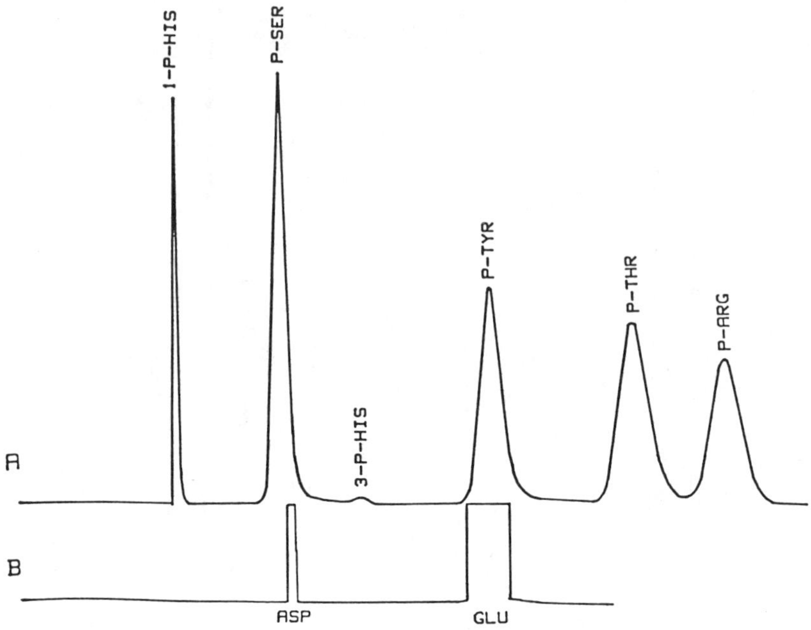

FIGURE LC 2. Separation of phosphoamino acids derivatized with *o*-phthalaldehyde on PRP-1, a polymer-based reversed-phase column. HPLC was performed with 88% buffer A, 12% buffer B. The phosphate buffer was 0.3 *M* NaH_2PO_4. Buffer A contained 100 mℓ phosphate buffer, 25 mℓ tetrahydrofuran, and 1875 mℓ water; buffer B contained 45 mℓ phosphate buffer, 1100 mℓ acetonitrile, and 855 mℓ water. The flow rate was 1.0 mℓ/min. (A) 800 pmol phosphoserine, phosphotyrosine, and phosphoarginine, 1200 pmol phosphothreonine, and 6 μg phosphohistidine-containing mixture in water were reacted with derivatizing reagent and injected onto the column. The graph shows fluorescence as a function of elution time. (B) 10 nmol aspartate, glutamate, and asparagine were derivatized and injected. The graph shows fluorescence as a function of elution on the same scale as in (A). Asparagine did not elute under these conditions and was removed from the column by a wash step. Reference: Carlomagno, L., Heubner, V. D., and Matthews, H. R., *Anal. Biochem.*, 149, 344, 1985. (Reproduced by permission of Academic Press.)

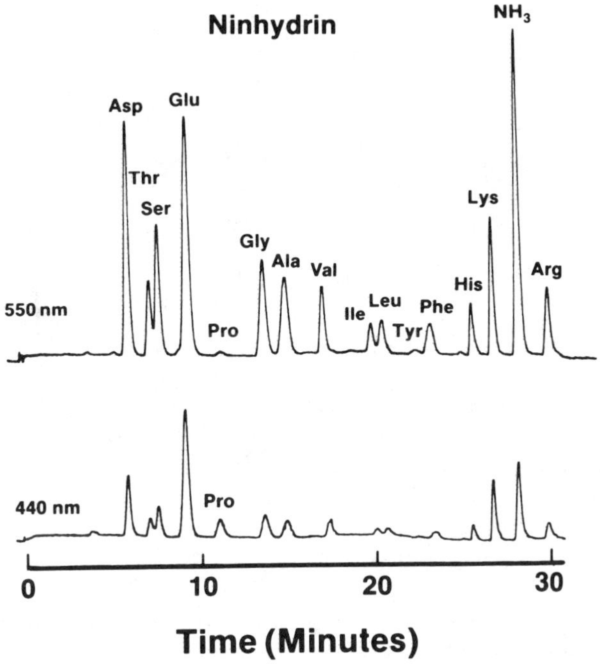

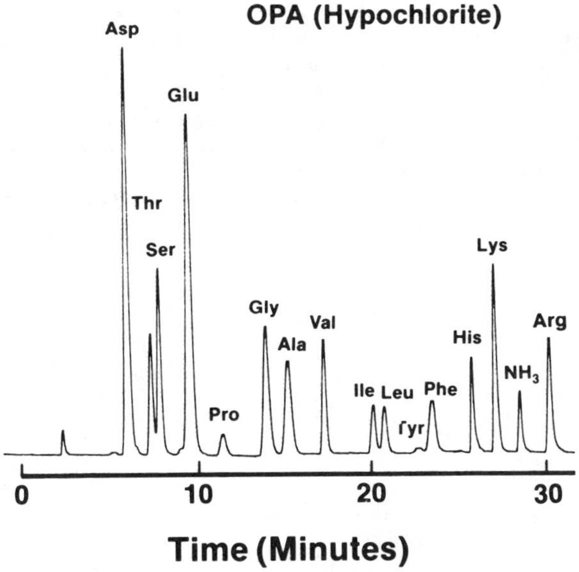

FIGURE LC 3. Comparative chromatograms of a soy protein hydrolyzate by the ninhydrin and sodium hypochlorite-*o*-phthalaldehyde methods. Chromatographic conditions: Perkin-Elmer high-speed amino acid analysis column, 120 × 4.6 mm I.D. packed with 6 μm sulfonated ion exchange resin. Temperature: 60°C. Chromatographic buffers: Pierce Buffelute, A (8 min), B (5 min), C (13 min), step gradient. Flow rate: 0.5 mℓ/min. Sodium hypochlorite-*o*-phthalaldehyde: detection, fluorescence, excitation filter at 360 nm, emission filter at 418—700 nm. Ninhydrin detection, adsorbance at 550 nm. Reference: Dong, M. W. and Gant, J. R., *J. Chromatogr.*, 327, 17, 1985. (Reproduced from Dong, M. W. and Gant, J. R., *J. Chromatogr.*, 327, 17, 1985. With permission.)

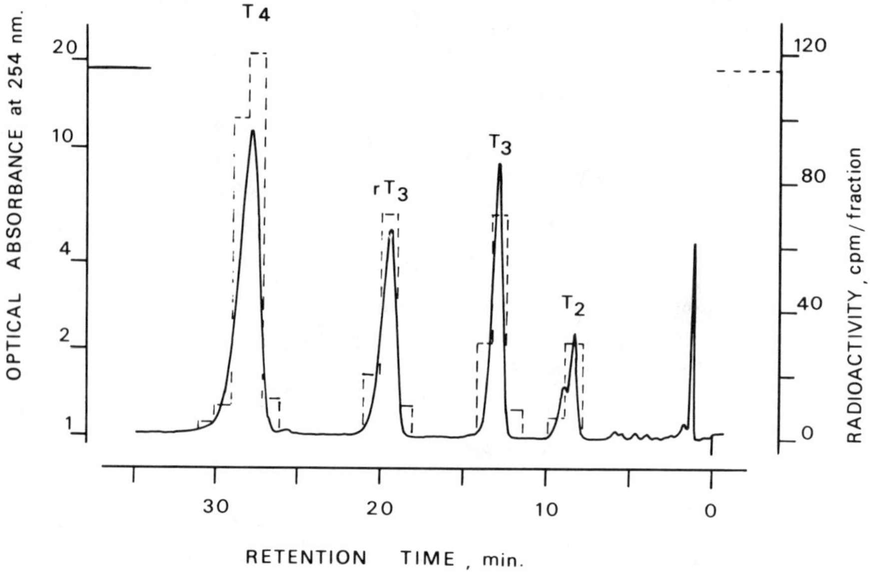

FIGURE LC 4. HPLC chromatogram of iodothyronines. After the injection of ^{125}I-T_4 in a human subject, radioiodothyronines were extracted from serum using a SEP-PAK® column (Waters Associates). Unlabeled iodothyronines were added as a carrier and as markers of optical absorbance. Chromatography was performed using an HPLC glass tube, 3 × 300 mm, packed with 10 μm C_{18} silica gel. Isocratic elution was carried out with methanol/water, 55:45, adjusted to pH 3 with H_3PO_4, at a flow rate of 1.2 mℓ/min. T_4 = thyroxine; rT_3 = reverse T_3, 3,3′,5′-triiodothyronine; T_3 = 3,5,3′-triiodothyronine; T_2 = 3,3′-diiodothyronine + 3′,5′-diiodothyronine. -------, optical absorbance profile at 254 nm; ———, radioactivity-counting profile (cpm per fraction). Reference: Bianchi, R., Molea, N., Cazzuola, F., Fusani, L., Lotti, M., Bertelli, P., Ferdeghini, M., and Mariani, G., *J. Chromatogr.*, 297, 393, 1984. (Reproduced from Bianchi, R., Molea, N., Cazzuola, F., Fusani, L., Lotti, M., Bertelli, P., Ferdeghini, M., and Mariani, G., *J. Chromatogr.*, 297, 393, 1984. With permission.)

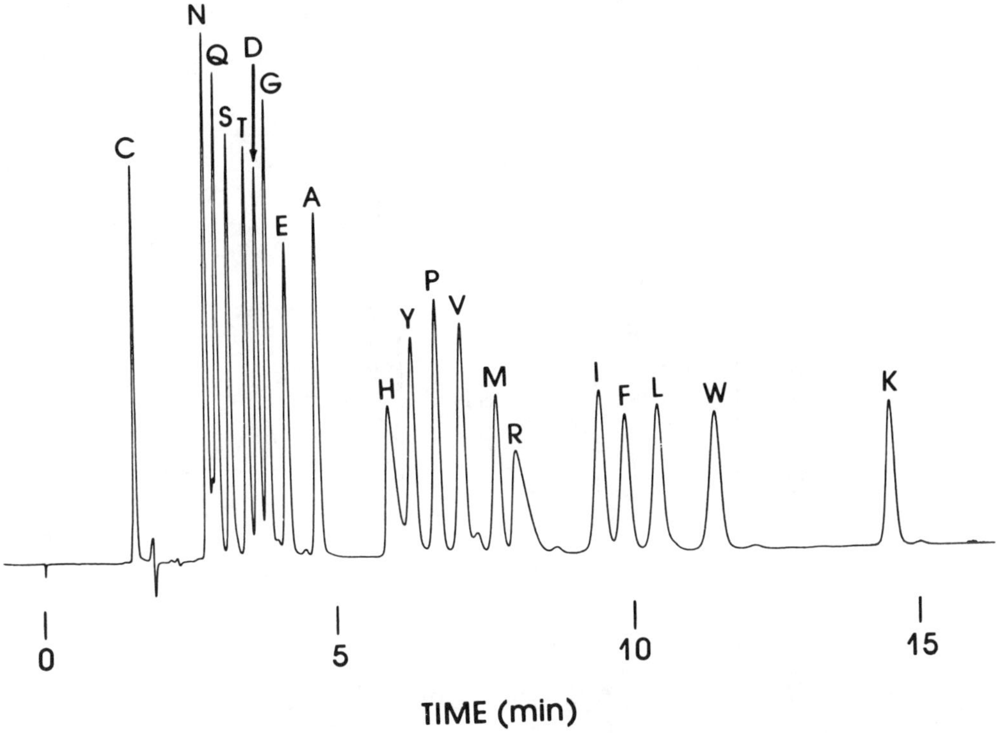

FIGURE LC 5. Isocratic separation of PTH amino acids. Column: Zorbax® PTH, 4.6 mm I.D. × 25 cm. Mobile phase: 5.8 m*M* H_3PO_4 brought to pH 3.15 with NaOH/acetonitrile/tetrahydrofuran, 66:18:16. Flow rate: 15 mℓ/min. Temperature: 35°C. Detection: UV at 254 nm. Peak identities; PTH derivatives. A, alanine; C, cysteine; D, aspartic acid; E, glutamic acid; F, phenylalanine; G, glycine; H, histidine; I, isoleucine; K, lysine; L, leucine; M, methionine; N, asparagine; P, proline; Q, glutamine; R, arginine; S, serine; T, threonine; V, valine; W, tryptophan; Y, tryosine. Reference: DuPont Zorbax Bio Series PTH Product Bulletin, Figure 1.

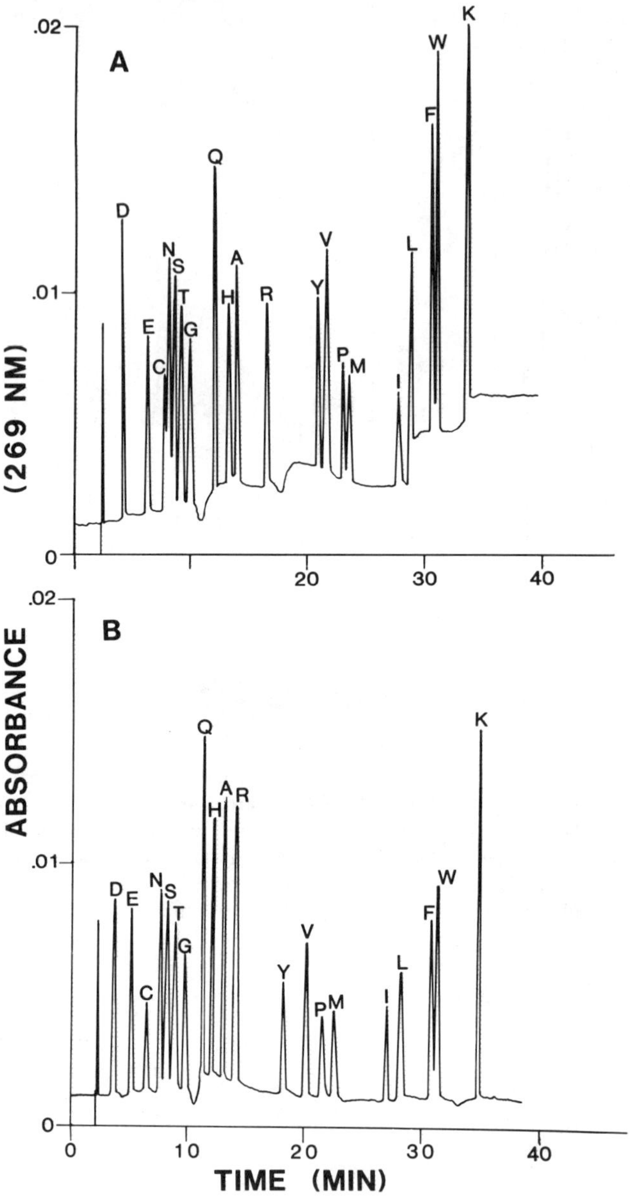

FIGURE LC 6. Separation of PTH amino acids on Bakerbond® wide-pore diphenyl columns. Each peak represents 300 pmol of the respective amino acid. The flow rate was 1.5 mℓ/min. (A) Buffer A is 66 m*M* trifluoroacetate/4 m*M* acetate, pH 5.6 and buffer B is acetonitrile/35 m*M* trifluoroacetate, pH 3.6, 75:25. The program (time in minutes) is (0) 5% B, (7) 18% B, (14) 28% B, (25) 45% B, (30) 50% B, (32) 100% B, (36) 5% B. The column temperature is 30°C. (B) Buffer A is 66 m*M* trifluoroacetate/4 m*M* acetate, pH 5.8 and buffer B is acetonitrile/35 m*M* trifluoroacetate, pH 3.6, 75:25. The program (time in minutes) is (0) 8% B, (7) 26% B, (20) 32% B, (25) 40% B, (30) 50% B, (36) 100% B, (40) 8% B. The column temperature is 27°C. Peaks D = aspartic acid; E = glutamic acid; C = carboxymethyl cysteine; N = asparagine; S = serine; T = threonine; G = glycine; Q = glutamine; H = histidine; A = alanine; R = arginine; Y = tyrosine; V = valine; P = proline; M = methionine; I = isoleucine; L = leucine; F = phenylalanine; W = tryptophan; K = lysine. Reference: Kruggel, W. G. and Lewis, R. V., *J. Chromatogr.*, 342, 376, 1985. (Reproduced by permission of Elsevier Science Publishers.)

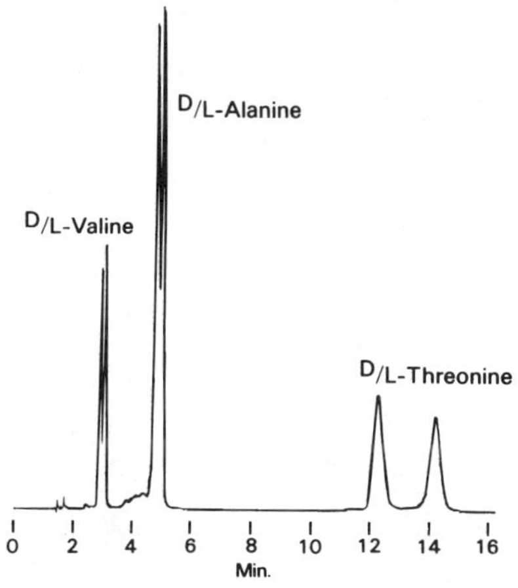

FIGURE LC 7. Separation of the enantiomers of PTH amino acids on a Supelcosil® LC-(R)-Urea column. Column: 25 cm × 4.6 mm, with a 5-μm packing. Mobile phase, *n*-hexane/1,2-dichloroethane/ethanol, 50:10:1. Flow rate: 2.0 mℓ/min. Temperature: ambient. Detection: UV at 254 nm. Reference: *Supelco Reporter,* 4(2), 1, 1985, Figure C. (Reprinted with permission of Supelco Inc., Bellefonte, PA 16823.)

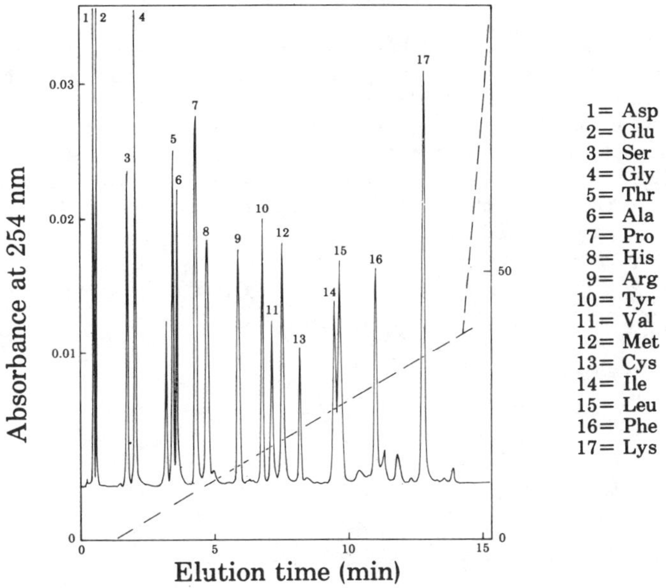

FIGURE LC 8. The separation of phenylthiocarbamyl amino acids. Column: SuperPac Cartridge Spherisorb® ODS2 reversed-phase column, 3-μm particles. Solvent: A, 12 m*M* sodium phosphate buffer, pH 6.4; B, 500 mℓ acetonitrile, 500 mℓ, 24 m*M* sodium phosphate buffer, pH 6.4. Gradient: 0—45% B in 12 min, 45—100% B in 1 min. Flow rate: 1.0 mℓ/min. Temperature: 40°C. Detection at 254 nm. Reference: LKB Liquid Chromatography Application Note 442. (Reproduced by permission of LKB Produkter AB.)

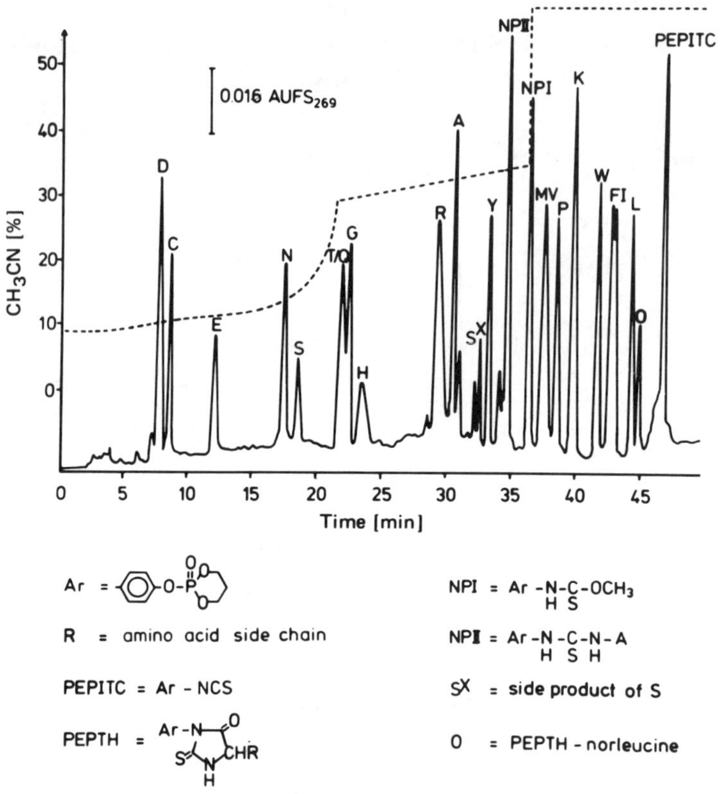

FIGURE LC 9. HPLC of PEPTH-amino acids. The amino acid derivatives (phosphoric esters of phenylthiohydantoin derivatives of amino acids) were separated at 52°C on a reverse-phase C-18 column (Hypersil® or LiChrosphere®, 45 × 250 mm) using a concave gradient of 20 m*M* sodium acetate, pH 5.0, containing 12—24% of acetonitrile. The acetonitrile gradient is indicated by the dotted line. Extinction was measured at 269 nm. Amino acids are abbreviated in the one-letter code. Other abbreviations are shown above. Reference: Ender, B. E., Gassen, H.-G., and Machleidt, W., *Hoppe Seylers Z. Physiol. Chem.*, 365, 839, 1984. (Reproduced by permission of Verlag Walter De Gruyter.)

Table TLC 1
AMINO ACIDS ON CELLULOSE AND SILICA GEL LAYERS

Layer	L1	L2	L3	L4
Solvent	S1	S1	S2	S2
Detection	D1	D1	D1	D1
Technique	T1	T1	T1	T1
Amino acid	**$R_f \times 100$**			
Alanine	41.9	29.0	32.4	28.8
Arginine	25.6	11.0	12.9	10.0
Aspartic acid	26.3	14.8	25.3	21.8
Cystine	6.9	3.2	14.1	7.1
Glutamic acid	34.4	22.6	30.0	28.2
Glycine	29.4	14.8	25.9	23.5
Histidine	20.0	7.1	11.7	7.1
Isoleucine	73.1	60.0	49.4	47.1
Leucine	75.0	63.9	51.8	48.8
Lysine	18.1	7.1	10.0	7.1
Methionine	41.0	22.5	47.3	43.5
Phenylalanine	67.5	54.8	52.4	50.0
Proline	43.8	33.5	24.1	21.2
Serine	26.9	16.1	26.4	24.1
Threonine	32.5	21.3	30.0	27.6
Tryptophan	55.6	36.1	54.1	51.8
Tyrosine	50.0	36.1	49.4	45.9
Valine	63.1	48.4	43.5	41.2

Layer: L1 = Baker-Flex® 20 × 20-cm cellulose sheets
L2 = Baker-Flex® 20 × 20-cm microcrystalline cellulose sheets
L3 = Whatman® 20 × 20-cm K6 silica gel plates
L4 = Whatman® 10 × 10-cm high performance silica gel plates

Solvent: S1 = 2-butanol/acetic acid/water, 3:1:1
S2 = *n*-butanol/acetic acid/water, 3:1:1

Detection: D1 = ninhydrin

Technique: T1 = silica gel and cellulose layers were developed in paper-lined glass tanks that had been preequilibrated with solvent for 20 min; chromatograms were dried at 100°C and the cooled plates sprayed with ninhydrin reagent; heating at 90—100°C for 5 min produced blue to purple zones of amino acids

REFERENCE

1. **Sleckman, B. P. and Sherma, J.,** *J. Liquid Chromatogr.*, 5, 1051, 1982.

Table TLC 2
AMINO ACIDS ON STRONG ACID ION EXCHANGE SHEETS

Layer	L1	L1	L1
Solvent	S1	S1	S1
Detection	D1	D1	D1
Technique	T1	T2	T3
Amino acid		**$R_f \times 100$**	
Alanine	50.9	51.2	53.6
Arginine	1.8	2.2	2.2
Aspartic acid	71.5	68.2	68.6
Cystine	55.9	50.0	57.9
Glutamic acid	34.5	29.4	30.6
Glycine	55.6	52.4	53.6
Histidine	10.6	8.9	10.0
Isoleucine	27.8	22.2	23.3
Leucine	21.8	17.8	19.4
Lysine	7.5	5.0	5.6
Methionine	28.0	27.2	25.0
Phenylalanine	14.4	11.1	11.7
Serine	67.1	64.7	67.1
Threonine	67.1	60.6	57.2
Tryptophan	1.8	2.2	2.2
Tyrosine	11.9	13.9	15.5
Valine	42.5	35.0	34.4

Layer: L1 = Fixion® 20 × 20-cm strong acid ion exchange sheets (Na^+ form)

Solvent: S1 = run buffer; 84 g citric acid monohydrate + 16.0 g NaOH + 5.8 g NaCl + 54.0 g ethylene glycol + ca. 4 mℓ of concentrated HCl added dropwise to give pH 3.3; equilibration buffer, run buffer diluted 30 times (pH 3.8)

Detection: D1 = chromatograms are completely dried at 100°C and the cooled plates sprayed with ninhydrin reagent; heating at 90—100°C for 5 min then produces blue to purple zones of amino acids

Technique: T1 = a 1-cm-wide lane on each side of the sheet is scraped free of resin, and the layer developed without prior treatment

T2 = the layer is preequilibrated for 16 hr with the equilibration buffer by attaching filter paper to the top of the layer to allow a continuous flow of solution; after this period a 1-cm strip is cut from the bottom edge of the layer in direct contact with the equilibration buffer, the layer is dried, spotted, and developed in the run buffer

T3 = the plate is preequilibrated, dried, spotted, and developed in the run buffer at 45°C rather than at ambient temperature; all the layers are developed for a distance of 17 cm beyond the origin

REFERENCE

1. **Sleckman, B. P. and Sherma, J.,** *J. Liquid Chromatogr.*, 5, 1051, 1982.

Table TLC 3
AMINO ACIDS ON REVERSED-PHASE THIN LAYERS

Layer	L1	L2	L3	L4	L5	L6
Solvent	S1	S1	S1	S1	S1	S1
Detection	D1	D1	D1	D1	D1	D1
Technique	T1	T1	T1	T1	T1	T1
Amino acid			**$R_f \times 100$**			
Alanine	50	32	36	51	42	41
Arginine	0.6	13	3.6	17	19	16
Asparagine	31	ND	31	47	24	21
Aspartic acid	53	25	41	44	29	29
Cysteine	NV	ND	18	5.5	12	13
Cystine	NV	14	NV	NV	NV	NV
Glutamic acid	61	30	41	52	42	40
Glutamine	44	ND	33	50	33	35
Histidine	6.3	12	21	5.5	15	11
Isoleucine	71	49	47	50	63	60
Leucine	72	52	49	52	77	72
Methionine	69	47	46	50	75	71
Phenylalanine	72	52	50	49	77	71
Proline	59	24	27	42	55	51
Serine	30	26	37	52	35	36
Threonine	47	30	37	49	47	45
Tryptophan	74	54	54	51	69	70
Tyrosine	72	49	50	59	71	66
Valine	66	44	42	51	75	71

Abbreviations: ND = no data; NV = not visualized

Layer: L1 = Whatman® KC_{18} chemically bonded reversed-phase plate
L2 = Whatman® K6 silica gel
L3 = Analtech® reversed-phase plate containing a long-chain hydrocarbon-impregnated support layer
L4 = OPTI-UP plate containing a layer of silica gel chemically bonded with C_{12}
L5 = Baker Flex® 10% acetylated cellulose (plastic backed)
L6 = Analtech® 20% acetylated cellulose (glass backed)

Solvent: S1 = *n*-propanol/water, 7:3

Detection: D1 = chromatograms are oven dried at 100°C, sprayed with 0.1% ninhydrin in acetone, and heated again for 5 min or longer to detect the amino acids as colored spots on a white background; the reagent produced purple, red, and tan spots on silica gel and KC_{18} chemically bonded silica gel plates, while all spots appear some shade of purple on cellulose

Technique: T1 = plates are developed in glass chambers that are lined with paper and preequilibrated with the mobile phase for at least 10 min; development is carried out to a point 15 cm above the origin line

REFERENCE

1. **Sherma, J., Sleckman, B. P., and Armstrong, D. W.,** *J. Liquid Chromatogr.*, 6, 95, 1983.

Table TLC 4
AMINO ACIDS ON REVERSED-PHASE THIN LAYERS USING MICELLAR MOBILE PHASES

Layer	L1	L1	L2	L3	L3	L4
Solvent	S1	S2	S3	S4	S5	S6
Detection	D1	D1	D1	D1	D1	D1
Technique	T1	T1	T1	T1	T1	T1
Amino acid				$R_f \times 100$		
Alanine	85	78	64	80	48	47
Arginine	67	54	79	21	ND	17
Asparagine	NV	78	NV	80	ND	ND
Aspartic acid	85	78	79	80	ND	23
Cysteine	NV	78	NV	62	ND	ND
Cystine	NV	NV	NV	NV	ND	10
Glutamic acid	83	78	79	80	ND	44
Glutamine	83	78	71	80	59	ND
Histidine	43	33	36	32	25	11
Isoleucine	79	77	50	45	40	54
Leucine	79	71	64	40	40	58
Methionine	79	78	64	52	44	58
Phenylalanine	62	60	71	35	40	54
Proline	83	78	NV	66	NV	ND
Serine	83	78	71	80	59	44
Threonine	85	78	64	80	56	47
Tryptophan	47	46	71	29	40	50
Tyrosine	79	78	64	61	50	58
Valine	79	78	50	62	44	54

Abbreviations: ND = no data; NV = not visualized

Layer: L1 = KC_{18} plate
L2 = C_{18} layer
L3 = Whatman® C_{18} plate impregnated by dipping in a 4% solution of dodecylbenzene sulfonic acid in 96% ethanol
L4 = KC_{18} layer impregnated with cetyltrimethylammonium bromide from a 10% solution of this compound in 96% ethanol

Solvent: S1 = 0.015 *M* sodium dodecylsulfate + 0.5 *M* NaCl
S2 = saturated aqueous cetyltrimethylammonium bromide/H_2O, 1:19, also containing 0.5 *M* NaCl
S3 = 1.3 *M* sodium dioctylsulfosuccinate in cyclohexane/water, 50:4
S4 = 1 *M* acetic acid + 0.2 *M* HCl in methanol/water, 1:1
S5 = 1 *M* acetic acid + 0.2 *M* HCl in methanol/water, 7:3
S6 = methanol/water, 9:1, + 1 *M* acetic acid + 0.2 *M* HCl

Detection: D1 = chromatograms are dried at 100°C, sprayed with 0.1% ninhydrin in acetone, and heated again for 5 min or longer to detect amino acids as colored spots on a white background

Technique: T1 = plates are developed in glass chambers that are lined with paper and preequilibrated with mobile phase for at least 10 min; development is carried out to a point 15 cm above the origin line

REFERENCE

1. **Sherma, J., Sleckman, B. P., and Armstrong, D. W.,** *J. Liquid Chromatogr.*, 6, 95, 1983.

Table TLC 5
AMINO ACIDS ON STANNIC TUNGSTATE LAYERS

Layer	L1	L1	L1	L1	L1	L1	L1	L1
Solvent	S1	S2	S3	S4	S5	S6	S7	S8
Detection	D1	D1	D1	D1	D1	D1	D1	D1
Technique	T1	T1	T1	T1	T1	T1	T1	T1
Amino acid				**R_f × 100**				
DL-Alanine	86		53	87	70	62	61	62
Glycine	52	52	55	69	86	67	86	84
DL-Isoleucine	53	45	63	84	24	76	68	86
L-Leucine	86	77	52	40	86	49	45	59
DL-Methionine	57	24	71	52	92		82	78
DL-Phenylalanine	62	77			47	69	86	61
L-Proline	95	59	32	45	60	20	74	78
DL-Tryptophan	80	45	50		85	45	55	71
DL-Valine	79	72	63	62	77	86	89	91
L-Cystine HC1		72	54	73	61	80	72	87
L-Cystine	75	76	48		84	85		67
DL-2-Amino-*n*-butyric acid	72	66	74	70	84	85	87	85
DL-Norleucine		56	85	62	56	58	77	42
DL-Serine	48	85	82	52	59	53	38	86
DL-Threonine		60	88	84		72	78	56
L-Hydroxyproline	68	44	87	85		86	91	
L-Tyrosine	45		53	25	45	76		
DL-3,4-Dihydroxyphenylalanine	39	33		20	95	92	96	95
DL-Aspartic acid	55	85	86	85	78	41	50	67
L-Glutamic acid	85	98	51	56	77	29	63	67
L-Arginine mono HC1	48	50	78	38	41	69	80	53
L-Ornithine HC1		53	39	23	85	94	47	35
L-Histidine HC1		60	95	67	61	65	74	20
L-Lysine HC1	92	85	48	85	85	84	33	76

Layer: L1 = stannic tungstate spread on glass plates 20 × 20 cm to give a layer 0.1 mm thick
Solvent: S1 = *n*-butanol/acetic acid/water, 5:4:1
S2 = ethyl acetate/formic acid, 6:4
S3 = *n*-butanol saturated with water/acetic acid, 3:1
S4 = acetone/formic acid/water, 2:2:1
S5 = ethyl acetate/pyridine/water, 2:1:2
S6 = acetone/ethanol/water, 6:1:3
S7 = acetic acid/formic acid/water, 4:3:2
S8 = ethanol/ethyl acetate/*n*-butanol, 3:4:2
Detection: D1 = 0.2% ninhydrin in *n*-butanol saturated with water
Technique: T1 = the solvents were allowed to ascend 12 cm from the point of application

REFERENCE

1. **Nabi, S. A., Farooqui, W. U., Siddiqui, Z. M., and Rao, R. A. K.,** *J. Liquid Chromatogr.*, 6, 109, 1983.

Table TLC 6
ENANTIOMERIC AMINO ACIDS

Amino acid	R_f × 100[a]	Solvent
Glutamine	37(S)	S1
	53(R)	
Isoleucine	37(2R,3R)	S1
	44(2S,3S)	
Phenylalanine	38(R)	S1
	45(S)	
Proline	40(R)	S2
	59(S)	
3-Thiazolidine-4-carboxylic acid	42(S)	S1
	52(R)	
Tyrosine	26(S)	S2
	34(R)	
Tryptophan	39(R)	S1
	45(S)	

[a] The configuration of the enantiomer is given in brackets.

Conditions: Layer: A glass plate covered with hydrophobic silicic acid (RP18-TLC) is immersed for 1 min in a 0.25% Cu (II) acetate solution in methanol/water, 1:9, and is then dried; the plate is then immersed in a 0.8% methanolic solution of the chiral selector, (2S,4R,2′RS)-4-hydroxy-1-(2′-hydroxydodecyl)-proline for 1 min; after being allowed to dry in air the plate is ready for chromatography

Solvent: S1 = methanol/water/acetonitrile, 50:50:200
S2 = methanol/water/acetonitrile, 50:50:30

Detection: 0.1% ninhydrin

Technique: Chromatography is carried out in a chamber saturated with solvent for 30 — 45 min, the development distance being 14 cm

REFERENCE

1. **Günther, K., Martens, J., and Schickedanz, M.,** *Angew. Chem. Int. Ed. Engl.*, 23, 506, 1984.

Table TLC 7
ENANTIOMERIC RESOLUTION OF RACEMIC AMINO ACIDS

Layer	L1
Solvent	S1
Dectection	D1
Technique	T1

Compound	**$R_f \times 100$**	
Valine	0.54(D)[a]	0.62(L)
Methionine	0.54(D)	0.59(L)
allo-Isoleucine	0.51(D)	0.61(L)
Norleucine	0.53(D)	0.62(L)
2-Aminobutyric acid	0.48	0.52
O-Benzylserine	0.54(D)	0.65(L)
3-Chloroalanine	0.57	0.64
S-(2-Chlorobenzyl)-cysteine	0.45	0.58
S-(3-Thiabutyl)-cysteine	0.53	0.64
S-(2-Thiapropyl)-cysteine	0.53	0.64
cis-4-Hydroxyproline	0.41(L)	0.59(D)
Phenylglycine	0.57	0.67
3-Cyclopentylalanine	0.46	0.56
Homophenylalanine	0.49(D)	0.58(L)
4-Methoxyphenylalanine	0.52	0.64
4-Aminophenylalanine	0.33	0.47
4-Bromophenylalanine	0.44	0.58
4-Chlorophenylalanine	0.46	0.59
2-Fluorophenylalanine	0.55	0.61
4-Iodophenylalanine	0.45(D)	0.61(L)
4-Nitrophenylalanine	0.52	0.61
O-Benzyltyrosine	0.48(D)	0.64(L)
3-Fluorotyrosine	0.64	0.71
4-Methyltryptophan	0.50	0.58
5-Methyltryptophan	0.52	0.63
6-Methyltryptophan	0.52	0.64
7-Methyltryptophan	0.51	0.64
5-Bromotryptophan	0.46	0.58
5-Methoxytryptophan	0.55	0.66
2-(1-Methylcyclopropyl)-glycine	0.59	0.57
N-Methylphenylalanine	0.50(D)	0.61(L)
N-Formyl-*tert*-leucine	0.48(+)	0.61(−)
3-Amino-3,5,5-trimethyl-butyrolactone HCl	0.50	0.59

[a] The configuration of the compounds is given in brackets.

Layer: L1 = a glass plate covered with silicic acid (RP 18-TLC) is immersed for 1 min in a 0.25% solution of Cu (II) acetate in methanol/water, 1:9; it is then immersed in a 0.8% methanolic solution of the chiral selector, (2S, 4R, 2′RS)-4-hydroxy-l-(2′-hydroxydodecyl)-proline; after drying in air the plate is ready for chromatography

Solvent: S1 = methanol/water/acetonitrile, 50:50:200

Detection: D1 = 0.1% ninhydrin reagent

Table TLC 7 (continued)
ENANTIOMERIC RESOLUTION OF RACEMIC AMINO ACIDS

Technique: T1 = 2 μℓ of a 1% solution of the racemic compound is applied to the TLC plate; after elution with solvent for 30 — 90 min with a development distance of 13 cm in a saturated chamber, the plates are dried

REFERENCE

1. **Günther, K., Schickedanz, M., and Martens, J.,** *Naturwissenschaften,* 72, 149, 1985.

Table TLC 8
ENANTIOMERS OF DANSYL AMINO ACIDS

Layer	L1
Solvent	S1
Detection	D1
Technique	T1
Dansyl amino acid	**α**
Alanine	2.1
Aspartic acid	1.3
Methionine	1.3
Phenylalanine	2.0
Serine	1.4

Layer: L1 = RP-18 plates, 10 × 20 cm (Merck)

Solvent: S1 = 8 m*M* *N*,*N*-di-*n*-propyl-L-alanine and 4 m*M* Cu (II) acetate dissolved in 0.3 *M* sodium acetate in water/acetonitrile, 70:30, apparent pH 7

Detection: D1 = UV light at 366 nm

Technique: T1 = the plates were developed in a special chromatographic chamber using a temperature gradient of 6.2°C/cm

REFERENCE

1. **Grinberg, N. and Weinstein, S.,** *J. Chromatogr.*, 303, 251, 1984.

Table TLC 9
DANSYL AMINO ACIDS

Layer	L1	L1
Solvent	S1	S2
Detection	D1	D1
Technique	T1	T1
Dansyl amino acid	**R_f × 100**	
Alanine	19.5	29.4
Arginine	20.3	30.0
Aspartic acid	53.1	62.9
Cysteic acid	53.9	63.5
Dansylic acid	35.1	47.0
Glutamic acid	48.4	59.4
Glycine	23.4	37.0
Histidine	22.5	38.8
Isoleucine	7.0	15.8
Leucine	7.8	16.8
Lysine	8.5	13.5
Methionine	11.7	17.6
Phenylalanine	7.0	11.7
Proline	16.4	25.2
Serine	29.6	42.9
Threonine	24.2	34.7
Tyrosine	3.9	5.8
Tryptophan	5.4	12.3
Valine	12.5	20.5

Layer: L1 = RP-18 plates, 10 × 20 cm (Merck)

Solvent: S1 = an elution gradient was applied; the mobile phase A was 0.3 *M* sodium acetate in water/acetonitrile, 80:20, adjusted to an apparent pH of 6.3 with glacial acetic acid; to this was added 0.3 *M* sodium acetate in water/acetonitrile, 70:30, apparent pH 6.8, at a flow rate of 0.5 mℓ/min, to give a final acetonitrile concentration of 38%; a convex gradient elution was obtained

S2 = as S1, but with a final acetonitrile concentration of 47%

Detection: D1 = UV light at 366 nm

Technique: T1 = the elution gradient was performed in special chromatographic chambers; the plates were equilibrated before application of the dansyl amino acids by development in the mobile phase A; samples were spotted 2.5 cm from the bottom of the plates and 1 cm from the lateral edge; the plates were immersed about 0.3 cm in the mobile phase and developed

REFERENCE

1. **Grinberg, N. and Weinstein, S.,** *J. Chromatogr.*, 303, 251, 1984.

Table TLC 10
DANSYL AMINO ACIDS

Layer	L1	L1
Solvent	S1	S2
Detection	D1	D1
Technique	T1	T1
Amino acid	R_x[a] × 100	
Alanine	92	155
β-Alanine	100	170
DL-α-Aminoadipic acid	100	80
α-Amino dansylhistidine	202	5
γ-Aminoisobutyric acid	75	182
DL-β-Aminoisobutyric acid	80	200
Arginine	208	24
Arginosuccinic acid	180	0
Asparagine	172	42
Aspartic acid	93	32
Citrulline	145	59
Cysteic acid	38	0
Cysteine(bis-dansyl)	11	17
Cystine (bis-dansyl)	11	17
α- or β-Dansylcysteine	200	0
Glutamic acid	92	53
Glutamine	157	58
Glycine	100	100
Histidine (bis-dansyl)	19	186
Homocysteine	11	17
Homocystine	11	17
Hydroxyproline	119	73
Isoleucine	29	209
Leucine	25	195
Lysine (bis-dansyl)	13	132
Methionine	49	154
L-Methionine sulfoxide	150	108
DL-Methionine sulfone	133	66
1-Methylhistidine	202	30
3-Methylhistidine	202	80
Norvaline	42	191
Ornithine	32	112
Phenylalanine	38	164
Proline	88	209
Serine	138	46
Taurine	30	0
Taurocholic acid	30	0
Threonine	135	67
Tryptophan	13	68
Tyrosine (bis-dansyl)	0	175
Tyrosine *O*-dansyltyrosine)	39	27
Valine	56	200

$$R_x = \frac{\text{Migration distance of dansyl amino acid}}{\text{Migration distance of dansyl glycine}}$$

Layer: L1 = micropolyamide plates, Schleicher & Schuell

Solvent: S1 = formic acid/water, 1.5:98.5
S2 = benzene/acetic acid, 4.5:1

Table TLC 10 (continued)
DANSYL AMINO ACIDS

Detection: D1 = the yellow fluorescent spots are detected under UV light at 254 nm

Technique: T1 = solutions are applied 5 mm from the side of the plate; the plates are then introduced into flat presaturated tanks, solvents being removed after each analysis; after migration for 5 min in solvent 1, the plates are dried with hot air and then cooled to room temperature; they are then placed in solvent 2 and migration allowed to proceed for 5 min in a direction perpendicular to the first; interpretation is carried out by comparison with control chromatograms; the stability of the fluorescence in the absence of light allows easy storage of chromatograms and comparison with later samples

REFERENCE

1. **Biou, D., Queyrel, N., Visseaux, M. N., Collignon, I., and Pays, M.,** *J. Chromatogr.*, 226, 477, 1981.

Table TLC 11
ENANTIOMERIC SEPARATION OF DANSYL AMINO ACIDS

	Eluent		$R_f \times 100$	
Amino acid	**Acetonitrile (%)**	**pH**	L	D
Aspartic acid	33	6.8	9	21
Glutamic acid	33	6.8	24	40
Leucine	50	7.5	47	38
Methionine[a]	50	7.5	34	40
Norleucine	50	7.5	42	33
Norvaline[a]	50	7.5	42	36
Phenylalanine	50	7.5	35	41
Serine[a]	50	7.5	41	51
Threonine[a]	50	7.5	47	52

[a] With the complex (2 m*M*) in the eluent.

Conditions: RP-18 F_{254}S HPTLC plates (Merck), 10 × 20 cm, are immersed for 1 hr in a 0.3 *M* solution of sodium acetate in water/acetonitrile, 60:40, adjusted to pH 7.5 with acetic acid; after drying at 100°C in an oven for 1 hr and cooling to room temperature the plates are immersed for at least 2 hr in water/acetonitrile, 10:90, containing the ligand Phe-NN-2 (4 m*M*) and copper acetate (4 m*M*); the plates are dried at 60°C for 1 hr and are ready for use or stored for later use; the ligand Phe-NN-2 consists of two L-phenylalanine residues joined via amide bonds with an ethylene bridge

Table TLC 11 (continued)
ENANTIOMERIC SEPARATION OF DANSYL AMINO ACIDS

Technique: Dansyl amino acids in aqueous solution are applied to the plates with a micro-syringe and developed in water/acetonitrile at different pHs, the percentage of acetonitrile and the pH depending on the types of amino acids to be separated

Detection: The dansyl amino acids are detected with a fluorescent lamp (365 nm)

REFERENCE

1. **Marchelli, R., Virgili, R., Armani, E., and Dossena, A.,** *J. Chromatogr.*, 355, 354, 1986.

Table TLC 12
PHENYLTHIOHYDANTOIN DERIVATIVES OF APOLAR AMINO ACIDS

Layer	L1
Solvent	S1
Detection	D1
Technique	T1

Derivative of	**$R_f \times 100$**
Alanine	47.9
Glycine	35.9
Isoleucine	81.7
Leucine	84.5
Lysine	28.9
Methionine	57.7
Phenylalanine	62.0
Proline	64.8
Serine	12.7
Threonine	21.1
Tryptophan	49.3
Tyrosine	40.8
Valine	71.1

Layer: L1 = high-performance thin-layer precoated silica gel plates F_{254} (Merck)

Solvent: S1 = chloroform/butyl acetate, 1:1

Detection: D1 = derivatives are visible under UV light due to their fluorescence quenching effect

Technique: T1 = the chromatographic plate is used in a horizontal position, a buffer trough being installed at the side of the plate

Table TLC 12 (continued)
PHENYLTHIOHYDANTOIN DERIVATIVES OF APOLAR AMINO ACIDS

REFERENCE

1. **Yang, C. Y.,** *Hoppe Seylers Z. Physiol. Chem.*, 361, 1599, 1980.

Table TLC 13
PHENYLTHIOHYDANTOIN DERIVATIVES OF POLAR AMINO ACIDS

Layer	L1
Solvent	S1
Detection	D1
Technique	T1

Derivative of	**$R_f \times 100$**
Arginine	0
Asparagine	65.2
Aspartic acid	7.6
Cysteic acid	34.8
Glutamic acid	16.6
Glutamine	71.2
Histidine	15.2

Layer: L1 = high-performance thin-layer precoated silica gel plates F_{254} (Merck)

Solvent: S1 = chloroform/butyl acetate/methanol/ethanol, 7:7:3:3

Dectection: D1 = derivatives are visible under UV light due to their fluorescence quenching effect

Technique: T1 = the chromatographic plate is used in a horizontal position, a buffer trough being installed at the side of the plate

REFERENCE

1. **Yang, C. Y.,** *Hoppe Seylers Z. Physiol. Chem.*, 361, 1599, 1980.

Table TLC 14
2,4-DINITROPHENYL(DNP)-AMINO ACIDS

Layer	L1	L2	L2	L2	L2	L2	L2
Solvent	S1	S1	S2	S3	S4	S5	S6
Dectection	D1	D1	D1	D1	D1	D1	D1
Technique	T1	T1	T1	T1	T1	T1	T1
Compound				**$R_f \times 100$**			
DNP-Gly	62	57	86	58	68	71	56
DNP-Ala	49	43	85	47	60	86	73
DNP-Ser	74	70	89	71	78	37	25
DNP-Thr	64	62	87	63	69	48	35
DNP-Val	32	27	78	27	40	96	87
DNP-Leu	22	17	68	18	26	99	94
DNP-Ile	22	17	69	19	27	99	94
DNP-Pro	52	53	87	52	62	83	72
DNP-Met-O_2	74	69	91	72	79	15	8
DNP-Trp	33	30	83	30	43	85	71
DNP-Phe	27	22	72	22	32	92	86
Di-DNP-Tyr	16	10	48	10	12	66	51
DNP-Asp	70	68	91	71	88	41	29
DNP-Glu	65	62	91	64	88	49	36
DNP-$CySO_3Na$	94	91	92	85	93	0	0
Di-DNP-Lys	24	18	64	20	26	57	38
α-*N*-DNP-Arg	70	68	59	80	71	0	0
Di-DNP-His	48	43	85	35	54	0	0
DNP-OH	53	51	91	52	75	99	96
DNP-NH_2	48	48	35	51	41	92	75

Layer:	L1	=	silanized silica gel RP-8 (Merck)
	L2	=	silanized silica gel RP-18 (Merck)
Solvent:	S1	=	1 *M* acetic acid in 60% methanol
	S2	=	1 *M* NH_3 in 60% methanol
	S3	=	3% KCl + 1 *M* acetic acid in 60% methanol
	S4	=	3% KCl + 1 *M* NH_3 in 60% methanol
	S5	=	hexane/ethyl acetate/acetic acid, 75:23:2
	S6	=	hexane/ethyl acetate/acetic acid, 80:18:2
Detection:	D1	=	exposure to UV light (360 nm with dried plates or 254 nm with wet ones)
Technique:	T1	=	the migration distance was 6 cm; the experiments were carried out at 25°C using a thermostatic chamber

REFERENCE

1. **Lepri, L., Desideri, P. G., and Heimler, D.,** *J. Chromatogr.*, 235, 411, 1982.

Table TLC 15
AMINO ACID ESTERS

Layer	L1	L1
Solvent	S1	S2
Detection	D1	D1

Amino acid ester	**$R_f \times 100$**	
1	61	70
2	57	66
3	55	77
4	50	65
5	55	70
6	54	70
7	42	70
8	32	22

Note: The formulas of the compounds are shown below.

Abbreviations: But = *tert*-butyl ether; OBut = *tert*-butyl ester; Z = benzyloxy-carbonyl

Layer: L1 = silica gel 60 F_{254} plates, 20 × 20 cm (Merck)

Solvent: S1 = chloroform/methanol/acetic acid, 95:5:5

S2 = *n*-butanol/pyridine/acetic acid/water, 42:24:4:30

Detection: D1 = 0.2% ninhydrin in acetone; for compounds with blocked amino groups the chromatograms were first sprayed with 40% HBr in glacial acetic acid, heated for 15 min at 100°C, and then sprayed with ninhydrin

1 HCl·H-DL-Phe(p-Cl)-OEt

2 HCl·H-DL-Phe(p-F)-OEt

3 Z-DL-Phe(p-Cl)-OEt

4 Z-DL-Phe(p-F)-OEt

OH
5 Z-Tyr-OMe

Table TLC 15 (continued)
AMINO ACID ESTERS

6 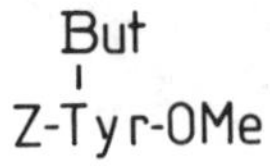

7 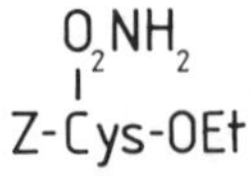

8 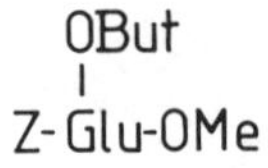

REFERENCE

1. **Aleksiev, B., Schamlian, P., Widenov, G., Stoev, S., Zachariev, S., and Golovinski, E.,** *Hoppe Seylers Z. Physiol. Chem.*, 362, 1323, 1981.

Table TLC 16
PHOSPHORYLATED AMINO ACIDS

Layer	L1	L1	L1	L1	L1
Solvent	S1	S2	S3	S4	S5
Detection	D1	D1	D1	D1	D1
Phosphoamino acid			**$R_f \times 100$**		
Phosphoserine	77	71	71	75	73
Phosphothreonine	75	71	70	73	75
Phosphotyrosine	66	55	53	56	59

Layer: L1 = polyamide plates, 5 × 15 cm

Solvent: S1, S2, S3, S4, and S5 = 5% propionic acid containing 0, 0.0063, 0.013, 0.025, and 0.05% sodium dodecyl sulfate, respectively

Detection: D1 = the dried plates are sprayed with 0.2% ninhydrin and dried in an oven at approximately 50°C for 1 min; the spots should be located within 5 — 10 min, otherwise they cannot be differentiated from the background color

REFERENCE

1. **Chang, W. C., Lee, M. L., Chou, C. K., and Lee, S. C.,** *Anal. Biochem.*, 132, 342, 1983.

Table TLC 17
UNUSUAL t-BUTYLOXYCARBONYL AMINO ACIDS

Amino acid	$R_f \times 100$	Solvent
L-α-Aminobutyric acid	59	S3
	75	S1
D-Alanine	70	S1
β-Alanine	69	S3
L-Azetidine-2-carboxylic acid	56	S1
	71	S2
L-Cyclohexylalanine	76	S3
L-4-Hydroxyproline	22	S1
	39	S2
D-Methionine	71	S1
L-Norleucine	71	S3
L-4-Chlorophenylalanine	60	S3
L-4-Nitrophenylalanine	68	S1
L-Pipecolic acid	73	S3
D-Proline	51	S3
	77	S2
DL-3,4-Dehydroproline	71	S1
Sarcosine	65	S1
D-Allothreonine	40	S1
DL-Allothreonine	40	S1
L-Thiazolidine-4-carboxylic acid	47	S3
	78	S1
D-Tyrosine	65	S1

Conditions:

Layer:	L1	=	silica gel precoated plates, 60 F_{254} (Merck), 0.25 mm thickness, 20 cm length
Solvent:	S1	=	benzene/ethyl acetate/acetic acid/water, 100:100:20:10 (upper phase)
	S2	=	benzene/ethyl acetate/acetic acid/water, 100:100:40:15 (upper phase)
	S3	=	ethyl acetate/methyl alcohol/water, 80:20:5
Detection:	D1	=	hydrobromic acid/ninhydrin and chlorine

REFERENCE

1. **Perseo, G., Piani, S., and De Castiglione, R.,** *Int. J. Pept. Protein Res.,* 21, 227, 1983.

Table TLC 18
*N*α-TRITYL AMINO ACID 1-HYDROXYBENZOTRIAZOLE ESTERS

Layer	L1		L1	
Solvent	S1		S2	
Detection	D1		D1	
	$R_f \times 100$			
Amino acid	**Ester form**	**Amide form**	**Ester form**	**Amide form**
Alanine	70	61	44	29
Glycine	a	a	a	a
Isoleucine	71	66	47	36
Leucine	71	66	46	37
Methionine	73	63	43	29
Phenylalanine	72	64	44	31
Proline	61	54	40	27
Tyrosine	61	50	37	23
Valine	73	66	49	36

Note: The derivatives occur in three isomeric forms, an ester I and two amides II and III.

[a] Considerable hydrolysis on TLC plates.

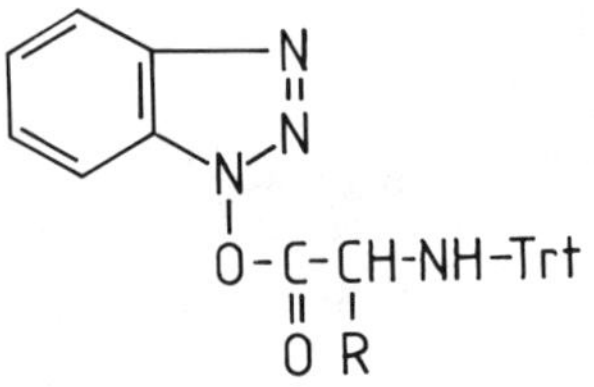

I

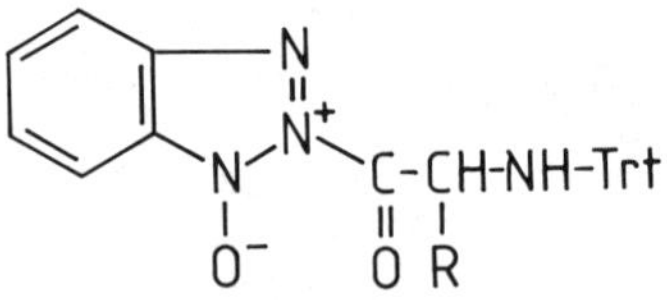

II

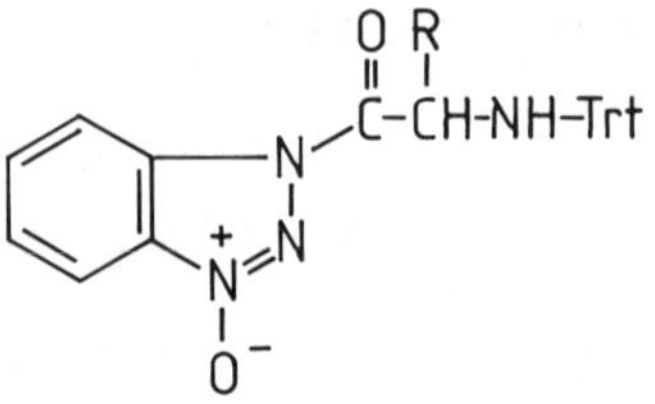

III

Table TLC 18 (continued)
$N\alpha$-TRITYL AMINO ACID 1-HYDROXYBENZOTRIAZOLE ESTERS

Layer:	L1	=	silica gel 60 F_{254}, 0.25-mm-layer thickness precoated on glass plates (Merck)
Solvent:	S1	=	ethyl acetate/benzene, 4:6
	S2	=	cyclohexane/ethyl acetate, 7:3
Detection:	D1	=	UV light at 254 nm, ninhydrin, and chlorine-tolidine reagents

REFERENCE

1. **Barlos, K., Papaioannou, D., and Theodoropoulos, D.,** *Int. J. Pept. Protein Res.*, 23, 300, 1984.

Table TLC 19
BENZYLOXYCARBONYL AMINO ACIDS

Layer	L1	L1	L1
Solvent	S1	S2	S3
Detection	D1	D1	D1
Amino acid		**$R_f \times 100$**	
Pipecolic acid	78	65	40
D-Pipecolic acid	80	65	40
L-2-Pyrrolidineacetic acid dicyclohexylamine salt	75	85	65
L-2-Pyrrolidineacetic acid	75	85	65
L-2-Pyrrolidineacetic acid amide	65	48	35
L-3-Amino-4-phenylbutyric acid	79	65	40
N-Methylphenylalanine dicyclohexylamine salt	82	16	6

Layer:	L1	=	silica gel precoated glass plates 60 F_{254} (Merck)
Solvent:	S1	=	*n*-butylalcohol/acetic acid/water, 4:1:1
	S2	=	a mixture of ethyl acetate and a stock solution of pyridine/acetic acid/water, 20:6:11, in the proportion ethyl acetate/stock, 9:1
	S3	=	chloroform/methyl alcohol, 9:1
Detection:	D1	=	chlorination, toluidine, or TMD (*N,N,N′,N′*-tetramethyl-4,4′-diaminodiphenylmethane) reagents

REFERENCE

1. **Balaspiri, L., Toth, M. V., Somlai, C., and Kovacs, K.,** *Int. J. Pept. Protein Res.*, 26, 1, 1985.

Table TLC 20
t-BUTYLOXYCARBONYL AMINO ACIDS

Layer	L1	L1	L1
Solvent	S1	S2	S3
Detection	D1	D1	D1
Amino acid		**$R_f \times 100$**	
Pipecolic acid	82	70	68
D-Pipecolic acid	81	70	69
L-2-Pyrrolidineacetic acid	81	72	67
D-2-Pyrrolidineacetic acid	79	72	68
L-3-Amino-4-phenylbutyric acid	84	78	65
D-3-Amino-4-phenylbutyric acid	84	76	66
L-3,4-Dehydroproline	81	48	58
D-3,4-Dehydroproline	81	48	60
γ-Aminobutyric acid	81		78
L-4-Chlorophenylalanine	72	66	
D-Proline	69	60	39

Layer: L1 = silica gel precoated glass plates 60 F_{254} (Merck)

Solvent: S1 = *n*-butyl alcohol/acetic acid/water, 4:1:1
S2 = a mixture of ethyl acetate and a stock solution of pyridine/acetic acid/water, 20:6:11, in the proportion ethyl acetate/stock, 9:1
S3 = chloroform/methyl alcohol, 9:1

Detection: D1 = chlorination, toluidine, or TMD (*N,N,N′,N′*-tetramethyl-4,4′-diaminodiphenylmethane) reagents

REFERENCE

1. **Balaspiri, L., Toth, M. V., Somlai, C., and Kovacs, K.,** *Int. J. Pept. Protein Res.*, 26, 1, 1985.

Table TLC 21
9-FLUORENYLMETHYLOXYCARBONYL AMINO ACIDS

Layer	L1	L1	L1
Solvent	S1	S2	S3
Detection	D1	D1	D1
Amino acid		**$R_f \times 100$**	
D-Proline	74	78	33
Pipecolic acid	83	93	48
D-Pipecolic acid	84	94	48
β-Alanine	80	92	37
4-L-Hydroxyproline	66	37	7

Table TLC 21 (continued)
9-FLUORENYLMETHYLOXYCARBONYL AMINO ACIDS

Layer:	L1	=	silica gel precoated glass plates, 60 F_{254} (Merck)
Solvent:	S1	=	*n*-butyl alcohol/acetic acid/water, 4:1:1
	S2	=	a mixture of ethyl acetate and a stock solution of pyridine/acetic acid/water, 20:6:11, in the proportion ethyl acetate/stock, 9:1
	S3	=	chloroform/methyl alcohol, 9:1
Detection:	D1	=	chlorination, toluidine, or TMD (*N,N,N′,N′*-tetramethyl-4,4′-diaminodiphenylmethane) reagents

REFERENCE

1. **Balaspiri, L., Toth, M. V., Somlai, C., and Kovacs, K.,** *Int. J. Pept. Protein Res.*, 26, 1, 1985.

Table TLC 22
N^{α}-9-FLUORENYLMETHYLOXYCARBONYL AMINO ACIDS

Layer	L1	L1	L1
Solvent	S1	S2	S3
Compound		**$R_f \times 100$**	
Fmoc-Ala-OH		57	34
Fmoc-Asn-OH		8	
Fmoc-Gln-OH		10	
Fmoc-Gly-OH		46	26
Fmoc-Ile-OH	33	77	41
Fmoc-Leu-OH	17	76	40
Fmoc-Met-OH	22	68	40
Fmoc-Phe-OH	17	65	41
Fmoc-Pro-OH	24	80	36
Fmoc-Trp-OH	15	55	31
Fmoc-Val-OH	26	77	38
Fmoc-Arg(Boc)-OH		12	
Fmoc-Asp(OtBu)-OH		54	35
Fmoc-Cys(tBu)-OH	30	70	36
Fmoc-Cys(StBu)-OH		68	
Fmoc-Glu(OtBu)-OH	17	73	
Fmoc-His(N^{im}-Boc-Tf)-OH		22	
Fmoc-Lys(Boc)-OH	14	72	38
Fmoc-Ser(tBu)-OH	30	80	36
Fmoc-Thr(tBu)-OH	40	90	41
Fmoc-Tyr(tBu)-OH	30	68	38
Fmoc-Ser-OH		15	
Fmoc-Ser-OBzl	85		
Fmoc-Ser(tBu)-OBzl	95		
Fmoc-Thr-OBzl	90		
Fmoc-Thr(tBu)-OBzl	95		

Table TLC 22 (continued)
N^{α}-9-FLUORENYLMETHYLOXYCARBONYL AMINO ACIDS

Abbreviations: Boc = *tert*-butyloxycarbonyl;
Boc-Tf = 2,2,2-trifluoro-l-*tert*-butyloxycarbonylaminoethyl;
Fmoc = 9-fluorenylmethyloxycarbonyl;
OtBu = *tert*-butyl ester;
StBu = *tert*-butyl sulfide;
tBu = *tert*-butyl ether

Layer: L1 = Precoated silica gel G-60 (F_{254}) plates, 10 cm
Solvent: S1 = EtOH/hexane, 3:1
S2 = EtOH/$CHCl_3$, 1:3
S3 = $CHCl_3$/MeOH/AcOH, 9:10

REFERENCE

1. **Chang, C. D., Waki, M., Ahmad, M., Meienhofer, J., Lundell, E. O., and Haug, J. D.,** *Int. J. Pept. Protein Res.,* 15, 59, 1980.

Table TLC 23
SUBSTITUTED THIAZOLE AND THIAZOLINE-AMINO ACID DERIVATIVES

Layer	L1
Solvent	S1
Detection	D1

R_a	$R_f \times 100$
Compounds of Type A	
-Gly	51
-L-Ala	80
-DL-Val	63
-L-Leu	45
-DL-Ser	35
-Gly-OMe	76
-L-Ala-OMe	80
-L-Val-OMe	68
-L-Leu-OMe	45
-DL-Ser-OMe	90

Table TLC 23 (continued)
SUBSTITUTED THIAZOLE AND THIAZOLINE-AMINO ACID DERIVATIVES

R_a	$R_f \times 100$
Compounds of Type B	
Pht-Gly-	60
Pht-L-Ala-	50
Pht-β-Ala-	54
Pht-L-Leu	71
Pht-L-Phe-	59
Tos-Gly-	41
Tos-DL-Ala-	51
Tos-β-Ala-	46
Tos-L-Val-	83
Tos-DL-Val-	79
L-Phe-	59
L-Ala-	69
Tos-L-Val-L-Leu-	66

[a] R in compounds of type A and type B.

Abbreviations: Tos = *p*-toluene sulfonyl; Pht = phthalyl

TYPE A TYPE B

Layer: L1 = silica gel G
Solvent: S1 = benzene/ethyl acetate, 1:1
Detection: D1 = iodine-potassium iodide (20%)

REFERENCE

1. **El-Naggar, A. M., Ahmed, F. S. M., Abd El-Salam, A. M., Haroun, B. M., and Latif, M. S. A.,** *Int. J. Pept. Protein Res.,* 19, 408, 1982.

Table TLC 24
3,6-DINITRO-1,8-NAPHTHALOYL- AND 3,6-DIAMINO-1,8-NAPHTHALOYL AMINO ACIDS

Layer	L1
Solvent	S1
Detection	D1

Table TLC 24 (continued)
3,6-DINITRO-1,8-NAPHTHALOYL- AND 3,6-DIAMINO-1,8-NAPHTHALOYL AMINO ACIDS

R[a]	$R_f \times 100$
Type A Compounds	
-Gly	72
-β-Ala	68
-L-Val	83
-L-Leu	82
-DL-Phe	69
-L-Ser	54
-L-Tyr	70
-DL-Trp	66
-Gly-OMe	80
-β-Ala-OMe	75
-L-Val-OMe	93
-L-Leu-OMe	91
-L-Ser-OMe	70
-DL-Phe-OMe	81
-L-Tyr-OMe	83
Type B Compounds	
-Gly	85
-β-Ala	80
-L-Val	90
-L-Leu	92
-L-Ser	79
-DL-Phe	86
-L-Tyr	91
-DL-Trp	78

[a] R in compounds of type A and type B.

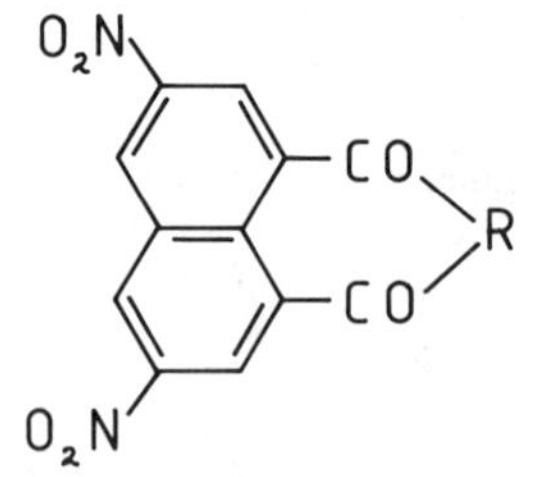

TYPE A

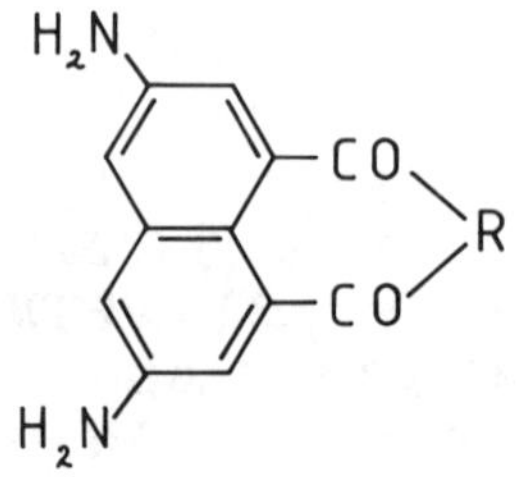

TYPE B

Table TLC 24 (continued)
3,6-DINITRO-1,8-NAPHTHALOYL- AND 3,6-DIAMINO-1,8-NAPHTHALOYL AMINO ACIDS

Layer:	L1	=	silica gel G
Solvent:	S1	=	benzene/ethyl acetate, 1:1
Detection:	D1	=	iodine-potassium iodide (20%)

REFERENCE

1. **El-Naggar, A. M., Zaher, M. R., and Salem, A. A.,** *Int. J. Pept. Protein Res.*, 20, 1, 1982.

Table TLC 25
3-HYDROXYNAPHTHALENE-2-CARBONYLAMINO ACID METHYL ESTERS AND HYDRAZIDES

Layer	L1
Solvent	S1
Detection	D1
R[a]	**R_f × 100**
Gly-OMe	66
L-Ala-OMe	74
β-Ala-OMe	74
L-Val-OMe	77
L-Leu-OMe	80
DL-Ser-OMe	84
L-Met-OMe	86
L-Phe-OMe	76
L-Tyr-OMe	86
L-Trp-OMe	88
Gly-N_2H_3	68
L-Ala-N_2H_3	78
β-Ala-N_2H_3	77
L-Val-N_2H_3	79
L-Leu-N_2H_3	80
DL-Ser-N_2H_3	82
L-Met-N_2H_3	72
L-Phe-N_2H_3	81
L-Tyr-N_2H_3	79
L-Trp-N_2H_3	83

[a] R in the compound.

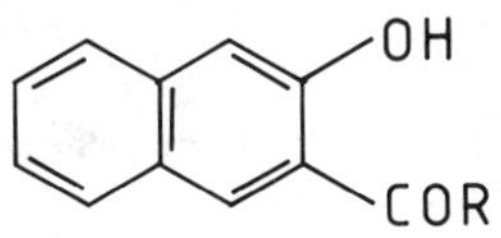

Table TLC 25 (continued)
3-HYDROXYNAPHTHALENE-2-CARBONYLAMINO ACID METHYL ESTERS AND HYDRAZIDES

Layer:	L1	=	silica gel G
Solvent:	S1	=	benzene/ethyl acetate, 1:1
Detection:	D1	=	iodine-potassium iodide (20%) or chlorosulfonic acid/acetic acid, 1:3

REFERENCE

1. **El-Naggar, A. M., Ahmed, F. S. M., Badie, M. F., and Kamel, K. M.,** *Int. J. Pept. Protein Res.*, 22, 251, 1983.

Table TLC 26
2-(2′-PYRIDYL)-ETHYL ESTER (PET ESTER) DERIVATIVES OF FUNCTIONAL AMINO ACIDS

Layer	L1	L1	L1
Solvent	S1	S2	S3
Derivative		$R_f \times 100$	
Boc-Gly-OPet	42	44	77
Boc-Ala-OPet	27	48	40
Boc-Phe-OPet	55	57	39
Boc-Pro-OPet	42	49	64
Boc-Gln-OPet	60	61	77
Boc-Ser-OPet	21	21	43
Boc-Thr-OPet	37	26	43
Boc-Tyr-OPet	50	34	73
Boc-Tyr(BZl)-OPet	44	62	
Boc-Trp-OPet	56	42	75
Boc-Asp(OBzl)OPet	62	62	82
Boc-Asp-OPet DCHA	36	20	38
Boc-Asp-OPet	36	14	51
Boc-Glu(OBzl)OPet	66	87	97
Boc-Glu-OPet	55	75	0
Z-Pro-OPet	15	53	44
Z-Val-OPet	27	57	50
Z-Lys(Boc)-OPet	58	50	74
Boc-Lys(Z)-OPet	94	66	56
Z-Ser-OPet	44	37	70

Abbreviations: Pet = 2-(2′-pyridyl)-ethyl-; DCHA = dicyclohexylamine

Layer:	L1	=	TLC plates precoated with 254 nm fluorescent indicator
Solvent:	S1	=	*n*-butanol/water/acetic acid, 3:1:1
	S2	=	chloroform/methanol/acetic acid, 95:5:3
	S3	=	pyridine/*n*-butanol/water/ethyl acetate, 20:10:5:3

Table TLC 26 (continued) 2-(2′-PYRIDYL)-ETHYL ESTER (PET ESTER) DERIVATIVES OF FUNCTIONAL AMINO ACIDS

REFERENCE

1. **Kessler, H., Becker, G., Kogler, H., Friesa, J., and Kerssebaum, R.,** *Int. J. Pept. Protein Res.*, 28, 342, 1986.

Table TLC 27 AMINO ACIDS AND DERIVATIVES

Layer	L1	L1	L1	L2
Solvent	S1	S2	S3	S4
Detection	D1	D1	D1	D1
Technique	T1	T1	T1	T1
Compound		**$R_f \times 100$**		
Glu	60	72	33	96
Glu-NH_2	49	58	16	95
Glu-γ-NHOH	40	57	4	95
Glu-γ-$NHNH_2$	57	64	21	95
Glu-γ-OMe	46	55	18	90
Glu-γ-OEt	46	55	18	79
Glu-$(OMe)_2$	27	35	8	43
Glu-$(OEt)_2$	27	35	8	34
Asp	67	74	40	96
Asp-β-NHOH	61	68	31	95
Asp-$(OMe)_2$	37	45	10	52
Gly	66	74	40	96
Gly-NHOH	56	65	25	90
Gly-OMe	51	60	21	70
Ala	69	77	49	96
Ala-NHOH	63	72	35	n.d.[a]
Ala-OMe	52	60	21	66
β-Ala	52	60	23	96
β-Ala-NHOH	47	56	16	91
β-Ala-OMe	41	50	12	79
Met	50	58	21	84
Val	67	78	46	90
Tyr	57	64	27	76
Phe	53	58	19	65
Trp	26	36	7	43
N-Methyl-Trp	15	19	1	45
7-Methyl-Trp	23	34	7	28
6-Methyl-Trp	32	41	11	28
5-Methyl-Trp	21	29	4	28
5-Methoxy-Trp	17	25	2	35
5-Hydroxy-Trp	32	43	11	61
Trm	6	11	2	30
N-Methyl-Trm	5	9	0	31
5 Methyl-Trm	7	15	0	20
5-Methoxy-Trm	4	7	0	27

[a] n.d. = not determined.

Table TLC 27 (continued)
AMINO ACIDS AND DERIVATIVES

Abbreviations: -NHOH = hydroxamate
-$NHNH_2$ = hydrazide
-OMe = methyl ester
-OEt = ethyl ester
-$(OME)_2$ = dimethyl ester
-$(OEt)_2$ = diethyl ester; Trm = tryptamine

Layer:	L1	=	ammonium tungstophosphate (AWP) + $CaSO_4 \cdot ^1/_2 H_2O$ (4:2)
	L2	=	Sil C_{18}-50 (Macherey, Nagel & Co.)
Solvent:	S1	=	1 *M* NH_4NO_3
	S2	=	2 *M* NH_4NO_3
	S3	=	0.5 *M* HNO_3
	S4	=	acetic acid/methanol/water (1:20:79)
Detection:	D1	=	the layers were sprayed with 1% ninhydrin in pyridine/acetic acid, 5:1, and the plates were heated for 5 min at 100°C; *N*-methyltryptamine and *N*-methyltryptophan were detected by spraying with 1% *p*-dimethylaminobenzaldehyde in concentrated HCl/methanol, 1:1, and heating the layers at 50°C for 20 min
Technique:	T1	=	the migration distance was 6 cm for the ready-for-use plates and 10 cm for the inorganic exchanger; chromatography was carried out at 25°C

REFERENCE

1. **Lepri, L., Desideri, P. G., and Heimler, D.,** *J. Chromatogr.*, 268, 493, 1983.

Table TLC 28
ENANTIOMERIC AMINO ACID DERIVATIVES

Compound	$R_f \times 100$	Solvent	Detection
Dansyl-D-leucine	49	S1	D1
Dansyl-L-leucine	66	S1	D1
Dansyl-D-methionine	28	S2	D1
Dansyl-L-methionine	43	S2	D1
Dansyl-D-alanine	25	S2	D1
Dansyl-L-alanine	33	S2	D1
Dansyl-D-valine	31	S2	D1
Dansyl-L-valine	42	S2	D1
D-Alanine-β-naphthylamide	16	S3	D2
L-Alanine-β-naphthylamide	25	S3	D2
D-Methionine-β-naphthylamide	16	S3	D2
L-Methionine-β-naphthylamide	24	S3	D2

Table TLC 28 (continued)
ENANTIOMERIC AMINO ACID DERIVATIVES

Conditions:	Layer:	L1	=	β-cyclodextrin bonded through a spacer to Macherey Nagel® silica gel plus binder (AS-TEC "all solvent binder", Advanced Separation Technologies Inc.)
	Solvent:	S1	=	methanol/1% triethylammonium acetate, pH 4.1, 40:60
		S2	=	methanol/1% triethylammonium acetate, pH 4.1, 25:75
		S3	=	methanol/1% triethylammonium acetate, pH 4.1, 30:70
	Detection:	D1	=	fluorescence
		D2	=	ninhydrin
	Technique:	TLC plates, 5 × 20 cm, are prepared by mixing 1.5 g of β-cyclodextrin-bonded silica gel in 15 mℓ of 50% aqueous methanol with 0.002 g of binder; the slurry is spread to a thickness of 3 mm on a clean glass plate and left to air dry; the plate is heated in an oven to 75°C for 15 min before use; development is conducted at 20°C in a developing chamber		

REFERENCE

1. **Alak, A. and Armstrong, D. W.,** *Anal. Chem.,* 58, 582, 1986.

Table TLC 29
PROTECTED DERIVATIVES OF DIPROPYL GLYCINE (2-AMINO-2-PROPYLPENTANOIC ACID, Dpg)

Derivative	$R_f \times 100$	Solvent
Dpg	70	S5
Dpg-OMe	50	S5
Dpg-OEt	60	S5
Dpg-OBu^t	60	S5
Dpg-OPh	60	S4
Z-Dpg	60	S3
Z-Dpg-OPh	75	S1
Boc-Dpg	50	S3
Nps-Dpg	30	S2
Tfa-Dpg	70	S3
Z-Dpg-Cl	70	S3

Abbreviations: OMe = methyl ester; OEt = ethyl ester; OBu^t = *t*-butyl ester; OPh = phenyl ester; Z = benzyloxycarbonyl; Boc = *t*-butyloxycarbonyl; Nps = *o*-nitrophenylsulfenyl; Tfa = trifluoroacetyl

Conditions:	Layer:	L1	=	Merck F254 plates
	Solvent:	S1	=	$CHCl_3$
		S2	=	$CHCl_3$/EtOH, 20:1
		S3	=	$CHCl_3$/EtOH, 10:1
		S4	=	$CHCl_3$/EtOH, 4:1
		S5	=	*n*-BuOH/AcOH/H_2O, 4:1:1

REFERENCE

1. **Hardy, P. M. and Lingham, I. N.,** *Int. J. Pept. Protein Res.,* 21, 392, 1983.

Table TLC 30
L-t-BUTYLGLYCINE AND DERIVATIVES

Layer	L1	L1	L1	L1
Solvent	S1	S2	S3	S4
Detection	D1	D1	D1	D1
Compound		$R_f \times 100$		
t-Butylglycine		21	23	
t-Butylglycine amide		28	48	50
5-Butylglycine methyl ester	52	38	64	
t-Butoxycarbonyl butyl glycine	56	78		

Layer: L1 = silica gel plates F_{254} (Merck)
Solvent: S1 = chloroform/methanol, 9:1
S2 = 2-butanol/acetic acid/water, 72:7:21
S3 = 1-butanol/aqueous ammonia (25%), 10:3
S4 = 2-propanol/water/pyridine, 36:32:32
Detection: D1 = I_2, ninhydrin, Reindel-Hoppe reagent, or fluorescence quenching in UV

REFERENCE

1. **Fauchere, J. L. and Petermann, C.,** *Helv. Chim. Acta,* 63, 824, 1980.

Table TLC 31
γ-METHYL-L-LEUCINE (NEOPENTYLGLYCINE, Neo) AND DERIVATIVES

Layer	L1	L1	L1	L1	L1
Solvent	S1	S2	S3	S4	S5
Detection	D1	D1	D1	D1	D1
Compound			$R_f \times 100$		
4,4-Dimethyl-2-([S]-α-methylbenzylimino)-pentanonitrile, hydrochloride		90	82	72	
Mbz-Neo-NH_2HCl	55		54	79	
H-Neo-NH_2	68		33	58	
H-Neo-OH			32	19	61
H-Neo-OMe	41	12	40	75	
Boc-Neo-OH	30	72	73	38	

Abbreviation: Mbz = (*S*)-α-(methylbenzyl)

Layer: L1 = silica gel plates F_{254} (Merck) (Reference 2)
Solvent: S1 = chloroform/methanol, 1:1
S2 = chloroform/methanol/acetic acid, 95:5:3
S3 = 2-butanol/acetic acid/water, 72:7:1
S4 = l-butanol/aqueous ammonia (25%), 10:3
S5 = isopropyl alcohol/water/pyridine, 36:32:32
Detection: D1 = I_2, ninhydrin, Reindel-Hoppe reagent, or fluorescence quenching in UV (Reference 2)

REFERENCES

1. **Fauchere, J.-L. and Petermann, C.,** *Int. J. Pept. Protein Res.,* 18, 249, 1981.
2. **Fauchere, J.-L. and Petermann, C.,** *Helv. Chim. Acta,* 63, 824, 1980.

Table TLC 32
L-(+)-ADAMANTYL ALANINE (Ada) AND DERIVATIVES

Layer	L1	L1	L1	L1	L1	L1
Solvent	S1	S2	S3	S4	S5	S6
Detection	D1	D1	D1	D1	D1	D1
Compound				**R_f × 100**		
Ada, HCl			28			38
Boc-Ada-OH	8	42		22		
H.Ada OtBu.HCl					9	55

Layer: L1 = silica gel plates F_{254} (Merck)
Solvent: S1 = hexane/ethyl acetate, 6:4
S2 = chloroform/methanol, 9:1
S3 = chloroform/methanol, 1:1
S4 = chloroform/acetone, 7:3
S5 = chloroform/methanol/acetic acid, 95:5:3
S6 = 1-butanol/acetic acid/water, 10:1:3
Detection: D1 = I_2, ninhydrin, Reindel-Hoppe reagent, or fluorescence quenching in UV

REFERENCE

1. **Do, K. Q., Thanei, P., Caviezel, M., and Schwyzer, R.,** *Helv. Chim. Acta,* 62, 956, 1979.

Table TLC 33
PHENYLALANINE DERIVATIVES

Compound	R_f × 100	Solvent
H-Phe	47	S4
	50	S5
Boc-Phe	44	S6
H-(4-SO_3H) Phe	12	S1
	26	S2
	12	S3
N-Boc-(4′-SO_3^-)Phe	48	S5
H-(4′-SO_2NH_2)Phe	26	S1
	13	S2
	26	S3
N^α-Boc(4′-SO_2NH_2)Phe	22	S6
H-(4′-SH)Phe	46	S4
	55	S5
N-Boc-(4′-SH)Phe	40	S6
N-Boc-(4′-S-Boc)Phe	43	S6
(H-4′-Phe)$_2$S$_2$	40	S4
	38	S5
(N-Boc-4′-Phe)$_2$S$_2$	26	S6
H-(4′-S-S-butyl)Phe	64	S4
N-Boc(4′-S-S-butyl)Phe	50	S6
H-(4′-S-S-tolyl)Phe	62	S4
N-Boc-(4′-S-S-tolyl)Phe	56	S6
H-(4′-S-Acm)Phe	38	S4
	46	S5
N^α-Boc-(4′-S-Acm)Phe	15	S6
H-(4′-S-Bzl)Phe	59	S4
	61	S5

Table TLC 33 (continued)
PHENYLALANINE DERIVATIVES

Compound	R_f × 100	Solvent
N-Boc-(4′-S-Bzl)Phe	49	S6
H-(4′-S-BrBzl)Phe	59	S4
	61	S5
N-Boc-(4′-S-BrBzl)Phe	49	S6
H-(4′-S-CH_3)Phe	54	S4
N-Boc-(4′-S-CH_3)Phe	47	S6
H-Cys	31	S4

Abbreviations: BrBzl = *m*-bromobenzyl; Boc = t-butoxycarbonyl; Acm = acetamidomethyl

Conditions: Layer: Merck precoated silica gel plates, G60-F254

Solvent:
- S1 = 2-propanol/concentrated NH_4OH, 3:1
- S2 = *n*-butanol/0.05 *M* ammonium acetate, pH 7.0, 2:1
- S3 = *n*-butanol/acetic acid/water, 4:1:1
- S4 = *n*-butanol/acetic acid/water, 5:2:3
- S5 = *n*-butanol/acetic acid/water/pyridine, 30:6:20:12
- S6 = chloroform/acetic acid/methanol, 95:5:3

Detection: Spots were visualized with fluorescence or ninhydrin

REFERENCE

1. **Escher, E., Bernier, M., and Parent, P.,** *Helv. Chim. Acta,* 66, 1355, 1983.

Table TLC 34
TRYPTOPHAN ANALOGS

Layer	L1
Solvent	S1
Detection	D1
Analog	**R_f × 100**
Benzofurylalanine	54
Benzothienylalanine	54
5-Fluorotryptophan	43
6-Fluorotryptophan	46
4,5,6,7-Tetrafluorotryptophan	46

Layer: L1 = silica gel TS (Merck F_{254})
Solvent: S1 = 1-butanol/acetic acid/water, 4:1:1
Detection: D1 = ninhydrin, TDM reagent, and Ehrlich's reagent

REFERENCE

1. **Rajh, H. M., Uitzetter, J. H., Westerhuis, L. W., Van Den Dries, C. L., and Tesser, G. I.,** *Int. J. Pept. Protein Res.,* 14, 68, 1979.

Table TLC 35
2-THIOETHER DERIVATIVES OF TRYPTOPHAN

Layer	L1	L1	L2
Solvent	S1	S2	S3
Detection	D1	D1	D1

Amino acid	**R_f × 100**			**Ninhydrin color (cellulose)**
α-Hpi	47	30	52	Lilac
β-Hpi	41	30	47	Lilac
Trp(SCH_3)	36	47	64	Inidgo
Trp(SCH_2CH_3)	42	59	69	Indigo
Trp(SCH_2CH_2OH)	18	29	57	Indigo
Trp(SCH_2CH_2COOH)	25	37	63	Indigo
Trpytathionine	5	5	25	Indigo
Oxindolylalanine	30	33	56	Blue-mauve

Abbreviation: Hpi = L-3a-hydroxy-1,2,3,3a,8,8a-hexahydropyrrolo-(2,3-b) indole-2-carboxylic acid

Layer: L1 = cellulose/aluminum foil sheets length 10 cm, layer thickness 0.1 mm
L2 = silica gel
Solvent: S1 = methanol/water, 9:1
S2 = ethyl acetate/acetic acid/water, 4:1:1
S3 = *n*-propanol/acetic acid, 3:1
Detection: D1 = dried sheets were sprayed with a solution of 0.2% ninhydrin and 1% lutidine in acetone and heated at 70°C for development

REFERENCE

1. **Savige, W. E. and Fontana, A.,** *Int. J. Pept. Protein Res.,* 15, 102, 1980.

Table TLC 36
BASIC AMINO ACIDS DERIVATIZED WITH NBD-Cl (7-CHLORO-4-NITROBENZO-2-OXA-1,3-DIAZOLE)

Layer	L1
Solvent	S1
Detection	D1
Technique	T1

Amino acid	**R_f × 100**
γ-Aminobutyric acid	32
Arginine	4
Histidine	12
Hydroxylysine	1
Lysine	2
1-Methylhistidine	16
3-Methylhistidine	18
Ornithine	1

Table TLC 36 (continued)
BASIC AMINO ACIDS DERIVATIZED WITH NBD-Cl (7-CHLORO-4-NITROBENZO-2-OXA-1,3-DIAZOLE)

Layer:	L1	=	Merck silica gel plates without fluorescent indicator
Solvent:	S1	=	chloroform/methanol/ethyl acetate/acetone/triethylamine, 75:15:10:10:5
Detection:	D1	=	the fluorescent spots are recorded with a spectrofluorometer equipped with a thin-layer recording device; the excitation light is set at 340 nm with an additional violet filter absorbing light over 500 nm; the emitted light is measured at 525 nm with an additional yellow filter to absorb radiation under 450 nm
Technique:	T1	=	samples of the derivatized solution are deposited in triplicate on a starting line 1.5 cm from the lower edge of a 20 × 20-cm plate; the spots are dried with a stream of cold air and predeveloped in methanol at the bottom of a chromatographic glass jar; this predevelopment, which is stopped when the spots have traveled 5 mm, produces thin spots all deposited 2 cm from the lower edge; the plate is dried in an oven at 65°C for 5 min and transferred to a second chromatographic jar previously saturated with the solvent; development takes 75 min and the solvent front is 16 cm from the bottom of the plate; the plate is dried at 65°C for 5 min and stored in the dark before spectrofluorometric measurement

REFERENCE

1. **Monboisse, J. C., Pierrelee, P., Bisker, A., Pailler, V., Randoux, A., and Borel, J. P.,** *J. Chromatogr.*, 233, 355, 1982.

Table TLC 37
2-METHOXYETHOXYMETHYL (MEM) DERIVATIVES OF HYDROXYPROLINE AND SERINE

Layer	L1	L1	L1
Solvent	S1	S2	S3
Protected amino acid		**$R_f \times 100$**	
Boc-Ser(Mem)-ONb		91	83
Z-Ser(Mem)-OMe	85	87	78
Boc-Hyp(Mem)-ONb	82	88	75
Z-Hyp(Mem)-OMe	80	92	72
Bpoc-Hyp(Mem)-ONb	84	90	84
Nps-Nyp(Mem)-ONb	88	96	91
Ddz-Hyp(Mem)-ONb	83	86	77

Abbreviations: Boc = t-butoxycarbonyl; Bpoc = 2-(*p*-biphenylyl)isopropyloxycarbonyl; Ddz = α, α-dimethyl-3,5-dimethoxybenzyloxycarbonyl; ONb = 4-nitrobenzyl ester; Nps = 2-nitrophenylthio

Layer: L1 = silica gel 60 F_{254} (Merck)
Solvent: S1 = chloroform/methanol, 95:5
S2 = chloroform/methanol, 8:3
S3 = acetic acid/*n*-butanol/water, 4:4:1

REFERENCE

1. **Vadolas, D., Germann, H. P., Thakur, S., Keller, W., and Heidemann, E.,** *Int. J. Pept. Protein Res.*, 25, 554, 1985.

Table TLC 38
HYDROXYPROLINE AND SERINE WITH PROTECTED HYDROXYL FUNCTIONS

Layer	L1	L1	L1
Solvent	S1	S2	S3
Protected hydroxyamino acid		**$R_f \times 100$**	
Z-Hyp(Mem)-OMe	80	92	72
Z-Ser(Mem)-OMe	85	87	78
Boc-Hyp(Mem)-ONb	82	88	75
Boc-Ser(Mem)-ONb		91	83
NpS-Hyp(Mem)-ONb	88	96	91
Ddz-Hyp(Mem)-ONb	83	86	77
Bpoc-Hyp(Mem)ONb	84	90	84

Abbreviations: Mem = 2-methoxyethoxymethyl
Boc = t-butoxycarbonyl
ONb = 4-nitrobenzyl ester
Ddz = α,α-dimethyl-3,5-dimethoxybenzyloxy-carbonyl
Bpoc = 2-(p-biphenylyl) isopropyloxycarbonyl
NpS = 2-nitrophenylthio

Layer: L1 = silica gel Merck 60 F-254
Solvent: S1 = chloroform/methanol, 95:5
S2 = chloroform/methanol, 8:3
S3 = acetic acid/*n*-butanol/water, 4:4:1

REFERENCE

1. **Vadolas, D., Germann, H. P., Thakur, S., Keller, W., and Heidemann, E.,** *Int. J. Pept. Protein Res.*, 25, 554, 1985.

Table TLC 39
HYDROXYPROLINE ISOMERS AFTER DERIVATIZATION WITH 7-CHLORO-4-NITROBENZO-2-OXA-1,3-DIAZOLE (NBD-Cl)

Layer	L1	L1	L1	L1	L1	L1
Solvent	S1	S2	S3	S4	S5	S6
Detection	D1	D1	D1	D1	D1	D1
Technique	T1	T1	T1	T1	T1	T1
Amino acid			**$R_f \times 100$**			
Asparagine	25	13	12	18	27	33
Histidine	23	14	12	16	25	23
3-Hydroxyproline	38	19	28	36	52	48
4-Hydroxyproline	25	12	20	26	41	36
Lysine	10	7	2.5	2.5	5	4.5
Methionine	37	27	31	36	48	48
Proline	43	29	56	49	60	55
Serine	32	23	33	37	49	48
Threonine	41	31	51	48	56	55

Layer: L1 = silica gel plates, 0.2 mm thickness (Merck)
Solvent: S1 = methanol/toluene/acetone/triethylamine, 20:40:35:5
S2 = methanol/toluene/acetone/triethylamine, 15:40:40:5
S3 = methanol/chloroform/toluene/acetone/tributylamine, 5:60:20:15
S4 = methanol/chloroform/toluene/acetone/tributylamine, 12.5:25:25:25:12.5
S5 = methanol/chloroform/toluene/acetone/tributylamine, 15:20:40:20:5
S6 = methanol/toluene/acetone/tributylamine, 15:40:40:5
Detection: D1 = the dried plates are stored in the dark prior to measurement of fluorescence; the plate is positioned in a spectrofluorometer equipped with a thin-layer plate recording attachment; the excitation light is set at 340 nm with an additional green filter absorbing light over 500 nm; the emitted light is read at 525 nm with an additional violet filter to absorb radiation under 450 nm; alternatively, the fluorescent spots may be eluted with ethanol/water, 50:50, and the fluroescence read in this solution, but the sensitivity is lower
Technique: T1 = a solution of the NBD-amino acids is spotted linearly on a length of 5 mm at 1 cm from the inferior edge of a silica gel plate; after drying under a stream of nitrogen, the plate is predeveloped in methanol to obtain a thinner deposit, so that the starting line is 2 cm from the lower edge; prior to the development, the plates are activated by heating at 65°C for 10 min; they are developed in glass tanks previously saturated with the appropriate solvent; development is stopped when the front has moved up 10 cm from the starting line; the plates are then dried in an oven at 65°C for 5 min

REFERENCE

1. **Bisker, A., Pailler, V., Randoux, A., and Borel, J. P.,** *Anal. Biochem.*, 122, 52, 1982.

Table TLC 40
HYDROXYPROLINE ISOMERS AS DERIVATIVES WITH 7-CHLORO-4-NITROBENZO-2-OXA-1,3-DIAZOLE (NBD-Cl)

Layer	L1	L1	L1	L1	L1
Solvent	S1	S2	S3	S4	S5
Detection	D1	D1	D1	D1	D1
Technique	T1	T1	T1	T1	T1
Amino acid			**$R_f \times 100$**		
trans-4-L-Hydroxyproline	11	10	6	7	38
cis-4-L-Hydroxyproline	21	18	12	13	47
trans-3-L-Hydroxyproline	29	30	32	26	55
cis-3-DL-Hydroxyproline	27	25	27	23	54
L-Proline	29	37	39	41	56

Layer: L1 = silica gel
Solvent: S1 = acetone/toluene/methanol/triethylamine, 40:40:15:5
S2 = acetone/chloroform/methanol/triethylamine/ethyl acetate, 10:70:15:5:10
S3 = acetone/chloroform/methanol/tributylamine, 20:60:5:15
S4 = chloroform/methanol/triethylamine, 80:11:11
S5 = acetone/toluene/methanol/tributylamine, 40:40:15:5
Detection: D1 = the plate is dried at 65°C and scanned in a recording fluorometer equipped with a plate-scanning device
Technique: T1 = the primary amino acids are reacted with *o*-phthalaldehyde and hydroxyprolines and proline are reacted with NBD-Cl

REFERENCE

1. **Bellon, G., Berg, R., Chastang, F., Malgras, A., and Borel, J. P.,** *Anal. Biochem.,* 137, 151, 1984.

Table TLC 41
DIMETHYLAMINOAZOBENZENE-THIOHYDANTOIN (DABTH) DERIVATIVES OF IODOTYROSINES AND IODOTHYRONINES

Layer	L1	L1
Solvent	S1	S2
Detection	D1	D1

DABTH derivative	**R_f × 100**	
3,3′-Diiodothyronine	0	30
3,5-Diiodothyronine	0	28
3,5,3′-Triiodothyronine	0	40
3,3′,5′-Triiodothyronine	0	59
Thyroxine	0	66
3-Iodotyrosine	12	24
3,5-Diiodotyrosine	8	48

Layer: L1 = polyamide sheets (Schleicher & Schuell)

Solvent: S1 = acetic acid/water, 1:2

S2 = toluene/*n*-hexane/acetic acid, 2:1:1

Detection: D1 = the plates were dried and exposed to HCl vapor for some seconds when the spots of DABTH-iodotyrosines and DABTH-iodothyronines turned from yellow to red

REFERENCE

1. **Marriq, C., Rolland, M., and Lissitzky, S.,** *Anal. Biochem.*, 116, 89, 1981.

Table TLC 42
AROMATIC AMINES

Layer	L1
Solvent	S1
Detection	D1
Technique	T1

Amine	$R_f \times 100$	Spot color
4-Methyl-3-nitroaniline	66	Violet
2-Nitroaniline	63	Violet
4-Methoxy-3-nitroaniline	58	Blue
3-Nitroaniline	58	Red-purple
4-Chloro-2-nitroaniline	56	Violet
3,3′-Dichlorobenzidine	55	Blue
2-Methoxy-4-nitroaniline	52	Purple
2-Methoxy-5-nitroaniline	52	Red-purple
2-Chloro-4-nitroaniline	48	Pink
4-Chloroaniline	43	Purple
2-Methyl-4-nitroaniline	38	Red-purple
4-Nitroaniline	34	Red-purple
2,4-Dinitroaniline	30	Pink
3,3′-Dimethoxybenzidine	25	Blue

Layer: L1 = glass TLC plates, 20 × 20 cm, coated with silica gel G (Anasil G, 250 μm)

Solvent: S1 = benzene/ethyl acetate, 80:20

Detection: D1 = the dried plates are exposed to nitrogen oxide fumes, prepared by the reaction of HCl with $NaNO_2$ in a closed TLC tank for 10 — 30 sec, and are then sprayed with a 0.1% solution of NEDA (*N*-[1-naphthyl]ethylene diamine dihydrochloride) in methanol

The primary aromatic amines listed in this table give deep-colored spots which remain sharp with no diffusion

Technique: T1 = bands (0.5 cm) of the amine solution are applied to the plate 2.5 cm from the bottom with a micro glass-capillary tube; the plates are developed in a vapor-saturated chromatographic tank until the solvent has migrated 15 cm from the origin; the plates are then air dried in a hood for 15 — 20 min

REFERENCE

1. **Narang, A. S., Choudhury, D. R., and Richards, A.,** *J. Chromatogr. Sci.*, 20, 235, 1982.

Table TLC 43
DIAMINES AND TRIAMINES

Layer	L1	L1
Solvent	S1	S2
Detection	D1	D1
Amine	**R_f × 100**	
$H_2N(CH_2)_3NH_2$	40	72
$H_2N(CH_2)_4NH_2$ (putrescine)	43	62
$H_2N(CH_2)_5NH_2$ (cadaverine)	49	58
$H_2N(CH_2)_6NH_2$	57	66
$H_2N(CH_2)_7NH_2$	66	73
$H_2N(CH_2)_8NH_2$	76	82
$H_2N(CH_2)_3NH(CH_2)_3NH_2$ (sym-norspermidine)	15	67
$H_2N(CH_2)_3N(CH_3)(CH_2)_3NH_2$	14	33
$H_2N(CH_2)_3NH(CH_2)_4NH_2$ (spermidine)	18	50
$H_2N(CH_2)_3NH(CH_2)_5NH_2$ (aminopropylcadaverine)	22	43
$H_2N(CH_2)_3NH(CH_2)_6NH_2$	26	51
$H_2N(CH_2)_3NH(CH_2)_8NH_2$	42	72
$H_2N(CH_2)_4NH(CH_2)_4NH_2$ (sym-homospermidine)	20	52

Layer: L1 = precoated silica gel plates, silica gel 60 F_{254} (Merck)
Solvent: S1 = *n*-butanol/acetic acid/pyridine/water, 3:3:2:1
S2 = *n*-butanol/acetic acid/pyridine/formaldehyde, 3:3:2:1
Detection: D1 = ninhydrin

REFERENCE

1. **Shirahata, A., Takeda, Y., Kawase, M., and Samejima, K.,** *J. Chromatogr.*, 262, 451, 1983.

Table TLC 44
TETRAAMINES

Layer	L1	L1	L2
Solvent	S1	S2	S3
Detection	D1	D1	D1
Amine		**R_f × 100**	
$H_2N(CH_2)_2NH(CH_2)_2NH(CH_2)_2NH_2$	5	36	10
$H_2N(CH_2)_3NH(CH_2)_5NH(CH_2)_3NH_2$	6	27	19
$H_2N(CH_2)_3NH(CH_2)_6NH(CH_2)_3NH_2$	8	36	25
$H_2N(CH_2)_3NH(CH_2)_3NH(CH_2)_3NH_2$ (sym-norspermine)	4	27	13
$H_2N(CH_2)_3NH(CH_2)_4NH(CH_2)_3NH_2$ (spermine)	6	26	17
$H_2N(CH_2)_3NH(CH_2)_3NH(CH_2)_4NH_2$ (thermospermine)	7	40	18
$H_2N(CH_2)_4NH(CH_2)_3NH(CH_2)_4NH_2$ (canavalmine)	7	34	22
$H_2N(CH_2)_3NH(CH_2)_4NH(CH_2)_4NH_2$	8	24	22
$H_2N(CH_2)_4NH(CH_2)_4NH(CH_2)_4NH_2$	8	23	26

Layer: L1 = precoated silica gel plates, silica gel 60 F_{254} (Merck)
L2 = precoated cellulose plates, Avicel®SF
Solvent: S1 = *n*-butanol/acetic acid/pyridine/water, 3:3:2:1
S2 = *n*-butanol/acetic acid/pyridine/formaldehyde, 3:3:2:1
S3 = isopropanol/concentrated HCl/water, 8:3:2
Detection: D1 = ninhydrin

REFERENCE

1. **Shirahata, A., Takeda, Y., Kawase, M., and Samejima, K.,** *J. Chromatogr.*, 262, 451, 1983.

Table TLC 45
POLYAMINES

Layer	L1
Solvent	S1
Detection	D1
Technique	T1

Polyamine	$R_f \times 100$
Arginine	38
Agmatine	8
Methylamine	75
Ornithine	66
Putrescine	26
Spermidine	11
Spermine	5

Layer: L1 = Fixion® 50 × 8 ion exchange thin-layer chromato-sheets (Na^+) (Chinoin, Budapest, Hungary)

Solvent: S1 = 200 mmol/ℓ of KH_2PO_4 and 2 mol/ℓ of NaCl, the pH of the solution being adjusted to 7.5 with NaOH

Detection: D1 = the dried sheets are developed with ninhydrin reagent containing cadmium acetate

Technique: T1 = the chromato-sheets are run for about 4 hr at room temperature; the concentration of polyamines and amino acids can be determined by measuring the density of the ninhydrin spots in a video densitometer; resolution can be improved by overrun chromatography, i.e., by placing a filter paper strip horizontally on the top of the plate over its whole length; care has to be taken for smooth contact of the paper and the plate

REFERENCE

1. **Bardocz, S. and Karsai, T.,** *J. Chromatogr.*, 223, 198, 1981.

Table TLC 46
PRIMARY MONO- AND DIAMINES AND AMINO ACIDS

Layer	L1	L2	L2	L3	L4	L4
Solvent	S1	S1	S2	S3	S4	S5
Detection	D1	D1	D1	D1	D1	D1
Technique	T1	T1	T1	T1	T1	T1
Compound			$R_f \times 100$			
Methylamine	75	18	68	92	30	69
Ethylamine	76	16	66	85	42	75
n-Butylamine	75	10	50	55	23	65
Isobutylamine	76	11	53	62	34	70
n-Amylamine	70	7	39	34	20	60
Isoamylamine	70	8	41	45	19	62
n-Hexylamine	67	5	30	24	16	53
n-Heptylamine	65	4	23	14	2	e.s.[a]
n-Octylamine	59	3	17	7	0	3
n-Decylamine	e.s.	3	9	1	0	0
n-Dodecylamine	e.s.	1	3	0	0	0
n-Tetradecylamine	e.s.	0	0	0	0	0
1-Phenylethylamine	75	11	41	37	7	41
2-Phenylethylamine	75	9	37	28	6	36
1,2-Diaminoethane	50	2	51	82	5	49
1,2-Diaminopropane	54	2	50	81	7	58
1,3-Diaminopropane	54	2	51	84	2	36
1,4-Diaminobutane	57	1	51	84	2	32
1,5-Diaminopentane	59	2	52	85	1	26
1,6-Diaminohexane	62	3	51	79	1	19
1,7-Diaminoheptane	63	4	47	72	0	17
1,8-Diaminooctane	62	4	42	56	0	16
Spermidine	29	2	34	81	0	10
Spermine	12	0	23	72	0	2
2-Aminoisobutyric acid	92	27	76	81	63	84
2-Aminobutyric acid	92	23	75	82	50	82
3-Aminoisobutyric acid	87	20	75	84	29	68
4-Aminobutyric acid	88	24	79	89	20	60
2,4-Diaminobutyric acid	e.s.	3	61	91	4	55
n-Leucine	84	9	54	49	39	74
ε-Amino-*n*-caproic acid	85	20	73	77	14	52
Aminomethanesulfonic acid	94	24	77	82	70	80
2-Aminoethanesulfonic acid	94	86	92	95	56	71

[a] e.s. = elongated spot.

Layer: L1 = silanized silica gel RP-18 (Merck)
L2 = RP-18 plates impregnated with 4% dodecylbenzenesulfonic acid (HBDS)
L3 = SIL C_{18}-50 plates (Macherey, Nagel & Co.) impregnated with 4% HBDS
L4 = ammonium tungstophosphate (AWP) + $CaSO_4 \cdot {}^1/_2\ H_2O$ (4:2)

Solvent: S1 = 1 *M* acetic acid in 60% methanol
S2 = 1 *M* acetic acid + 1 *M* HCl in 60% methanol
S3 = 1 *M* acetic acid + 1 *M* HCl in 30% methanol
S4 = 1 *M* HNO_3
S5 = 2 *M* NH_4NO_3

Detection: D1 = the layers are sprayed with 1% ninhydrin in pyridine/acetic acid, 5:1, and the plates are heated for 5 min at 100°C

Technique: T1 = the migration distance is 6 cm for the ready-to-use plates and 10 cm for the inorganic exchanger; chromatography is carried out at 25°C

Table TLC 46 (continued)
PRIMARY MONO- AND DIAMINES AND AMINO ACIDS

REFERENCE

1. **Lepri, L., Desideri, P. G., Heimler, D., and Giannessi, S.,** *J. Chromatogr.*, 245, 297, 1982.

Table TLC 47
4′-DIMETHYLAMINOAZOBENZENE-4′-SULFONYL (DABSYL) AMINES[1]

Layer	L1	L1	L1	L1	L1	L1
Solvent	S1	S2	S3	S4	S5	S6
Technique	T1	T1	T1	T1	T1	T1
Amine			**$R_f \times 100$**			
Diethylamine	27	34	75	89	95	100
Diisopropylamine	26	32	72	85	100	
Diisobutylamine	38	51	72	85	100	
Di-*n*-hexylamine	49	62	91	95	100	
Di-*n*-decylamine	61	80	95	100		
Octylamine	24	41	65	93	100	
Nonylamine	21	36	64	86	95	
Decylamine	25	40	69	88	100	
Dodecylamine	19	38	74	89	100	

Layer: L1 = silica (Si60 GF 254, Merck)
Solvent: S1 to S6 = *n*-heptane containing 10, 20, 40, 50, 70, and 90% methyl ethyl ketone, respectively
Technique: Sandwich BN chambers with glass distributors were used[2]

REFERENCES

1. **Jusiak, J. and Soczewinski, E.,** *J. Chromatogr.*, 248, 263, 1982.
2. **Soczewinski, E. and Wawrzynowicz, T.,** *Chromatographia*, 11, 446, 1978.

Table TLC 48
4′-DIMETHYLAMINOAZOBENZENE-4′-SULFONYL (DABSYL) AMINES[1]

Layer	L1	L1	L1	L1	L1	L1
Solvent	S1	S2	S3	S4	S5	S6
Technique	T1	T1	T1	T1	T1	T1
Amine			**$R_f \times 100$**			
Diethylamine	25	48	82	86	98	100
Diisopropylamine	31	47	81	87	100	
Diisobutylamine	49	63	91	100		
Di-*n*-hexylamine	59	70	93	100		
Di-*n*-decylamine	66	82	93	100		
Octylamine	27	52	83	90	95	100
Nonylamine	26	51	92	95	100	
Decylamine	31	52	94	100		
Dodecylamine	27	48	90	95	100	

Layer: L1 = silica (Si 60 GF 254, Merck)

Solvent: S1 to S6 = *n*-heptane containing 10, 20, 40, 50, 70, and 90% dioxane, respectively

Technique: T1 = sandwich BN chambers with glass distributors were used[2]

REFERENCES

1. **Jusiak, J. and Soczewinski, E.,** *J. Chromatogr.*, 248, 263, 1982.
2. **Soczewinski, E. and Wawrzynowicz, T.,** *Chromatographia,* 11, 446, 1978.

Table TLC 49
4′-DIMETHYLAMINOAZOBENZENE-4′-SULFONYL (DABSYL) AMINES[1]

Layer	L1	L1	L1	L1	L1	L1	L1
Solvent	S1	S2	S3	S4	S5	S6	S7
Technique	T1	T1	T1	T1	T1	T1	T1
Amine				**R_f × 100**			
Diethylamine	7	16	35	39	54	64	67
Diisopropylamine	12	21	40	44	56	65	68
Diisobutylamine	19	35	55	60	68	78	82
Di-*n*-hexylamine	25	37	60	70	78	84	87
Di-*n*-decylamine	30	42	66	70	78	87	89
Octylamine	4	10	25	31	49	59	66
Nonylamine	5	9	26	34	54	69	72
Decylamine	6	8	31	39	55	74	77
Dodecylamine	7	9	30	39	56	77	80

Layer:	L1	=	silica (Si 60 GF 254, Merck)
Solvent:	S1 to S7	=	*n*-heptane containing 10, 20, 40, 50, 70, 90, and 100%, respectively, of diisopropyl ether
Technique:	T1	=	sandwich BN chambers with glass distributors were used[2]

REFERENCES

1. **Jusiak, J. and Soczewinski, E.,** *J. Chromatogr.,* 248, 263, 1982.
2. **Soczewinski, E. and Wawrzynowicz, T.,** *Chromatographia,* 11, 446, 1978.

Table TLC 50
4′-DIMETHYLAMINOAZOBENZENE-4′-SULFONYL (DABSYL) AMINES[1]

Layer	L1	L1	L1	L1
Solvent	S1	S2	S3	S4
Technique	T1	T1	T1	T1
Amine		**$R_f \times 100$**		
Diethylamine	61	46	28	23
Diisopropylamine	69	58	42	33
Diisobutylamine	51	29	14	12
Di-*n*-hexylamine	37	17	11	
Di-*n*-decylamine	14	5		
Octylamine	58	36	21	13
Nonylamine	52	28	14	10
Decylamine	50	25	11	
Dodecylamine	43	16	7	

Layer: L1 = HPTLC precoated plates, RP-18, Merck

Solvent: S1, S2, S3, S4 = water/methanol mixtures containing 100, 95, 90, and 85%, respectively, of methanol

Technique: T1 = sandwich BN chambers with glass distributors were used[2]

REFERENCES

1. **Jusiak, J. and Soczewinski, E.,** *J. Chromatogr.*, 248, 263, 1982.
2. **Soczewinski, E. and Wawrzynowicz, T.,** *Chromatographia,* 11, 446, 1978.

Table TLC 51
PRIMARY ALIPHATIC AMINES DERIVATIZED WITH 1,2-NAPHTHOYLENEBENZIMIDAZOLE-6-SULFOCHLORIDE

Layer	L1	L1	L1	L1	L1	L1	L1	L1
Solvent	S1	S2	S3	S4	S5	S6	S7	S8
Detection	D1	D1	D1	D1	D1	D1	D1	D1
	$R_f \times 100$							
Amine								
Ammonia	23	14	15	15	11	12	29	51
Methylamine	28	30	22	28	17	18	41	60
Ethylamine	37	47	29	38	27	25	48	66
n-Propylamine	47	55	38	43	35	31	58	69
n-Butylamine	52	60	48	51	44	35	63	74
n-Hexylamine	57	67	56	57	60	40	68	78
n-Heptylamine	60	70	50	60	67	43	71	81
n-Octylamine	63	72	62	62	73	46	73	83
n-Decylamine	68	76	70	68	78	49	78	88

Layer: L1 = silica gel thin-layer plates (Silufol®, 150 × 150 mm)
Solvent: S1 = *n*-hexane/methylethyl ketone, 7:5
S2 = *n*-hexane/ethylacetate, 1:1
S3 = *n*-hexane/cyclohexane, 2:1
S4 = *n*-hexane/tetrahydrofuran, 7:5
S5 = *n*-hexane/dioxan, 2:1
S6 = benzene/ethyl acetate, 8:3
S7 = benzene/dioxan, 7:3
S8 = benzene/ethyl acetate/acetic acid, 10:5:0.1
Detection: D1 = the spots are fluorescent; at the end of chromatography, the plate is dried and the spot of the separated derivative extracted with acetone; the volume of the acetone extract is adjusted to 10 mℓ and the fluorescence of the solution measured for quantitative determination; the fluorescence reading is corrected for a blank value obtained in the absence of amine derivatives

REFERENCE

1. **Jandera, P., Pechova, H., Tocksteinova, D., Churacek, J., and Kralovsky, J.,** *Chromatographia,* 16, 275, 1982.

Table TLC 52
SECONDARY ALIPHATIC AMINES DERIVATIZED WITH 1,2-NAPHTHOYLENE BENZIMIDAZOLE-6-SULFOCHLORIDE

Layer	L1	L1	L1	L1	L1	L1	L1	L1
Solvent	S1	S2	S3	S4	S5	S6	S7	S8
Detection	D1	D1	D1	D1	D1	D1	D1	D1
Amine				**$R_f \times 100$**				
Dimethylamine	27	18	45	29	28	48	64	68
Diethylamine	37	29	61	43	40	52	75	79
Di-*n*-propylamine	44	37	70	55	51	60	81	84
Di-*n*-butylamine	53	45	78	62	60	66	88	88
Di-*n*-octylamine	74	67	93	79	83	73	94	91

Layer: L1 = silica gel thin-layer plates (Silufol®, 150 × 150 mm)
Solvent: S1 = *n*-hexane/methylethylketone, 3:1
S2 = *n*-hexane/ethyl acetate, 7:3
S3 = *n*-hexane/tetrahydrofuran, 7:5
S4 = *n*-hexane/cyclohexanone, 7:3
S5 = *n*-hexane/dioxan, 7:3
S6 = *n*-hexane/acetone, 9:5
S7 = benzene/dioxan, 7:3
S8 = benzene/ethyl acetate, 7:3
Detection: D1 = the spots are fluorescent; at the end of chromatography the plate is dried and the spot of the separated derivative extracted with acetone; the volume of the acetone extract is adjusted to 10 mℓ and the fluorescence of the solution measured for quantitative determination; the fluorescence reading is corrected for a blank value obtained in the absence of amine derivatives

REFERENCE

1. **Jandera, P., Pechova, H., Tocksteinova, D., Churacek, J., and Kralovsky, J.,** *Chromatographia,* 16, 275, 1982.

Table TLC 53
DABSYL (4-DIMETHYLAMINOAZOBENZENE-4′-SULFONYL) DERIVATIVES OF POLYAMINES

Layer	L1	L1	L1	L1	L1
Solvent	S1	S2	S3	S4	S5
Detection	D1	D1	D1	D1	D1
Technique	T1	T1	T1	T1	T1
Derivative			**$R_f \times 100$**		
Dabsyl ammonia	62	60	55	35	46
Dabsyl chloride	92	94	97	99	87
Dabsyl hydroxide	2	1	2	3	0
Dabsyl ornithine	37	39	33	4	25
Dabsyl putrescine	66	72	61	27	48
Dabsyl spermidine	70	87	71	58	51
Dabsyl spermine	87	94	82	78	78

Layer: L1 = thin-layer plastic sheets precoated with silica gel 60 (0.25 mm thick) (Merck)

Solvent: S1 = chloroform/dichloromethane/acetone/95% ethanol, 5:2:0.5:1

S2 = chloroform/dichloromethane/acetone/absolute ethanol, 5:2:0.5:1

S3 = chloroform/dichloromethane/absolute ethanol, 5:2:1

S4 = chloroform/trimethylamine, 5:1

S5 = chloroform/chloromethane/acetone/absolute ethanol/*n*-hexane, 5:2:1:0.5:3

Detection: D1 = the dabsyl derivatives are colored

Technique: T1 = chromatography is performed in a closed cylinder, 10 cm high, 6 cm in diameter, with a ground glass cover

REFERENCE

1. **Lin, J. K. and Wang, C. C.,** *J. Chromatogr.*, 227, 369, 1982.

Table TLC 54
AMINES DERIVATIZED WITH BENOXAPROFEN CHLORIDE

Layer	L1	L1	L1	L2
Solvent	S1	S2	S3	S4
Detection	D1	D1	D1	D1
Compound		**$R_f \times 100$**		
Amphetamine	14	53	81	29
	21	59		
Benoxaprofen	9	5	68	43
Benoxaprofen chloride	64	84	85	
Benzylamine	19	44	77	37
Maprotiline	25	69	82	13
Methamphetamine	27	65	79	23
	33			
α-Methylbenzylamine	16	45	81	31
	28	56		
Phenylbutylamine	18	48	80	26
β-Phenylethylamine	16	49	79	33
Procaine	8	14	30	4
Tolylethylamine	18	49	79	30
Tranylcypromine	16	39	76	32
	21	45		

Note: Two spots with differet R_f values represent diastereomers obtained by reaction of the optically active compound with racemic benoxaprofen chloride.

Layer: L1 = silica gel 60 (Merck)
L2 = reversed-phase material RP-18 (Merck)
Solvent: S1 = toluene/dichloromethane/tetrahydrofuran, 5:1:1, ammonia atmosphere
S2 = toluene/chloroform/tetrahydrofuran, 5:4:1, ammonia atmosphere
S3 = chloroform/methanol/formic acid/tetrahydrofuran, 110:20:5:2
S4 = methanol/water, 9:1
Detection: D1 = densitometric measurement of the intensity of fluorescence using a chromatogram spectrophotometer

REFERENCE

1. **Spahn, H., Weber, H., Mutschler, E., and Möhrke, W.,** *J. Chromatogr.*, 310, 167, 1984.

Table TLC 55
CHIRAL AMINES AFTER DERIVATIZATION WITH (*S*)-(+)-BENOXAPROFEN CHLORIDE

Layer	L1	L1
Solvent	S1	S2
Detection	D1	D1
Technique	T1	T1
Derivative of	**R_f × 100**	
(*S*)-(+)-Amphetamine	14	53
(R)-(−)-Amphetamine	21	59
(*S*)-(+)-Methamphetamine	27	65
(R)-(−)-Methamphetamine	33	65
(R)-(+)-α-Methylbenzylamine	28	56
(*S*)-(−)-α-Methylbenzylamine	16	45
(R)-(+)-Tranylcypromine	16	39
(*S*)-(−)-Tranylcypromine	21	45

Layer: L1 = silica gel 60 without F_{254}
Solvent: S1 = toluene/dichloromethane/tetrahydrofuran, 5:1:1, ammonia atmosphere
S2 = toluene/chloroform/tetrahydrofuran, 5:4:1, ammonia atmosphere
Detection: D1 = densitometric measurement of fluorescence intensity using a chromatogram spectrometer, the excitation wavelength being 313 nm
Technique: T1 = volume of 10 μℓ applied, band width 5 mm

REFERENCE

1. **Weber, H., Spahn, H., Mutschler, E., and Möhrke, W.,** *J. Chromatogr.*, 307, 145, 1984.

Table TLC 56
PRIMARY AMINES AFTER CONVERSION TO SALICYLALDEHYDEAZOMETHINE-DIPHENYLBORON CHELATES

Layer	L1
Solvent	S1
Detection	D1
Technique	T1

Amine	**R_f × 100**
2-Aminobutane	52
2-(Aminomethyl)-furan	50
Benzylamine	51
n-Butylamine	52
Cyclohexylamine	54
Cyclopentylamine	47
Methoxypropylamine	8
n-Propylamine	49

Layer: L1 = Kieselgel® 60 plates, 0.25-mm layer, without fluorescent indicator (Merck)
Solvent: S1 = benzene
Detection: D1 = fluorescence measurement of spots *in situ*
Technique: T1 = prior equilibration of the chamber at 18% relative humidity for 20 min

REFERENCE

1. **Hohaus, E.,** *Fresenius Z. Anal. Chem.*, 310, 70, 1982.

Section II

Techniques

II.I. The Liquid Chromatographic Determination of Free Amino Acids

II.II. The Gas Chromatographic Determination of Amino Acids

II.III. The Chromatographic Separation of Amino Acid Enantiomers

II.IV. The Liquid Chromatographic Separation of *o*-Phthalaldehyde Amino Acid Derivatives

II.V. The Chromatographic Separation of Phenylthiohydantoin Amino Acids

II.VI. The Chromatographic Separation of Dansyl Amino Acids

II.VII. The Chromatographic Separation of Phenylthiocarbamyl Amino Acids

II.VIII. The Chromatographic Separation of Dimethylamnobenzenesulfonyl (DABS) -Amino Acids

II.IX. The Chromatographic Separation of 4-*N,N*-Dimethylaminobenzene-4′-Thiohydantoin (DABTH) -Amino Acids

II.X. The Determintion of Proline and Hydroxyproline by Derivatization with 4-Chloro- or 4-Fluoro-7-Nitrobenzofurazan (NBD-Cl and NBD-F)

Section II.I

THE LIQUID CHROMATOGRAPHIC DETERMINATION OF FREE AMINO ACIDS

Automated amino acid analysis was first introduced by Moore and Stein[1] in the 1950s. Together with Spackman[2] they developed an amino acid analyzer that automatically coupled the amino acid separation with quantitation based on the ninhydrin reaction. Essentially amino acids are separated by cation exchange on cross-linked sulfonated polystyrene using citrate buffers. The use of a "microbead" resin with a uniform diameter of about 10 μm or less improved buffer composition, and high pressure chromatographic techniques have decreased the analysis time from 24 hr to approximately 1 hr, while microbore columns and improved spectrophotometer design have increased sensitivity about 100-fold. The increased sensitivity of amino acid determination has, however, also depended greatly on the introduction of new postcolumn fluorogenic detection reagents.

POSTCOLUMN DETECTION REAGENTS

Ninhydrin

Three postcolumn derivatization reagents have found widespread use in amino acid analysis: ninhydrin, fluorescamine, and *o*-phthalaldehyde. Ninhydrin has been used the longest; noteworthy is the fact that the reagent reacts with both primary and secondary amino acids, yielding derivatives that absorb strongly at 570 and 440 nm. For a complete amino acid assay two detectors or one wavelength-programmable detector are required. The novel technique of Le Page and Rocha[3] takes advantage of the slow rate of reaction of ninhydrin or other reagent with amino acids, the reagent being added to the mobile phase prior to separation rather than postcolumn. After reversed-phase ion-pair chromatography of the amino acids the eluate passes to a heated reaction coil where color development occurs. Ninhydrin was found to be useful in connection with the separation of aliphatic amines, while 1,2,3-peri-naphthindanetrione hydrate was useful for the majority of amino acids. Advantages claimed for the approach are that hardware needs are reduced, only one pump is necessary, and a mixing tee is eliminated from the equipment. Additionally, solute bands are not diluted by the reagent stream, and in consequence concentration and absorbance signals are proportionally greater.

o-Phthalaldehyde

The high sensitivity needed for the structure determination of proteins and peptides that can be obtained only in small amount is available through the use of the fluorogenic reagent, *o*-phthalaldehyde.[4] This reagent, used in conjunction with a thiol at pH 9, gives an intense blue fluorescence with α-amino acids and a sensitivity five to ten times greater than that of ninhydrin.[5] Most routine amino acid analyses, however, require the separation and determination of proline and sometimes hydroxyproline, which do not yield fluorescent products with *o*-phthalaldehyde. One approach used to overcome this difficulty is to first convert these secondary amino acids to primary amino acids by alkaline hypochlorite oxidation.

In the system of Cunico and Schlabach,[6] one pump-coil arrangement is used to add hypochlorite to the column effluent, a second for the addition of *o*-phthalaldehyde. Conversion of proline to a primary amine by hypochlorite is temperature dependent. Heating the reaction coil greatly increases the amount of primary amine formed, but there is a decreased response of the other primary amino acids. A concentration of 1% NaOCl produced the best peak height response for proline conversion; the response of primary amino acids decreased 15 to 30% at room temperature when this concentration of hypochlorite was added to the column effluent. The relative cystine response, however, decreased about twofold, this amino acid probably being converted to cysteic acid by the hypochlorite. Satisfactory results were obtained when hypochlorite was added only during the elution of proline.

Dong and Gant[7] added sodium hypochlorite continuously during the analysis, claiming that baseline disruption is thereby avoided. However, the destruction of primary amino acids by hypochlorite requires precise reaction conditions to be maintained in order to obtain reproducible and quantitative analyses. Of special importance in this connection is the linearity of the sodium hypochlorite-*o*-phthalaldehyde reaction. However, the data showed excellent linear correlation of all amino acids at the concentration range 100 to 500 pmol and 0.25 to 10 nmol. Ishida et al.[8] considered the method of adding hypochlorite only during elution of proline to be costly, complicated, and not very easy to operate. In contrast, the continuous addition of hypochlorite ensured satisfactory results provided the operational conditions ensured that hypochlorite degradation of primary amino acids that was reproducible.

Postcolumn addition of chloramine-T prior to the introduction of *o*-phthalaldehyde has been used for the detection of cyclic imino acids of plant origin following automated ion exchange fractionation.[9] Each compound exhibits characteristic optima with respect to the pH of the chloramine-T solution, the concentration of chloramine-T, and the temperature at which the oxidation is performed. Böhlen et al.[10,11] split peptide samples into two aliquots, one being hydrolyzed under reducing conditions, the other being reacted with performic acid prior to hydrolysis and used for the determination of proline and cystine, the latter as cysteic acid. The sample hydrolyzed under reducing conditions is injected onto the analysis column first, the column is then reequilibrated with starting buffer, and the second aliquot is injected. Prior to the elution of proline, hypochlorite is added to the column effluent.

o-Phthalaldehyde gives greater sensitivity of detection than other derivatization reagents. It meets two important criteria: first, it is not itself fluorescent and therefore does not interfere with detection. Second, the reaction occurs rapidly at room temperature, eliminating the requirement for long delay times between the mixing tee and detector which can contribute to band broadening. The alternative fluorescent reagent, fluorescamine, was developed historically from the chemistry of the reaction of ninhydrin with amino acids.[12] Its drawbacks are that two postcolumn pumps are required and it does not react with secondary amino acids. In amino acid analyzers with microbore columns fluorescamine shows no greater sensitivity than ninhydrin. It now seems to be less popular than *o*-phthalaldehyde. In a direct comparison of postcolumn derivatization techniques, amino acids were separated on polystyrene-based strong cation exchangers and then derivatized with *o*-phthalaldehyde or ninhydrin.[6] MicroPak® amino acid columns in the sodium form were used for amino acids found in acid hydrolyzates of peptides and proteins; the lithium form was used for amino acids and their metabolites in physiological fluids such as serum and urine. The conclusion was reached that the single-pump *o*-phthalaldehyde system is more sensitive and presents fewer technical problems than the ninhydrin system; it should be chosen whenever the detection of secondary amino acids is not important. The ninhydrin system is the system of choice when primary and secondary amino acids have to be detected; it requires only one postcolumn pump and provides sensitivities for secondary amino acids that are comparable to dual-pump *o*-phthalaldehyde-sodium hypochlorite systems. On analyzing a soy protein hydrolyzate excellent correlation between the ninhydrin and hypochlorite *o*-phthalaldehyde techniques was found, with an average deviation of approximately 3%.[7]

AVOIDANCE OF CONTAMINATION

A considerable amount of information is available on subnanomole-range amino acid analysis, and it is not too difficult to analyze amino acid standard mixtures at the 10-pmol level. However, as Böhlen and Schroeder[11] pointed out in 1982, it was not easy to obtain accurate amino acid compositions at the 50-pmol level, often due to contamination in the course of peptide hydrolysis. Since amino acid analysis at the highest sensitivity has become increasingly necessary, suitable methodology for the routine determination of amino acid composition using only 20 to 100 pmol of peptide has been developed using fluorogenic reagents. Baseline shifts, artifacts, and unknown contaminants may, however, limit the sensitivity of the analysis. The sources of fluorescence contamination have been attributed to impurities in the water, buffer solutions, and HCl.

Buffers

Impurities in amino acid analyzer buffers derive from contaminated reagents used in buffer production. These impurities are responsible for the so-called buffer change peaks and other baseline artifacts which can affect the accurate integration of peaks eluting in their vicinity. There is some difference of opinion as to whether commercial buffers or those prepared in the laboratory are to be preferred in order to overcome this problem. Böhlen and Schroeder[11] observed that the use of Millipore-treated water and analytical grade buffer salts in the preparation of buffers gave consistently good results. Buffers were prepared directly in volume-calibrated buffer storage bottles used in the amino acid analyzer. Distilled *n*-propanol and HCl required for pH adjustment were added before final adjustment of the volume. Buffers were protected from atmospheric contamination during use by connecting their reservoirs to a sulfuric acid trap. In the opinion of Hughes et al.,[13] in the determination of amino acids below the 1 nmol level, impurities in commercial buffers require that laboratory preparation of buffers is essential. They formulated a series of formate buffers that permit short analysis times and quantification at the low picomole level.

The preparation of buffer solutions free from troublesome impurities, however, can be time consuming and expensive. The use of highly purified commercial buffer solutions might therefore be a preferable alternative. However, buffer formulations vary among manufacturers and no standard amino acid buffer system, e.g., for use with *o*-phthalaldehyde detection, has been established. It is, therefore, difficult to select the best buffer system for a particular type of analysis, and the use of commercial premixed buffers requires careful scrutiny and examination. Dong and Gant[7] evaluated several commercial buffers. The Pierce® Buffelutes were best in terms of resolution, analysis time, and buffer background. The Pierce® Pico II and Pickering buffers gave good resolution and sensitivity at longer analysis times (30 to 50 min). LKB buffers gave good separation of all amino acids except glutamic acid and proline.

Barbarash and Quarles[14] found that the Durrum® Pico-Buffer system II was unsuitable for *o*-phthalaldehyde amino acid analysis. The Beckman® buffer system showed a rising baseline around the basic amino acid region,while the Dionex® system showed a large buffer change peak between the first and second buffers and a flatter baseline in the basic amino acid region. Both systems contained impurities that coeluted with histidine and lysine. The differences between the buffers could only be attributed to differences in their manufacture. The use of a combination of buffers, Beckman® buffer 1 with isopropanol and Dionex® buffers 2 and 3, produced a reduced buffer change peak, a reduction of interfering contaminants, and a smoother overall baseline.

Hydrochloric Acid

A major source of contamination in amino acid determination is commercially available HCl, which contains amino acids and ammonia in sufficient amounts to interfere with cleanliness requirements during buffer preparation and peptide hydrolysis. Böhlen and Schroeder[11] studied the amounts of amino acids found in a sample blank after hydrolysis with purified HCl. When less HCl was used for hydrolysis, contamination was reduced

Table 1
AMINO ACID CONTENT OF A SAMPLE BLANK AFTER HYDROLYSIS WITH 50 AND 5 μℓ HYDROCHLORIC ACID

Amino acid	50 μℓ HCl (pmol)	5 μℓ HCl (pmol)
Alanine	1.9 ± 3.9	1.2 ± 1.5
Asparagine or aspartic acid	5.9 ± 4.8	3.0 ± 2.3
Glutamine or glutamic acid	1.9 ± 2.6	0.6 ± 0.9
Glycine	10.3 ± 8.9	3.1 ± 7.3
Serine	9.1 ± 7.7	4.4 ± 3.2
Threonine	1.0 ± 1.9	0.5 ± 1.1

Note: Values represent means ± standard deviation from nine analyses. Blank levels of amino acids not listed were below detection level (<0.5 pmol).

REFERENCE

1. **Böhlen, P. and Schroeder, R.,** *Anal. Biochem.,* 126, 144, 1982.

(Table 1). This indicated that even purified HCl was a major source of contamination. Hydrochloric acid used for peptide hydrolysis was therefore purified as follows: constant boiling HCl was prepared by diluting concentrated HCl, refluxing overnight, and distilling over a Vigreux® column in an apparatus protected from laboratory air by a trap containing 50% (v/v) H_2SO_4. The purified acid was stored in glass bottles with tightly sealing polypropylene screw caps and was exposed to laboratory air for as short a time as possible. Refluxing HCl prior to distillation reduced the level of contamination, most notably of glycine. The nature of the interfering material that is eliminated by this procedure is not known.

Knecht and Chang,[15] using the DABS-C1 precolumn labeling method of amino acid determination (Section II.VIII), considered the most likely source of amino acid contaminants to be aqueous 6 *N* HCl. Gas phase hydrolysis reduced the background generated by liquid phase hydrolysis to about 25%. A blank sample following gas phase hydrolysis yielded an average of 0.5 to 3 pmol of aspartic acid, glutamic acid, serine, and glycine. This reduced level of background permitted reliable analysis of low nanogram amounts of protein hydrolyzates.

HPLC EQUIPMENT FOR AMINO ACID DETERMINATION

Ion exchange techniques, particularly those involving automatic analyzers, have been the dominant methods of amino acid determination over several decades. Increasing interest, however, is now being shown in the use of HPLC equipment in place of dedicated analyzers and in the employment of different types of column. Hughes et al.[13] have described how an HPLC apparatus normally employed for peptide or protein separation can be used for amino acid analysis. The equipment commonly used for HPLC (Altex or Kontron) consists of a microprocessor controlling the flow rates of two pumps; gradient elution from two solvents is achieved by adjusting these rates. Peptide detection is carried out by postcolumn derivatization via a sampling valve with fluorescamine or *o*-phthalaldehyde or directly by absorption in the UV region. These elements, with a small number of additions, were used to construct an amino acid analyzer. DC-4A resin was packed in a stainless steel column, 250 × 4.6 mm I.D. Stainless steel columns permitted the use of higher pressure than glass columns and higher flow rates were thus achievable. Analysis time, including regeneration and equilibration for a standard hydrolyzate, not including tryptophan, was 60 min. Reproducibility of the peak elution time enabled peak-height or -area measurements to be used for quantitation in the range 10 pmol to 25 nmol. The reproducibility, speed of analysis, and sensitivity are claimed to be equal or better than those of most commercial purpose-built analyzers. An apparatus of this kind, with the advantages of lower cost and greater versatility than commercial analyzers, more readily allows the most effective use of equipment.

Schuster[16] has quantitatively determined free amino acids in i.v. solutions and beverages, separations being performed on a 5-μm LiChrosorb® NH_2 column with an acetonitrile-phosphate buffer gradient. Amino acids are separated within 30 min and high sensitivity is achieved by detection in the UV at 200 nm. Reversed-phase packings have been successfully applied to the HPLC of amino acids. Since free amino acids do not normally separate well on a reversed-phase column due to the charges they carry, ion-pairing conditions are used. The ion pairs of amino acids with dodecyl sulfate have been separated on a reversed-phase Beckman® Ultrasphere I.P. column using a sequence of eluents prepared by mixing 0.2 *M* H_3PO_4 containing 10 m*M* sodium dodecyl sulfate, 0.2 *M* sodium acetate containing 10 m*M* sodium dodecyl sulfate, and methanol.[17] Amino acids occurring in tissues, except taurine and related weakly basic amino acids, were determined in 95 min with a sensitivity of less than 50 pmol per amino acid. These complex amino acid mixtures required the complicated sequence of eluents for their separation. The procedure was considered suitable for fully automated routine amino acid analysis.

Despite the increasing attention being paid to procedures such as those described, it seems probable that the automatic analyzer essentially in its present form will have an important role in amino acid determination for several years to come.

REFERENCES

1. **Moore, S. and Stein, W. H.,** *J. Biol. Chem.*, 192, 663, 1951.
2. **Moore, S., Spackman, D. H., and Stein, W. H.,** *Anal. Chem.*, 30, 1185, 1958.
3. **Le Page, J. N. and Rocha, E. M.,** *Anal. Chem.*, 55, 1360, 1983.
4. **Roth, M. and Hampai, A.,** *J. Chromatogr.*, 83, 353, 1973.
5. **Benson, J. R. and Hare, P. E.,** *Proc. Natl. Acad. Sci. U.S.A.*, 72, 619, 1975.
6. **Cunico, R. L. and Schlabach, T.,** *J. Chromatogr.*, 266, 461, 1983.
7. **Dong, M. W. and Gant, J. R.,** *J. Chromatogr.*, 327, 17, 1985.
8. **Ishida, Y., Fujita, T., and Asai, K.,** *J. Chromatogr.*, 204, 143, 1981.
9. **Bleeker, A. B. and Romeo, J. T.,** *Anal. Biochem.*, 121, 285, 1982.
10. **Böhlen, P. and Mellet, M.,** *Anal. Biochem.*, 94, 313, 1979.
11. **Böhlen, P. and Schroeder, R.,** *Anal. Biochem.*, 126, 144, 1982.
12. **Udenfriend, S., Stein, S., Böhlen, P., Dairman, W., Leingruber, W., and Weigele, M.,** *Science*, 178, 871, 1972.
13. **Hughes, G. J., Winterhalter, K. G., Boller, E., and Wilson, K. J.,** *J. Chromatogr.*, 235, 417, 1982.
14. **Barbarash, G. R. and Quarles, R. H.,** *Anal. Biochem.*, 119, 177, 1982.
15. **Knecht, R. and Chang, J.-Y.,** *Anal. Chem.*, 58, 2375, 1986.
16. **Schuster, R.,** *Anal. Chem.*, 52, 617, 1980.
17. **Seiler, N. and Knodgen, B.,** *J. Chromatogr.*, 341, 11, 1985.

Section II.II

THE GAS CHROMATOGRAPHIC DETERMINATION OF AMINO ACIDS

A considerable number of papers dealing with the gas chromatographic separation of amino acids have appeared during the last few years; this section describes recent developments.

AMINO ACID DERIVATIVES

Discussion of the most suitable derivatives for GC estimation and variations on their methods of synthesis increase. The proliferation probably reflects the difficulty of ensuring reproducibility in procedures between laboratories and at different times. Complete derivatization of the reactive groups of an amino acid molecule is necessary for satisfactory gas chromatographic analysis. The *N*-perfluoroacyl amino acid alkyl esters are probably the most convenient derivatives, allowing the determination of all common protein amino acids. A small ester group may be used to reduce separation time or allow lower over temperatures. However, the high volatility of derivatives with a small ester group may cause losses during evaporation.

The *n*-propyl, isopropyl, *n*-butyl, and isobutyl esters of amino acids have been prepared by Gamerith and acylated with trifluoroacetic anhydride (TFAA) or heptafluorobutyric anhydride (HFBA).[1] The derivatives were separated on 2% OV-17-1% OV-210 coated on to Supelcoport®, the HFB derivatives also being analyzed on a mixed phase containing 0.31% Carbowax® 20M, 0.28% Silar® 5CP, and 0.06% Lexan® on Chromosorb® W AW, 120-140 mesh. The various esters were found to have different retention times, the isopropyl esters eluting first, then the *n*-propyl esters, the isobutyl esters, and the *n*-butyl esters. This order is the same on all the stationary phases tested as well as for the two acyl groups. The effect of different ester groups is more pronounced with small amino acids. Using a small ester group and/or the 2-alcoholic group in place of the 1-alcoholic analog, the peaks are spaced further apart. This result could be of use when analyzing complex amino acid mixtures.

The retention times of the TFA and HFB derivatives of amino acids were also compared by Gamerith.[1] The *n*-propyl esters were analyzed on 2% OV-17-1% OV-120 on Supelcoport®, the isopropyl esters on 0.65% EGA on Chromosorb® W AW, and the *n*-butyl esters on either SP 2100 or 5% Carbowax® 20M. On three of these packings the TFA derivatives have shorter retention times than their HFB analogs, while on Carbowax® 20M the reverse is the case. Retention times relative to norleucine are similar for the TFA and HFB derivatives, although the length of time between the first and last eluting amino acids for a given type of derivative is less after acylation with HFBA than after acylation with TFAA. The chromatograms of the HFB derivatives thus appear to be more ''compressed'' than those of the TFA derivatives.

The order of elution of a hydroxylated amino acid relative to its nonhydroxylated analog depends on the polarity of the stationary phase.[1] The greater the polarity the longer is the retention time of the hydroxylated derivative. Yoneda[2] has esterified protein amino acids to their *n*-propyl esters and then acylated these to the *N*-heptafluorobutyryl derivatives, the reactions being carried out in a single tube at 100°C. A simple steam-heating apparatus was constructed that heats only the bottom of the reaction vessel.

Yeung et al.[3] have converted amino acids to their *N*(*O*,*S*)-pentafluorobenzoyl isobutyl esters by acid-catalyzed esterification followed by aqueous extractive pentafluorobenzoylation. The frequently used esterification reagent anhydrous isobutanol-3 *M* HCl is conventionally prepared by bubbling HCl gas into the alcohol and then titrating to the desired normality. The amino acid carboxylic group, however, can be effectively esterified by adding a solution of one drop of concentrated HCl per milliliter of distilled isobutanol to the amino acids before heating. The equilibrium of the esterification reaction shifts to the right owing to the large excess of isobutanol relative to the trace of water present in a drop of concentrated HCl. The simplicity of the technique makes this an attractive alternative reagent. The extractive perfluorobenzoylation with pentafluorobenzoyl chloride procedure of these authors is stated to combine efficiently the extraction, derivatization, and consumption of the derivatizing reagent within a reasonable time suitable for routine analysis of amino acids. The high sensitivity of their procedure permitted analysis of minute amounts of sample, while the high stability of the derivatives allowed the use of an automatic sample injector at ambient temperatures.

N-Methyl-*N*-*tert*-butyldimethylsilyltrifluoroacetamide (MTBSTFA) is a relatively new silylation reagent that leads to the *tert*-butyldimethylsilyl (TBDMS) derivative.[4-6] TBDMS derivatives are stable enough to allow them to be concentrated if necessary for improved sensitivity, or they may be injected directly into a chromatograph with the reaction solvent and reagent. TBDMS derivatives of protein amino acids have been separated on a wide bore, bonded silicone phase, glass capillary column.[4] Due to the excellent resolution and linear response the method was recommended for the analysis of biological samples. TBDMS derivatives were also readily separable on SE-30 (wall-coated open tubular) and OV-17 (bonded) capillary columns, as well as on packed columns coated with the same liquid phases.[5] The preparation of TBDMS derivatives is a simple, one-step, and rapid process. The derivatives are particularly useful for gas chromatographic/mass spectrometric analysis involving the use of stable isotopes and selected ion monitoring due to the frequently prominent ion at 57 units less than the molecular ion.[6]

Husek and co-workers[7,8] have shown that substituted oxazolidinones are suitable derivatives or the determination of amino acids. The amino acids are condensed with 1,3-dichlorotetrafluoroacetone and those with reactive side chains are treated with acid anhydride in a subsequent step. However, problems were encountered in generalizing the reaction conditions due to the reactive side chain groups requiring different chemical treatments. A later study showed that it was possible to so modify the reaction conditions that the 20 protein amino acids could be successfully derivatized and quantitatively analyzed within 1 hr.[8] The characteristic amino acids of collagen, proline, 4-hydroxyproline, 3-hydroxyproline, lysine, hydroxylysine, and ε-hydroxynorleucine can be determined as their *N*-trifluoroacetyl *n*-butyl esters using a silanized glass column packed with Gas Chrom® P coated with 3% QF-1 mixed with Gas Chrom® P coated with 1% SE-30 in the ratio of 3:2.[9]

A number of amino acids exhibit characteristic difficulties during derivatization and techniques to overcome specific problems have been devised. Complete acylation of the butyl ester of arginine depends on the use of a high temperature, for example, 150°C for 10 min.[10] Precautions must be taken to ensure that sufficient acylating agent is present at this temperature to prevent complete vaporization, and acylation tubes must be immersed in the oil bath to the level of liquid in the tube to allow efficient refluxing of the reagent. Failure to use high-purity heptafluorobutyric anhydride for acylation produces a poor response for arginine, small quantities of heptafluorobutyric acid having an adverse effect. Heptafluorobutyric acid may arise when water formed during esterification and/or isobutanol remaining in the azeotroped residue reacts on addition of heptafluorobutyric anhydride to give the acid. Removal of excess esterification reagent and the water generated during the isobutyl ester formation must therefore be complete.

When analyzing mixtures of synthetic amino acids as their trifluoroacetyl and heptafluorobutyryl *n*-butyl esters, Schneider et al.[11] found it necessary to remove even traces of the acylating agent before chromatography, since it destroyed the chemically bonded OV-1701 phase of a capillary column. Removal of the acylating agent and acid derived from it is more difficult if synthetic mixtures are replaced by complex biological samples; evaporation then must be repeated several times after the addition of solvent.

Recovery of tyrosine during analysis of the amino acid in serum as their *N*-isobutyloxycarbonyl methyl esters was below 70%.[12] The incomplete derivatization was attributed to adsorption of tyrosine to some specific components that were not removed by the cleanup process. The problem was solved by the use of *N*-ethyloxycarbonyl methyl esters, which could be prepared by reaction with ethyl chloroformate in the same manner as with *N*-isobutyloxycarbonyl methyl esters. The recovery of tyrosine then improved.

COLUMNS

The separation of amino acid derivatives by gas chromatography is usually performed with packed columns, but more recently capillary columns have gained favor.[13] Glass capillary columns offer higher speeds of analysis, better resolution, and improved sensitivity of detection. A number of unknown peaks were found when biological samples from transfer RNA were analyzed using packed columns, these impurities making accurate quantification difficult.[14] With a capillary column separation of the individual peaks was much better and quantification of 16 amino acids was possible. Yoneda[2] has chromatographed *N*-heptafluorobutyryl amino acid *n*-propyl esters on columns packed with 1% silicone OV-1 on dimethylchlorosilane-treated Chromosorb® G, 80-100 mesh, or with 3% silicone SE-30 on Gas-Chrom® Q, 60-80 mesh. Better reproducibility was obtained using the 3% SE-30 column than the 1% OV-1 column, the effect apparently being due to the higher concentration of the silicone liquid phase. However, both columns gave satisfactory results in the analysis of protein hydrolyzates.

The gas chromatographic analysis of pentafluoropropionyl- or heptafluorobutyryl-amino acid oxazolidinones using electron capture detection gave results comparable with those obtained using heptafluorobutyryl amino acid isobutyl esters.[15] Amino acids with the perfluoroacylated side chain reactive groups gave responses about one order of magnitude greater than the others, but they tended to be adsorbed on the column filling more readily than those with only the oxazolidinone ring. Significant losses due to adsorption were found in packed columns at levels below 100 pmol per amino acid. When the chromatographic support was eliminated from the capillary columns analysis down to femtomole amounts was obtained without apparent loss of the troublesome amino acid derivatives; however, impurities began to cause serious problems at this level. The support material has an influence on the separation of acyl alkyl amino acid esters.[1] Using the same stationary phase, EGA coated on Chromosorb® W AW or on Supelcoport®, different elution patterns were obtained for the same sample. The elution pattern did not change appreciably on increasing the loading from 0.65 to 2% EGA on Supelcoport®.

DETECTORS

There have been several studies on the choice of detectors for the gas chromatographic analysis of amino acids. When analyzing serum using a flame ionization detector, difficulties may arise from unidentified contaminants in the prepared samples causing noisy baselines and possibly overlapping amino acid peaks.[13] Excessive drift of the baseline at high temperatures can occur as a result of accumulation of nonvolatile compounds from relatively dirty samples that gradually decompose in the hot injector. These problems are minimized by using an alkali flame ionization detector, with high selectivity and sensitivity for nitrogen-containing compounds.

When the amino acids in an insulin hydrolyzate were analyzed as the *N(O,S)*-heptafluorobutyryl esters on a support-coated open tubular (SCOT) column, the chromatogram obtained with the alkali flame ionization detector was "cleaner" than that obtained with the flame ionization detector.[13] The electron capture detector exhibited many extraneous peaks because of the high sensitivity of detection and the presence of derivatization products from the breakdown of amino acids during HCl hydrolysis. Serum and tissue extracts proved much cleaner to analyze by gas chromatography than the more complex urine samples. The samples giving the least satisfactory chromatograms were those obtained from urine and analyzed using electron capture detection.

MacKenzie has studied the application of a nitrogen-sensitive (alkali flame) detector to the analysis of *N*-heptafluorobutyryl amino acid isobutyl esters using a capillary column and with the emphasis on the low picomole range.[16] Baseline rise precluded the use of the detector in combination with a packed column for temperature-programmed chromatography. The precise mechanism of the detector response is not known, but the data indicate that the response is primarily to primary amino nitrogen, secondary nitrogen eliciting a lesser molar response. The results for sulfur-containing amino acids suggest that the simpler the derivative, the more likely is the relative molar response (RMR) to accurately reflect response only to nitrogen. The detector appeared to be appreciably more sensitive for amino acid detection than any other that had been described. In addition, the low response to carbon allowed direct analysis of complex samples while avoiding cleanup procedures such as ion exchange, during which recovery of individual compounds may be incomplete.

Significantly increased sensitivity (generally corresponding to more than three orders of magnitude) toward the electron capture detector as compared with the flame ionization detector was observed when amino acids were chromatographed as the *N*-heptafluorobutyryl isobutyl esters using a support-coated open-tube capillary column.[17] This was attributed to the high electron withdrawing capacity of the acyl group, which depends on the increased polarity of the carbonyl carbon atom induced by electronegativity of the fluorine atoms. The linearity of the electron capture detector was satisfactory over the 10- to 400-pg range, the detection limit being less than 1 pg for most amino acids. The detector was considered suitable for the simultaneous analysis of large numbers of amino acids in small quantities. It is, however, sensitive to bleed, baseline drift, and sample contamination.

Bonvell and Monheimer have studied the quantitative flame photometric detection of *N*-heptafluorobutyryl, isobutyl, and ethyl ester derivatives of cysteine, cystine, and methionine employing 3% OV-101 on Chromosorb® W HP, 80-100 mesh, and 3% SE-30 on Chromosorb® W HP, 100-120 mesh, as column packings.[18] The ester mixtures gave similar responses when applied to the two packings, except that cystine yielded no signal when applied to 3% SE-30. This artifact may have arisen from the derivatization procedure. The integrated peak areas of the chromatograms corresponded to the sulfur content of the amino acids, both columns giving good resolution and response down to the 10-ng level of each amino acid. The response of cysteine and methionine on the 3% SE-30 column was not satisfactorily comparable with that obtained using the 3% OV-101 column.

ENANTIOMER LABELING

While sample preparation for ion exchange chromatography is relatively simple, the preparation for gas chromatography includes additional steps such as sorption of the amino acids on a cation exchanger, evaporation of water, esterification of the carboxyl group, acylation of amino and other functional groups in trifunctional amino acids, evaporation of reagents, dissolution to an appropriate concentration, withdrawal of a sample, and injection into the chromatograph. These steps involve risks of incomplete recovery or yield and only by adherence to meticulously standardized conditions are the analyses as reproducible as ion exchange techniques. In principle these drawbacks can be overcome by the use of appropriate internal standards, but for amino acids with their greatly different chemical properties, internal standards serve principally to compensate for physical losses, without reflecting the different chemical behavior of the amino acids during the analysis procedure.[9]

A chiral stationary phase, Chirasil-Val®, has high thermal stability and separates all individual amino acids and resolves all their enantiomers. This column gave the possibility of determining the enantiomeric composition of the common amino acids, and by enantiomer labeling with the optical antipodes as internal standards, their precise amounts.[19] The accuracy and precision of amino acid determination in protein hydrolyzates, serum, urine, and cerebrospinal fluid by enantiomer labeling were equal, or for most amino acids superior, to those by ion exchange procedures. In the analysis of blood serum, enantiomer labeling produced considerable simplification of sample preparation and allowed quantitative determination of amino acids with 50 $\mu\ell$ or less of serum.[20] The analysis of human saliva using enantiomer labeling gave an accuracy claimed to be equal or superior to that of other methods.[21]

CONCLUSIONS

Despite prolonged and careful studies of the gas chromatographic technique, ion exchange chromatography is still the preferred method of analysis for amino acids in biological fluids and protein hydrolysates.[19] The equipment for ion exchange chromatography is an expensive, dedicated, liquid chromatograph, but, correctly operated, an instrument of this type yields accurate results and normally requires little sample preparation. Also, as a result of rapid progress in liquid chromatography procedures, analysis times have been considerably shortened. In routine analysis the time and reliability of results are decisive factors in selecting a particular method, these considerations being occasionally outweighed by requirements of high sensitivity or low cost. Gas chromatographic procedures are superior with regard to sensitivity and cost. The time required for chromatography is short, but sample preparation is laborious, particularly when seeking high accuracy, and overall gas chromatographic analysis is more time consuming. The accuracy and reliability of gas chromatographic methods are comparable with those of ion exchange methods only with precise adherence to standardized procedures at all stages, which demands a widely experienced analyst. Many potential users who obtained frustrating results with gas chromatography reverted to ion exchange chromatography.

Before derivatization of amino acids can be performed, biological samples require some form of purification procedure, the most widely adopted being ion exchange. A careful study of described procedures revealed a wide diversity both in reagents used and in the degree of experimental detail.[22] The authors have concluded that over the previous 10 years little progress was made in sample purification techniques, the inherent disadvantages of this methodology casting considerable doubt on the accuracy of the quantitative analysis of amino acids.

REFERENCES

1. **Gamerith, G.,** *J. Chromatogr.*, 268, 403, 1983.
2. **Yoneda, T.,** *Anal. Biochem.*, 104, 247, 1980.
3. **Yeung, J. M., Baker, G. B., and Coutts, R. T.,** *J. Chromatogr.*, 378, 293, 1986.
4. **Biermann, C. J., Kinoshita, E. M., Marlett, J. S., and Steel, R. D.,** *J. Chromatogr.*, 357, 330, 1986.
5. **Mawhinney, T. P., Robinett, R. S. R., Atalay, A., and Madson, M. A.,** *J. Chromatogr.*, 358, 231, 1986.

6. **Schwenk, W. F., Berg, P. J., Beaufrere, B., Miles, J. M., and Haymond, M. W.,** *Anal. Biochem.*, 141, 101, 1984.
7. **Husek, P., Felt, V., and Matucha, M.,** *J. Chromatogr.*, 180, 53, 1979.
8. **Husek, P.,** *J. Chromatogr.*, 234, 381, 1982.
9. **Perier, C., Ronziere, M. C., Rattner, A., and Frey, J.,** *J. Chromatogr.*, 182, 155, 1980.
10. **Moodie, I. M., Burger, J. A., Hough, B. J., Shephard, G. S., and Labadarios, D.,** *J. Chromatogr.*, 347, 179, 1985.
11. **Schneider, K., Neupert, M., Spiteller, G., Henning, H. V., Matthaei, D., and Scheler, F.,** *J. Chromatogr.*, 345, 19, 1985.
12. **Yamamoto, S., Kiyama, S., Watanabe, Y., and Makita, M.,** *J. Chromatogr.*, 233, 39, 1982.
13. **Chauhan, J., Darbre, A., and Carlyle, R. F.,** *J. Chromatogr.*, 227, 305, 1982.
14. **Lindqvist, L. and Maenpaa, P. H.,** *J. Chromatogr.*, 232, 225, 1982.
15. **Husek, P. and Felt, V.,** *J. Chromatogr.*, 305, 442, 1984.
16. **MacKenzie, S. L.,** *J. Chromatogr.*, 358, 219, 1986.
17. **Bengtsson, G., Odham, G., and Westerdahl, G.,** *Anal. Biochem.*, 111, 163, 1981.
18. **Bonvell, S. I. and Monheimer, R. H.,** *J. Chromatogr. Sci.*, 18, 18, 1980.
19. **Frank, H., Vujtovic-Ockenga, N., and Rettenmeier, A.,** *J. Chromatogr.*, 279, 507, 1983.
20. **Frank, H., Rettenmeier, A., Welcker, H., Nicholson, G. J., and Bayer, E.,** *Anal. Chem.*, 54, 715, 1982.
21. **Rotgans, J., Wodarz, R., Schoknecht, W., and Drysch, K.,** *Arch. Oral Biol.*, 28, 1121, 1983.
22. **Labadarios, D., Moodie, I. M., and Shephard, G. S.,** *J. Chromatogr.*, 310, 223, 1984.

Section II.III

THE CHROMATOGRAPHIC SEPARATION OF AMINO ACID ENANTIOMERS

The resolution of amino acid enantiomers is important in peptide synthesis and structure determinations. Chromatographic methods of analysis have the advantage of resolving not only a pair of enantiomers, but also a mixture of several amino acid enantiomers. The gas chromatographic separation of D and L amino acids using chiral stationary phases requires the derivatization of amino acids to volatile compounds. For this reason high performance liquid chromatography can be a more attractive procedure.

GAS CHROMATOGRAPHIC METHODS

Feibush et al.[1] have resolved the *N*-trifluoroacetyl isopropyl esters of α, β, and γ amino acids on the diamide stationary phases *N*-dodecanoyl-1-valine-6-undecylamide and *N*-dodecanoyl-L-valine t-butylamide. The α amino acid derivatives emerged throughout in the sequence D before L-isomer; the derivatives of the γ amino acids showed a "reversed" order of emergence, the retention of the D-isomers being larger than that of the L-isomers. The derivatives of the β amino acids exhibited a variable behavior depending on the diamide phase used. Chang et al.[2] suggested that chiral recognition by the diamide involves the formation of diastereomeric association complexes in which the solute is hydrogen bonded through the "C_5" or "C_7" side (Figure 1), considering that the isopropyl group of valine met the requirement that the substituent at the asymmetric center of the solvent molecule must be sufficiently bulky to show stereoselective effects, but not so large as to make association with the solute difficult. However, they then explored the nature of the R″ group, reporting on four new phases, I to IV, where R′ and R‴ are $C_{11}H_{23}$ and *tert*-butyl, respectively, and R″ is a nonpolar aliphatic or aromatic substituent: I (L-alanine phase), R″ = CH_3; II (L-leucine phase), R″ = iso-C_4H_9; III (D-phenylglycine phase), R″ = C_6H_5; IV (L-phenylalanine phase), R″ = $C_6H_5CH_2$.

The resolution of *N*-TFA-isopropyl esters of α, β, and γ amino acids was studied using stainless steel capillary columns coated with the different phases. For α amino acids the order of emergence throughout was the L- after the D-isomer on the L-stationary phases. The changes made in the structure of the diamide solvents did not essentially affect the nature of the resolution mechanism. Phases II and IV had the highest stereoselectivity, with only small differences in the retention times for a given α amino acid. The data for II and IV were also close to those obtained on the *N*-lauroyl-*tert*-butylamide of valine. Phase I had the lowest retention times throughout. For the resolution of individual amino acids, phases II and IV were equivalent or even superior to the diamide phases derived from valine. The increase in R″ allowed a higher operating temperature to be used than on the respective *N*-acyl-*tert*-butylamide derivatives of valine. Phases I and II, especially II, resolved proline isomers reasonably well, but the aromatic phases II and IV showed no chiral recognition for this imino acid.

β Amino acids had considerably lower retention times than α amino acids. Phase I showed no chiral recognition, while phase II gave relatively low resolution factors. The diamides III and IV were more selective in their interaction with β amino acids, this being particularly true for phase IV which was recommended for analytical applications. Phases I and II exhibited chiral differentiation for the *N*-TFA-isopropyl esters of branched γ-amino acids but not for γ-aminovaleric acids, while phases III and IV showed no stereoselectivity for these compounds.

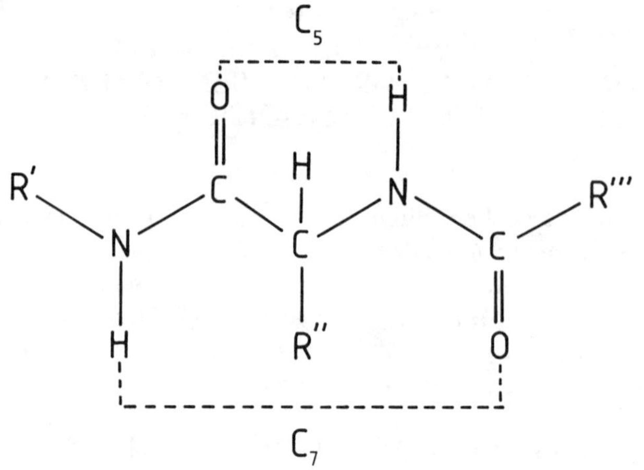

FIGURE 1. Diastereomeric association complex of a diamide.

Two diamide phases, *N*-lauroyl-L-valine *tert*-octylamide, employed at a maximum column temperature of 180°C, and *N*-docosanoyl-L-leucine *tert*-octylamide, at a maximum temperature of 220°C, showed good stereoselectivity for *N*-TFA esters of α and γ amino acids.[3] For α amino acids the order of elution throughout was the L- after the D-isomer, this order being reversed for γ amino acids. Using the *N*-lauroyl-L-valine *tert*-octylamide phase, the resolution factors on both stainless steel capillary columns and whisker-walled glass capillary columns were essentially the same. Most chiral stationary phases have poor thermal stability. The approach adopted by König and Benecke[4] to overcome this problem was to introduce chiral constituents into silicone OV-225. The cyano groups of OV-225 were reduced to amino groups using lithium aluminium hydride ($LiAlH_4$), and the amino groups were then coupled with benzyloxycarbonyl-L-valine and benzyloxycarbonyl-L-leucine. Both phases gave only poor separations of amino acid enantiomers; the separation factors for amino alcohols and amines, however, were surprisingly high. Operation of the columns at up to 130°C for several weeks did not decrease their separation efficiency.

The order of elution of amino acid enantiomers was not consistent. On the benzyloxycarbonyl-L-valine phase the derivatives of D-valine, D-isoleucine, and D-alloisoleucine had longer retention times than the corresponding L-enantiomers. With the benzyloxycarbonyl-L-leucine phase the derivatives of L-alanine and D-valine were retarded. For amines, the *S*-enantiomers had longer retention times in the case of 2-aminopentane and phenylethylamine.

König et al.[5] have used isocyanates as derivatizing reagents for amino acids and *N*-methylamino acids. α Amino acid esters are converted to *N*-ureido derivatives by isocyanates at room temperature or on brief heating in the presence of pyridine. When isopropyl isocyanate is used, a mixture of the mono- and bis-*N*-isopropylureido derivatives is obtained, while *tert*-butyl isocyanate yields the mono-derivatives exclusively. This can be explained by the steric requirements of the *tert*-butyl groups. *N-tert*-Butyl-ureido-DL-amino acid isopropyl esters were obtained in good yields and were separated into enantiomers on the chiral polysiloxane XE-60-L-valine-*(S)*-α-phenylethlamide. The reaction of isocyanates with *N*-methylamino acid esters yields ureido derivatives that are only partially separated on XE-60-L-valine-*(S)*- or -(R)-α-phenylethylamide. However, when isocyanates react with free *N*-methylamino acids at elevated temperatures, the carboxyl group is converted into an *N*-alkylamide group, while the *N*-methyl group is converted to *N*-alkylureido derivatives. The availability of an additional amide group increases enantioselective interaction with the stationary phase to provide complete enantiomer separation. The reaction of *N*-methylamino acids with isopropyl isocyanate occurs without noticeable racemization, and the corresponding *N*-isopropylureido/*N*-isopropylamides are formed in excellent yields from sample amounts as low as 100 μg. On XE-60-L-valine-(R)-α-phenylethylamide, the D-enantiomers are eluted after the L-enantiomers, in contrast to normal amino acids.

LIQUID CHROMATOGRAPHIC METHODS

Liquid chromatographic resolution of optical isomers is based either on the formation of diastereomers or on the application of chiral stationary or mobile phases.

THE SEPARATION OF DIASTEREOMERS

Marfey has synthesized N^2-(5-fluoro-2,4-dinitrophenyl)-L-alanine amide which contains a reactive fluorine atom.[6] Reaction of this compound with a mixture of D- and L-amino acids gives quantitative yields of diastereomers, which can be separated and estimated by HPLC, L-diastereomers being eluted before D-diastereomers. This behavior is probably due to stronger hydrogen bonding in the D-diastereomer producing a more hydrophobic molecule, which interacts more strongly with the reversed-phase column and therefore has a longer retention time than the L-diastereomer. The presence of a stable, highly absorbing chromophore allows the determination of diastereomers in the nanomole range. The amount of D-amino acid in the original mixture is ascertained by comparing the HPLC peak areas of the two diastereomers in the chromatogram. Aberhart et al.[7] have separated (3R)- and (3S)-β-leucine by reaction with Marfey's reagent and HPLC of the resulting diastereomers. Nimura et al.[8] have analyzed protein amino acid enantiomers by their derivatization with 2,3,4,6-tetra-*O*-acetyl-β-D-glucopyranosyl isothiocyanate to give diastereomeric thiourea derivatives. These derivatives are efficiently separated on an octadecyl silica gel column by gradient elution and are detected by their absorbance at 250 nm.

CHIRAL STATIONARY PHASES

The separation of racemates into enantiomers by means of column liquid chromatography is a rapidly developing field covering theoretical studies on the underlying retention mechanism and practical studies. Dalgleish[9] in 1952 suggested necessary criteria to enable chromatographic separation of enantiomers to occur. This suggestion indicated that for a unique spatial arrangement a chiral molecule must be associated with another chiral moiety via at least three "points". Diastereomers are thus formed by fixed positions in space, which are stereoselective. The bonds in these diastereomers need not be covalent, the interaction "points" being of a more labile nature, i.e., hydrogen bonds, hydrophobic interactions, electrostatic interactions, charge transfer, etc. The necessity for the three "points" contact has since been demonstrated experimentally.

One approach to the liquid chromatographic separation of enantiomers consists of the covalent binding of a chiral ligand to a silica support producing a stable and selective phase system. The system used by Davankov et al.[10] consisted of a reversed-phase packing coated with *N*-alkyl-L-hydroxyproline, where alkyl was *n*-heptyl, *n*-decyl, or *n*-hexadecyl, and an aqueous-organic eluent containing Cu(II) acetate. As the Cu(II) concentration was decreased at constant pH and methanol-water composition, there was a noticeable increase in retention. The enantioselectivity, however, remained unchanged. At constant Cu(II) and methanol-water composition the retention of amino acid enantiomers was increased by raising the pH above 5.5. Simultaneously, the enantioselectivity was enhanced. The high enantioselectivity of the system was attributed to a three-site sorbate-sorbent interaction involving bidentate coordination of two amino acids to a Cu(II) ion and hydrophobic attractions between hydrocarbon side chains of amino acids and the *n*-octadecyl groups of the support.

Roumeliotis et al.[11] synthesized chiral-bonded silica phases composed of L-proline and L-hydroxyproline amino linked to an *n*-propyl spacer anchored to the surface. Surface modification was brought about by reaction of the chiral silanizing reagent or by bonding the hydrocarbon spacer followed by fixation of the chiral ligand. LiChrosorb® Si 100 was reacted with L-prolylpropyltriethoxysilane (product I) and with L-hydroxyprolylpropyltriethoxysilane (product II). Product III was obtained by modifying LiChrosorb® Si 100 with 3-chloropropyltriethoxysilane and subsequent treatment with sodium iodide and L-hydroxyproline. It contained excess 3-(iodopropyl) groups. Using the three packings the separation of amino acid enantiomers was studied in eluents containing a constant concentration of Cu(II) acetate. The pH of the eluent was varied over the range 4 to 6 and the ammonium acetate concentration from 0.001 to 0.1 *M*. These were found to be the principal factors controlling the retention on and selectivity of the columns, the addition of organic solvents such as methanol, acetonitrile, or tetrahydrofuran producing only minor changes.

In later studies Roumeliotis et al.[12] examined the systematic variation of the hydrophobic spacer X in the $(Cu[L\text{-hydroxyprolyl}]X)^+$ complex bonded at the silica surface. Packings used were $(Cu[L\text{-hydroxyprolyl}]methyl)^+$ on LiChrosorb® Si 60 (1), $(Cu[L\text{-hydroxyprolyl}]methyl)^+$ on LiChrosorb® Si 100 (2), and $(Cu[L\text{-hydroxyprolyl}]n\text{-octyl})^+$ on LiChrosorb® Si 100 (3). They contained residual iodomethyl and ω-iodooctyl groups. In eluents containing 10^{-4} *M* copper acetate packing 1 exhibited a greater retention of amino acid enantiomers than packing 2, but the latter showed a superior selectivity. Retention of the enantiomers was determined by the eluent pH and methanol content as well as by the ammonium acetate concentration. Alterations in the hydrophobic spacer gave a sensitive control of retention and produced unique changes in selectivity, which were attributed to multiple solute-surface interactions, i.e., complexation, ion exchange, and hydrophobic interactions.

L-Proline-copper complexes display different affinities toward L- and D-enantiomers of amino acids depending on the method of fixation of the L-proline ligand to the stationary phase. This indicates the important role of the spacer between the fixed ligand and the sorbent matrix in the stereochemistry of ternary sorption complexes.[13] The nature of the matrix plays a subordinate role; asymmetric resins chelating Cu(II) ions to L-proline or L-hydroxyproline fixed in macroporous methacrylate copolymers loaded with Cu(II) gave chromatographic resolution of amino acid enantiomers and acceptable quantitative agreement with results obtained on silica gel sorbents with similar fixed chiral ligands.[14]

N-Decyl-L-histidine is another chiral coating that in conjunction with a Zorbax® C_8 packing or a Zorbax® ODS (C_{18}) packing resolves amino acid enantiomers.[15] Water containing trace amounts of Cu(II) ions was found to be the best eluent. With the exception of aspartic acid, L-enantiomers eluted with smaller k' values than D-enantiomers. The *n*-decyl-L-histidine coating had a markedly lower enantioselectivity than the *N*-decyl-L-hydroxyproline coating. This is not necessarily a disadvantage when determination of the enantiomeric composition of a solute is desired; an enantioselectivity exceeding $\alpha = 1.5$ decreases the precision in evaluating the ratio of two chromatographic peaks. In contrast, coatings with a higher enantioselectivity are preferable for preparative resolutions.

Berezin et al.[16] synthesized sorbents containing L-methionine-*d*-sulfoxide and L-methionine-l-sulfoxide groups on a macronet polystyrene matrix. The polymers were then loaded with copper or nickel ions. On the Cu(II)-loaded columns enantioselectivity was generally somewhat higher for the polymer containing L-l-sulfoxide groups. However, the differences in α values were mostly insignificant, the configuration at the sulfur atom in methionine sulfoxide apparently having little effect on enantioselectivity. The chromatographic results for histidine and aspartic acid using nickel ions indicated a significant influence of the sulfur atom configuration. As a rule, D-amino acid enantiomers were more strongly retained than the L-enantiomers, but histidine (copper- and nickel-loaded polymers) and aspartic acid (nickel-loaded polymers) showed the reverse elution order.

Gübitz et al.[17] have prepared stationary phases by treating silica gel with 3-glycidoxypropyltrimethoxysilane and bonding L-azetidine carboxylic acid, L-pipecolic acid, or L-phenylalanine to the product. When Cu(II) was used as the complexing agent, the elution order of amino acids was the D- before the L-enantiomer, with the exception of proline and hydroxyproline for which the order was reversed. On the phase incorporating phenylalanine, however, the elution order was the L- before the D-enantiomer.

N-2,4-Dinitrophenyl (DNP) amino acid esters can be resolved on silica gel coated with P-(+)-disodium hexahelicene-7,7′-dicarboxylate.[18] Except for DNP-phenylglycine esters, the D-isomer emerged from the column first. The enantiomers of *N*-benzoyl-DL-α-alanine have been separated on a column of albumin-agarose.[19] *N*-Aroyl-DL-amino acids have been resolved by affinity chromatography on a stationary phase of bovine serum albumin covalently bound to a 10-μm silica support.[20] In all instances k′ values increased with decreasing pH of the mobile phase. The *N*-benzoyl and *N*-naphthoyl derivatives of DL-alanine and DL-phenylalanine were completely resolved into the optical antipodes. The 2-substituent of the amino acid had a marked influence on the degree of resolution and also on the elution order of the enantiomers.

On a bonded cyclodextrin stationary phase the β-naphthylamide derivatives of DL-methionine and DL-alanine were separated, as were the β-naphthyl ester derivatives of amino acids.[21] The detection of as little as 0.2% of dansyl-D-leucine in the presence of the L-enantiomer constituted one of the more sensitive determinations of an optical impurity that had been reported. DL-Tryptophan was resolved on a column of crystalline cellulose (Avicel) using a sodium citrate buffer imidazole eluent.[22] The resolution was attributed to the "fair chirality" of crystalline Avicel®.

CHIRAL REAGENT IN THE MOBILE PHASE

To resolve optical isomers directly by chromatography, diastereomers must be produced *in situ* while the resolution is taking place. To form the diastereomers, a chiral reagent can be introduced to the mobile phase or to the stationary phase. Since the mobile phase is easier to manipulate, several workers have preferred to add the chiral reagent to it. Gilon et al.[23] chose the metal complex of the dipeptide L-aspartyl-L-phenylalanine methyl ester, available commercially as an artificial sweetener under the trade name Aspartame, as the chiral agent to add to the water-acetonitrile mixture constituting the mobile phase. Aspartame-Cu(II) or aspartame-Zn(II), in conjunction with an ODS column operated at 32°C, could be used to resolve mixtures of racemic aromatic amino acids. Gundlach et al.[24] separated the enantiomers of alkylamino acids using an aqueous eluent containing the Cu(II) complex of L-aspartyl-L-phenylalanine methyl ester at a concentration of 1 m*M*, the stationary phase being Nucleosil® 10 C-18.

In an extension of their studies, Gilon et al.[25] prepared L-aspartyl cyclohexamide (AspcHex). This compound has only one chiral center and lacks phenyl and methoxycarbonyl groups (the hydrophobic part consists of a cyclohexyl ring). Because of the low background absorbance of AspcHex, nonaromatic amino acids can be detected by virtue of the fact that they form a five-membered ring complex with copper that absorbs UV radiation at 230 nm. The presence of AspcHex-Cu(II) complex in the mobile phase thus allows the detection as well as the separation of amino acid enantiomers. A wide variety of enantiomers were separated using this reagent, and as with Aspartame, the L-enantiomer eluted before the D-enantiomer. A single asymmetric center in the chiral eluent was thus shown to be sufficient for the separation.

Gilon et al.[26] later used alkylamine derivatives of aspartic acid complexed with Cu(II) as resolving agents. The resolution obtained was a function of the chain length of the alkyl group of the agent; thus, mobile phases with L-aspartylethylamide Cu(II) exhibited lower resolution than those with L-aspartyl-*n*-hexylamide Cu(II) and k′ values and selectivity factors increase with increasing pH of the mobile phase. Most likely, optically active complexes containing a six-membered ring involving the β-carboxy and α amino groups of the aspartyl residue are formed. This six-membered ring is less stable than the five-membered ring formed by α amino and α-carboxy groups of amino acids. Thus, in a solution that contains $(AspcHex)_2Cu(II)$ an amino acid solute can replace one aspartyl group yielding a ternary complex (AspcHex)(amino acid)Cu(II). This ternary complex is essential for the separation of the amino acids. Molecular models suggest that hydrophobic interactions between the amino acid side chain and the aspartyl derivative form the third point contact and are necessary for separation. They are feasible with D-amino acids, but not with their L-isomers, explaining the observation that the D-isomers are retained longer on the column.

Ternary complexes containing Ni(II) are able to effect enantiomeric resolution, but their resolving power is less than that of similar complexes containing Cu(II). The chiral recognition system, however, is believed to be similar in both systems. Reversed-phase columns that have been loaded with chiral reagents can resolve enantiomers even when using mobile phases that do not contain chiral reagents.[27] This experiment indicates that when a chiral reagent is present in the mobile phase, it is adsorbed on the stationary phase and enantiomeric resolution takes place there. Resolution of enantiomers occurs under extreme conditions such as low pH or very low concentrations of the resolving agent, and the elution order is the same whether the chiral reagent is in the mobile phase or only in the stationary phase.

Kurganov and Davankov[28] have used Cu(II)-*N,N,N′,N′*-tetramethyl-(R)-propanediamine-1,2 (tpmn) complexes as chiral additives to the eluent for the resolution of amino acid enantiomers on reversed-phase LiChrosorb® RP-18 and normal-phase LiChrosorb® Si 100 silica gel packings. A 5×10^{-4} *M* solution of $Cu_2(tmpn)_2(OH)_2(ClO_4)_2$ in water/acetonitrile, 5:1, resolved the racemates of aromatic amino acids on reversed-phase silica gel, D-enantiomers being eluted after L-enantiomers. Almost all common racemic amino acids were resolved using the normal-phase system with water/acetonitrile, 1:9, containing the complex and additional free (R)-tpmn as eluent. Here the order of elution of the enantiomers was reversed, with aspartic and glutamic acids as exceptions. In both the reversed- and normal-phase systems the enantioselectivity values (α) were not very high, but were sufficient for the resolution of many amino acids.

The fact that in the reversed-phase system all the solutes resolved were aromatic amino acids suggests that the ternary sorption complexes of these solutes have an unusual conformation. Contrary to LiChrosorb® RP-18, unmodified LiChrosorb® Si 100 silica gel strongly retained the chiral resolving complex from its solution, which was also demonstrated by the intense blue color of the sorbent. The retention of amino acids in the normal-phase system differed considerably from that in the reversed-phase system, hydrophobic amino acids eluting first followed by hydrophilic amino acids. The overall results demonstrated the differences in the properties of ternary sorption complexes that exist on the sorbent surface and those of corresponding ternary complexes that exist in bulk solution.

Nimura et al.[29] separated D- and L-amino acids using a mobile phase containing *N*-(*p*-toluenesulfonyl)-L-phenylalanine Cu(II), (TosPhe-Cu[II]), and a column of *n*-octylsilyl-bonded silica gel, the amino acids being monitored with *o*-phthalaldehyde. Separation is dependent on the acetonitrile concentration in the mobile phase. Nimura et al. suggest that the free amino acid, TosPhe and Cu(II), form a ternary complex, the complex of the D-amino acid assuming a *trans* conformation around the Cu(II) ion and the L-isomer a *cis* conformation. This suggestion agrees with the fact that, except for acidic amino acids, the L-amino acid is eluted before the D-isomer. DL-Proline was resolved on an octadecylsilyl-bonded phase (Develosil® ODS) using TosPhe-Cu(II) as chiral additive, postcolumn monitoring being performed using 7-chloro-4-nitrobenzofurazan (NBD-Cl).[30] L-Proline was eluted before the D-isomer.

The Cu(II) complex of *N*-(*p*-toluenesulfonyl)-D-phenylglycine (TosPhG) resolved a wide range of amino acid enantiomers. Compared with the TosPhe-Cu(II) system, the α values of many amino acids increased and the elution orders for all pairs of enantiomers were reversed.[31] Nimura et al. suggested that the chiral additive is distributed on the surface of the alkyl-bonded stationary phase and may act as an immobilized chiral phase. In agreement with this view the reversed-phase column previously equilibrated by passing a tosyl amino acid-Cu(II) solution through it, retaining its resolving power for more than 5 hr when aqueous 0.5 m*M* copper sulfate was passed through it. Using TosPhG-Cu(II) as the chiral additive, simultaneous resolution of protein amino acid enantiomers was achieved by a column-switching technique using two ODS columns of different lengths and gradient elution with acetonitrile.[8] The column eluate was monitored fluorometrically following reaction with *o*-phthalaldehyde.

Weinstein et al.[32] have devised an HPLC technique for the enantiomeric analysis of the protein amino acids. The amino acid mixture is first separated into groups on a cation exchange resin using volatile aqueous pyridine buffers. The eluate is evaporated and amino acids within the groups are separated into D- and L-enantiomers on a reversed-phase column with a mobile phase containing *N,N*-di-*n*-propyl-L-alanine and cupric acetate. Postcolumn reaction with *o*-phthalaldehyde provides subnanomole sensitivity, while proline is detected using ninhydrin. Wernicke[33] has separated amino acid enantiomers by reversed-phase chromatography with an eluent containing Cu(II)-L-phenylalanine, *N*-methyl-L-phenylalanine, or *N,N*-dimethyl-L-phenylalanine complexes. The selectivity of the separation is greatly affected by pH value, the proportion of organic modifier in the mobile phase, and the ionic strength; using short columns chiral resolution is achieved in less than 20 sec.

Weinstein and Leiserowitz[34] have proposed a model for the separation of enantiomers by copper complexes of *N,N*-dialkyl-α amino acids. Molecular space filling models of the Cu(II) complex of *N,N*-di-*n*-propyl-L-alanine show a rigid molecule with little conformational freedom (Figure 2). Cu(II) complexes, in addition to having a square planar coordination, exhibit weaker interactions at the two axial positions. The model reveals that the molecule is hydrophobic on one side of the coordination plane, with the axial position on this side completely blocked by one *N*-alkyl group from each ligand molecule; on the other side of the plane the axial position is accessible to coordination. The molecule appears to be ideally shaped to be located in the interface between the mobile phase and the "organic" reversed-phase C-18 layer. Its hydrophobic side is embedded in the layer and its "cleft" containing the accessible copper atom points toward the aqueous phase. A large amount of complex is, indeed, adsorbed on the solid support.

Compounds that have been separated into enantiomers using Cu(II) complexes of *N,N*-dialkyl-α amino acids include α and β amino acids, α amino acid esters, dipeptides, α-hydroxy acids, and β-hydroxy acids. For any given *N,N*-dialkyl-α amino acid ligand the order of elution of the enantiomers is constant (e.g., D eluting before L for *N,N*-di-*n*-propyl-L-alanine) irrespective of the type of compound considered. All the compounds resolved contain the moiety 2 (Figure 3).

Association between 2 and the chiral complex 1 occurs. The carbonyl oxygen of 2 binds to the copper atom through the axial position, the NH (or OH) group is hydrogen bonded to the carboxyl oxygen of the *N,N*-dialkylamino-α amino acid, and the CHX chiral group of 2 makes van der Waals' contacts with the ligands. The association leads to diastereomeric intermediates $(CuL_2)L'$ and $(CuL_2)D'$, where L denotes ligand of the L configuration and D'- and L'-enantiomers being separated. This model explains the stability relationships between pairs $(CuL_2)L'$ and $(CuL_2)D'$; for example, with *N,N*-di-*n*-propyl-L-alanine as chiral additive, D-amino acids elute first, indicating that the L-amino acids interact more strongly with the complex immobilized on the reverse phase, and are thus retained longer.

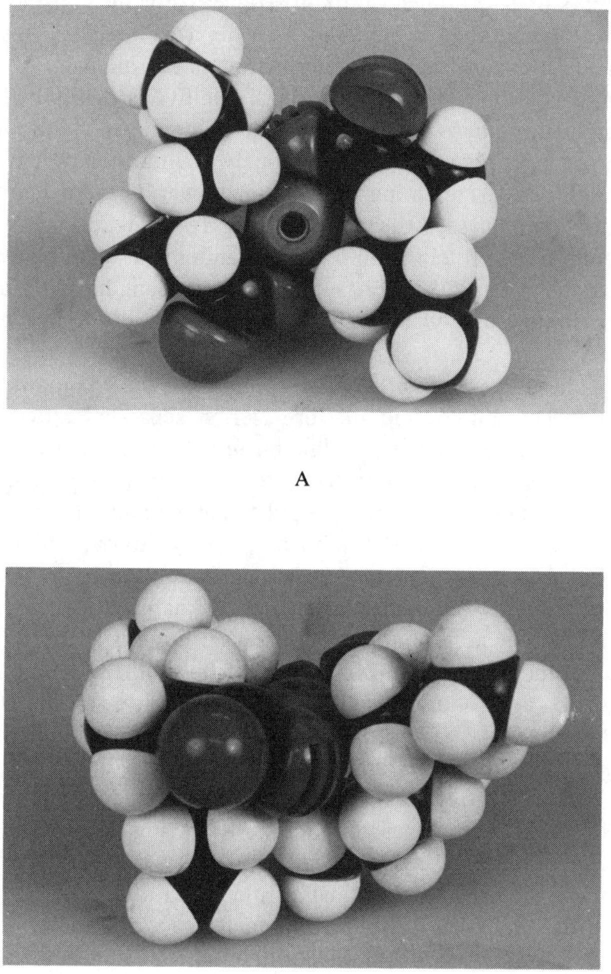

A

B

FIGURE 2. Molecular space-filling model of the Cu(II) complex of *N,N*-di-*n*-propyl-L-alanine. (A) View along axial position; (B) view perpendicular to axial position. Reference: Weinstein, S. and Leiserowitz, L., *Isr. J. Chem.*, 25, 334, 1985. (Reproduced by permission of the Weizmann Science Press of Israel.)

RESOLUTION OF THE ENANTIOMERS OF DANSYL AMINO ACIDS

Several authors have separated D- and L-dansyl amino acids using the Cu(II) complexes of amino acids in reversed-phase eluents. The use of dansyl derivatives rather than the free amino acids allows their chromatographic behavior to be studied by monitoring the fluorescence. Eluents containing Cu(II) complexes of L-proline, L-arginine, L-histidine, and L-histidine methyl ester separated many D- and L-dansyl amino acids.[35] Resolution of the optical isomers is attributed to mixed chelate complexation of the amino acid with the chiral ligand and the metal ion. The selectivity between D- and L-pairs depends on the alkyl substituent on the α-carbon atom of the amino acid; the higher the carbon content and the bulkier the alkyl group, the larger the selectivity factor. Resolution of optical isomers is dependent on pH, a change in pH of the mobile phase displacing the equilibrium distribution

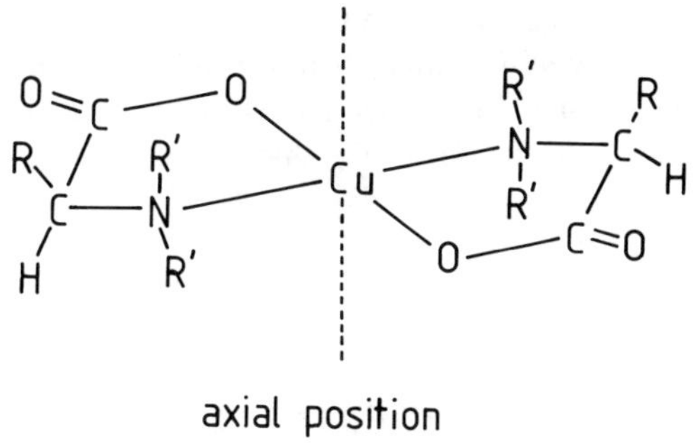

1

A

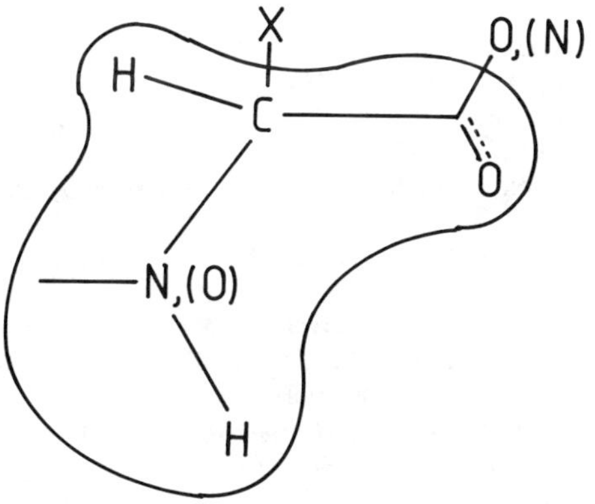

2

B

FIGURE 3. (1) Cu(II) complex of *N,N*-dialkyl-α-amino acid; (2) moiety common to compounds that are resolved using Cu(II) complexes of *N,N*-dialkyl-α amino acids.

of the metal species in solution. At pH 7 resolution of D- and L-isomers is close to unity. A gradient with a lower starting pH shows less resolution, but amino acid pairs like proline-alanine and serine-threonine are resolved. A one-column system with a controlled gradient of acetonitrile in buffer containing Cu(II) L-histidine methyl ester separated most of the dansyl amino acids except the derivatives of proline, hydroxyproline, alanine, threonine, and cysteine, and resolved their isomers in the same analysis.

Using a Cu(II)-L-proline mobile phase, the dansyl derivatives of several small polar amino acids were baseline resolved. While the employment of different column packing materials, LiChrosorb®-RP-8 and Spherisorb® C_{18}, did not affect the stereoselectivity between the D- and L-isomers, the retention and selectivity between the amino acids were changed. D-Serine eluted before the L-isomer, the reverse order of that generally observed for amino acids. The free hydroxyl group of serine may take part in coordination and alter the stability of the dansyl serine-Cu(II)-L-proline ternary complex. Almost all the enantiomers of the common D- and L-amino acids can be separated as the dansyl derivatives in a single analysis, using *N,N*-di-*n*-propyl-L-alanine and cupric acetate as the chiral additive.[36] The method is quantitative sensitive, fairly rapid, and can be adapted for routine automated analysis. By replacing *N,N*-di-*n*-propyl-L-alanine with *N,N*-di-*n*-propyl-D-alanine the order of elution can be reversed to reveal peaks not well separated with the L-ligand. In analysis of natural samples this procedure is useful in confirming that a peak is homogeneous and not contaminated by other compounds.

REFERENCES

1. **Feibush, B., Bolan, A., Altman, B., and Gil-Av, E.,** *J. Chem. Soc. Perkin Trans.*, 2, 1231, 1979.
2. **Chang, S.-C., Charles, R., and Gil-Av, E.,** *J. Chromatogr.*, 235, 87, 1982.
3. **Charles, R. and Watabe, K.,** *J. Chromatogr.*, 298, 253, 1984.
4. **König, W. A. and Benecke, I.,** *J. Chromatogr.*, 209, 91, 1981.
5. **König, W. A., Benecke, I., Lucht, N., Schmidt, E., Schulze, J., and Sievers, S.,** *J. Chromatogr.*, 279, 555, 1983.
6. **Marfey, P.,** *Carlsberg Res. Commun.*, 49, 591, 1984.
7. **Aberhart, D. J., Cotting, J.-A., and Lin, H.-J.,** *Anal. Biochem.*, 151, 88, 1985.
8. **Nimura, N., Toyama, A., and Kinoshita, T.,** *J. Chromatogr.*, 316, 547, 1982.
9. **Dalgleish, C.,** *J. Chem. Soc.*, 137, 1952.
10. **Davankov, V. A., Bochkov, A. S., Kuranov, A. A., Roumeliotis, P., and Unger, K. K.,** *Chromatographia*, 13, 677, 1980.
11. **Roumeliotis, P., Unger, K. K., Kurganov, A. A., and Davankov, V. A.,** *J. Chromatogr.*, 255, 51, 1983.
12. **Roumeliotis, P., Kurganov, A. A., and Davankov, V. A.,** *J. Chromatogr.*, 266, 439, 1983.
13. **Yamskov, I. A., Berezin, B. B., Davankov, V. A., Zolotarev, Yu. A., Dostavalov, I. N., and Myasoedov, N. F.,** *J. Chromatogr.*, 217, 539, 1981.
14. **Gübitz, G., Jellenz, N., and Santi, W.,** *J. Chromatogr.*, 203, 377, 1981.
15. **Davankov, V. A., Bochkov, A. S., and Belov, Yu. P.,** *J. Chromatogr.*, 218, 547, 1981.
16. **Berezin, B. B., Yamskov, I. A., and Davankov, V. A.,** *J. Chromatogr.*, 261, 301, 1983.
17. **Gübitz, G., Juffmann, F., and Jellenz, W.,** *Chromatographia*, 16, 103, 1982.
18. **Kim, Y. H., Balan, A., Tishbee, A., and Gil-Av, E.,** *J. Chem. Soc. Chem. Commun.*, p. 1336, 1982.
19. **Allenmark, S. and Bomgren, B.,** *J. Chromatogr.*, 252, 297, 1982.
20. **Allenmark, S., Bomgren, B., and Boren, H.,** *J. Chromatogr.*, 264, 63, 1983.
21. **Armstrong, D. W. and DeMond, W.,** *J. Chromatogr. Sci.*, 22, 411, 1984.
22. **Yuasa, S., Shimada, A., Kameyama, K., Yasul, M., and Adzuma, K.,** *J. Chromatogr. Sci.*, 18, 311, 1980.
23. **Gilon, C., Leshem, R., Tapuhi, Y., and Grushka, E.,** *J. Am. Chem. Soc.*, 101, 7612, 1979.
24. **Gundlach, G., Sattler, E. L., and Wagenbach, U.,** *Z. Anal. Chem.*, 311, 684, 1982.
25. **Gilon, C., Leshem, R., and Grushka, E.,** *Anal. Chem.*, 52, 1206, 1980.
26. **Gilon, C., Leshem, R., and Grushka, E.,** *J. Chromatogr.*, 203, 365, 1981.
27. **Grushka, E., Leshem, R., and Gilon, C.,** *J. Chromatogr.*, 255, 41, 1983.
28. **Kurganov, A. A. and Davankov, V. A.,** *J. Chromatogr.*, 218, 559, 1981.
29. **Nimura, N., Suzaki, T., Kasahara, Y., and Kinoshita, T.,** *Anal. Chem.*, 53, 1380, 1981.
30. **Nimura, N., Toyama, A., and Kinoshita, T.,** *J. Chromatogr.*, 234, 482, 1982.
31. **Nimura, N., Toyama, A., Kasahara, Y., and Kinoshita, T.,** *J. Chromatogr.*, 239, 67, 1982.
32. **Weinstein, S., Engel, M. H., and Hare, P. E.,** *Anal. Biochem.*, 121, 370, 1982.
33. **Wernicke, R.,** *J. Chromatogr. Sci.*, 23, 39, 1985.
34. **Weinstein, S. and Leiserowitz, L.,** *Isr. J. Chem.*, 25, 334, 1985.
35. **Lam, S.,** *J. Chromatogr. Sci.*, 22, 416, 1984.
36. **Weinstein, S. and Weiner, S.,** *J. Chromatogr.*, 303, 244, 1984.

Section II.IV

THE LIQUID CHROMATOGRAPHIC SEPARATION OF *o*-PHTHALALDEHYDE AMINO ACID DERIVATIVES

Ion exchange techniques of amino acid analysis have dominated the scene for many years, but are now being increasingly challenged by HPLC methods. The precolumn derivatization of amino acids using *o*-phthalaldehyde and a thiol and the rapid separation of the derivatives using reversed-phase HPLC have proved to be a precise and accurate technique of amino acid determination. The preparation of the derivatives has been extensively studied in order to determine the optimum conditions for the reaction.

PREPARATION OF THE DERIVATIVES

The reaction of *o*-phthalaldehyde with amino acids is represented by the equation

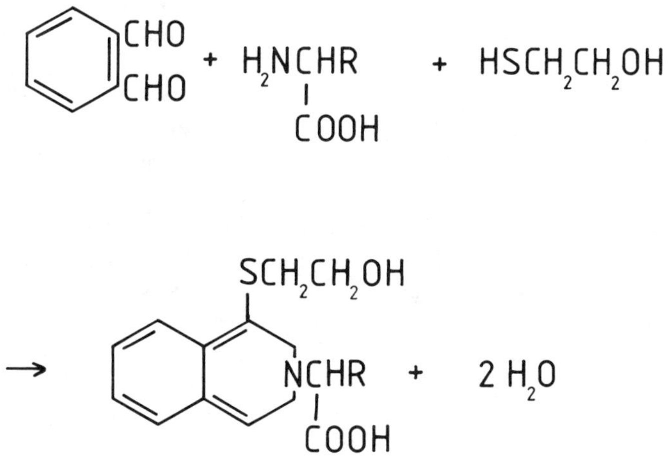

At pH values below 6 essentially no reaction occurs, while from pH 7.4 to 10.0 the yield of amino acid derivatives increases. The pH of 9.3 was chosen by Halfpenny and Brown[1] to provide satisfactory yields of acidic and neutral as well as basic amino acids, while Cooper et al.[2] found that the pH required to produce the maximum fluorescence response of the derivatives ranged from 9.5 to 11.0 for all the amino acids studied.

At pH 9.5 the rate of reaction of *o*-phthalaldehyde-2-mercaptoethanol and amino acids depends on the concentration of *o*-phthalaldehyde.[2] If the ratio of *o*-phthalaldehyde to amino acids is of the order of 10:1, immediate injection of the derivatives on to the column is necessary if losses in the fluorescence responses of glycine, ornithine, and lysine are to be avoided. If a concentration ratio of the order of 18:1 is employed, a time delay of 1 min before injection is required in order to obtain maximum derivative formation. A concentration ratio of 2-mercaptoethanol-*o*-phthalaldehyde of 2:1 is optimal for maximum fluorescence response. The molarity of the buffer used in the *o*-phthalaldehyde-2-mercaptoethanol reaction has no effect on the amino acid derivative formation. The ionic environment, however, plays an important role.[1] When the reaction is performed at pH 9 in phosphate buffer, the total fluorescence for a plasma sample is 45% of that found using sodium borate buffer.

For a fixed reaction time, the fluorescence responses of the amino acid derivatives are linear and almost independent of the composition of the derivatizing reagent as long as it is used in at least 200 times excess.[3] Nonlinear responses occur when a lower excess is employed. When physiological fluids such as urine or cerebrospinal fluid are analyzed, high concentrations of nonamino acid amines that consume reagent are often present. In these instances the sample should be greatly diluted before derivatization and should then be analyzed at a high detection sensitivity.

Lysine gives somewhat lower fluorescence than other amino acids with the *o*-phthalaldehyde reagent, but this can be improved by the inclusion of a detergent such as Brij or sodium dodecyl sulfate; the former was found to be superior to the latter.[4] Methanol or ethanol stabilizes the derivatives of all amino acids, methanol being preferred to ethanol, since the latter causes precipitation of borate buffer. One disadvantage of the use of *o*-phthalaldehyde is that cysteine produces a derivative with minimal fluorescence.[5] Oxidation of cystine or cysteine to cysteic acid or pretreatment of cysteine with iodoacetic acid produces a derivative that has a fluorescence intensity similar in magnitude to that of other amino acids on *o*-phthalaldehyde treatment.

STABILITY OF THE DERIVATIVES

The precolumn derivatization of amino acids using an *o*-phthalaldehyde reagent has many advantageous features, but the instability of the derivatives could hinder acceptance of the method. When amino acids are reacted with *o*-phthalaldehyde-2-mercaptoethanol for different times before injection onto the column, they display varying fluorescent responses as well as varying degrees of stability.[2] In particular, the derivatives of glycine, ornithine, and lysine are relatively unstable.[6] When the derivatives are injected onto the column immediately after reaction and the flow of solvent is stopped for different time intervals, most of the derivatives, including those of glycine, ornithine, and lysine, are stable. Only aspartic and glutamic acids show significant loss of fluorescence. As these diacidic amino acids are eluted within the first 4 min of chromatography, little loss of sensitivity is observed. Only the derivatives of glycine, lysine, and hydroxylysine show appreciable decay of fluorescence after 8 min of reaction, the lysine and hydroxylysine derivatives giving a very low response.[7]

CHOICE OF THE THIOL FOR THE DERIVATIZING REAGENT

When the preparation of *o*-phthalaldehyde amino acid derivatives was investigated using ethanethiol, 2-mercaptoethanol, or 3-mercaptopropionic acid as the thiol in the reagent mixture, maximum UV intensity was achieved after 130, 240, and 400 sec, respectively.[8] The best UV response and stability was obtained using 3-mercaptopropionic acid; degradation of amino acids are apparent when employing 2-mercaptoethanol. The relative fluorescence of individual amino acids using 3-mercaptopropionic acid was in accord with values derived using 2-mercaptoethanol. 3-Mercapto-1-propanol appeared to be a superior thiol to 2-mercaptoethanol; its use afforded detection limits of less than 200 fmol of amino acid.[9]

SEPARATION SYSTEMS

A number of chromatographic systems, often automated, have been developed for the separation of amino acid *o*-phthalaldehyde derivatives.

Columns

C_8 and C_{18} columns of the same dimensions, prepared using the same alkyl bonding and endcapping chemistries, showed the same relative retention of all amino acid derivatives and gave adequate resolution under correct conditions.[4] The C_8 column gave rather better resolution of glycine and threonine when tetrahydrofuran was absent from the elution buffers and was preferred for this reason. Other considerations also influenced the choice of the C_8 column; it could be reequilibrated to the starting conditions in less time than the C_{18} column and after 200 hr of analyses showed less peak shape degradation, peak tailing, and loss of peak intensity. Three different types of Altex Ultrasphere® columns — octadecylsilane (ODS), octadecylsilane/ion pairing (ODS IP), and cyanopropylsilane (CN) — were evaluated by Jones et al.[7] Although the amino acid derivatives were resolved on the CN column, variable peak heights and retention times were obtained and its use was discontinued. The ODS and ODS IP columns gave identical results.

A Waters Fatty Acid column gave a chromatograph showing 14 skewed peaks for 18 amino acids, while a LiChrosorb® RP-8 column gave 15 symmetrical peaks.[10] A Waters μBondapak® column showed 16 symmetrical peaks having two sets of amino acids not resolved. On this column the use of an acetonitrile-phosphate buffer gradient resolved the phenylalanine-isoleucine pair which was not resolved with an earlier methanol-pH 7.2 phosphate buffer system.

Difficulties that may be encountered by variations between columns are exemplified by the studies of Larsen and West.[11] The *o*-phthalaldehyde-amino acid derivatives were separated on a μBondapak® fatty acid analysis column using a linear gradient of acetonitrile. Resolution was sufficient to allow quantitation using peak height measurement, permitting quantitative analysis to be performed using a basic HPLC system. An Altex Ultrasphere® column did not give satisfactory resolution, while two different μBondapak® C_{18} columns exhibited different behavior. The authors suggested that the endcapping procedure used during production of the μBondapak® fatty acid analysis column enabled it to provide satisfactory results. To obtain the same separation using columns possessing "active" silica within the column may require a greatly different set of chromatographic conditions.

Particle Size of the Packing

A decrease in the length of a 5-μm particle size ODS column from 250 to 150 or 45 mm allowed amino acid derivative mixtures to be resolved in a shorter time and at low column pressures.[12] The decrease in column pressure permitted greater flexibility in selecting the column flow rate, thus making further decreases in analysis time feasible. An alternative approach to decreasing analysis time was to replace the 5-μm-particle-sized column with a 3-μm-particle-sized column. An Ultrasphere® ODS column (75 × 4.6 mm I.D.) and a Microsorb® C-18 (100 × 4.6 mm I.D.) column both gave analysis times of less than 20 min. The time could be reduced to less than 15 min by slight modifications of the gradient. Graser et al.[13] found that at a constant flow rate of 1.5 mℓ/min the use of a 3-μm-particle-sized column gave a marked improvement in resolution over a 4-μm-particle-size column. The excellent separation achieved between aspartic and glutamic acids and between arginine and tyrosine, as well as among isoleucine, phenylalanine, and leucine, was noteworthy.

Bhown and Bennett[14] demonstrated that it is feasible to use only one column to achieve all the different aspects of structural analysis of low molecular weight proteins. Using a mini ODS 5-μm Ultrasphere® column, 0.46 × 4.5 cm, and linear gradients of different solvents, they separated the α and β chains of hemoglobin and their tryptic peptides, then performed amino acid analysis by precolumn derivatization with *o*-phthalaldehyde, and finally identified PTH amino acids. Certain advantages are claimed for the method: the minicolumn is about half the price of long and medium-sized columns and does not require a precolumn, the operating pressure is low, the peaks are sharper with comparable resolution, and in most cases the separation time is reduced. Umagat et al.[15] have described an approach to total amino acid analysis whereby primary amino acids are treated with *o*-phthalaldehyde in the presence of mercaptoethanol and the derivatives separated on an ODS reversed-phase column. Cystine and cysteine are similarly determined after oxidation with performic acid to cysteic acid. Proline and hydroxyproline are reacted with 4-chloro-7-nitrobenzofurazan and their separation performed on the same column. Most amino acids showed detection limits below 1 pmol; for proline the limit was about 6 pmol.

The pH of the Eluent

The effect of pH on the separation of amino acid *o*-phthalaldehyde derivatives on a C-8 column can be stated quite simply: all amino acids elute at longer times as the pH is decreased, but the effect is much greater for amino acids with a carboxylic side chain.[4] The pH range over which these effects are observed is consistent with the acid dissociation of side chain carboxyl groups. Presumably, as the pH is decreased the acid side chains become protonated (unionized), thus strengthening interactions with the C-8 stationary phase and increasing retention time. At pH 5.7 using a C-8 (5-μm Adsorbosphere®) column and a linear methanol gradient, acidic and polar amino acids are eluted first, followed by short, alkyl side chain amino acids and finally by the more hydrophobic amino acids. As ionic strength of the eluent is increased, retention times increase for amino acids with an acidic side chain, increase to a lesser extent for neutral amino acids, and increase slightly or not at all for most basic amino acids. This suggests that while the separation is primarily reversed-phase, an ion-repulsion effect is also present. Since ionic strength affects arginine, histidine, and ammonia differently from other amino acids, this parameter is very useful when separations involving these compounds have to be improved.

Flow Rate

In an attempt to reduce analysis time, Graser et al.[13] employed a short 4-μm-particle size (LiChrospher®) reversed-phase column and an acetonitrile gradient, flow rate and gradient being varied. A flow rate of 1.2 mℓ/min was associated with a small reduction in analysis time, but resolution was also reduced compared with that obtained using a longer column. Increasing the flow rate to 1.5 mℓ/min with no change in the gradient did not decrease retention time or improve the resolution. However, a simultaneous increase in gradient slope and flow rate produced a marked decrease in analysis time with no appreciable change in resolution. By keeping the higher flow rate and further increasing the gradient the analysis time was shortened by 75% compared with the initial experiment. The additional increase in gradient appeared to be accompanied by improved resolution.

Temperature

Amino acid-*o*-phthalaldehyde derivatives eluted more rapidly as the temperature of a reversed-phase C-18 column was increased to 40°C.[1] The overall peak symmetry improved, giving better resolution for closely eluting peaks. Above 40°C the resolution of these amino acids decreased. An operating temperature of 40°C was therefore chosen.

Automated Systems

Winspear and Oaks[16] developed an automated procedure to prepare *o*-phthalaldehyde derivatives of amino acids as a preliminary to their HPLC analysis. A microprocessor-controlled pump supplies *o*-phthalaldehyde, while amino acid samples are pumped with an automatic sampler. Amino acids and reagent are combined in a mixing tee just before injection onto the column. In the opinion of these authors, the major cause of variability in the automated precolumn derivatization system probably arises from errors in the mixing ratios of sample and reagent. In the system of Buck and Krummen[17] the sample and reagent volumes are controlled very precisely by two loop injection valves, reaction occurring on the high pressure side of the HPLC system in a packed bed reactor. The reaction time is determined by the flow rate. The derivatization is very reproducible rendering the method very precise and making internal standards unnecessary. Similar results are obtained using 3-mercaptopropionic acid or ethanethiol instead of 2-mercaptoethanol in the derivatization reagent.

The on-line derivatization method of Fleury and Ashley[18] employs a Resolve® 5-μm spherical packing stainless steel column (Waters) and eliminates errors arising from spontaneous decay of the amino acid derivatives. The chromatographic run lasts 50 min and under the gradient conditions used gives excellent and reproducible separation of 21 amino acids found in serum and other physiological fluids. The automated device of Hodgin et al.[19] simultaneously mixes amino acids with *o*-phthalaldehyde reagent and at once injects the fluorogenic derivatives into a Spheri 5® RP-18 column. The derivatives are separated in under 20 min with a peak area precision of better than ±2.00% relative standard deviation. The fluorometric response is linear from 100 to 500 pmol of sample, the minimum detectable quantity being 100 fmol.

REFERENCES

1. **Halfpenny, A. P. and Brown, Ph. R.,** *J. High Res. Chromatogr. Chromatogr. Commun.*, 8, 243, 1985.
2. **Cooper, J. D. H., Ogden, G., McIntosh, J., and Turnell, D. C.,** *Anal. Biochem.*, 142, 98, 1984.
3. **Lindroth, P. and Mopper, K.,** *Anal. Chem.*, 51, 1667, 1979.
4. **Jarrett, H. W., Cooksy, K. D., Ellis, B., and Anderson, J. M.,** *Anal. Biochem.*, 153, 189, 1986.
5. **Cooper, J. D. H. and Turnell, D. C.,** *J. Chromatogr.*, 227, 158, 1982.
6. **Winspear, M. J. and Oaks, A.,** *J. Chromatogr.*, 270, 378, 1982.
7. **Jones, B. N., Paobo, S., and Stein, S.,** *J. Liquid Chromatogr.*, 4, 565, 1981.
8. **Godel, H., Graser, T., Földi, P., Pfaender, P., and Fürst, P.,** *J. Chromatogr.*, 297, 49, 1984.
9. **Stobaugh, J. F., Repta, A. J., Sternson, L. A., and Garren, K. W.,** *Anal. Biochem.*, 135, 495, 1983.
10. **Hill, D. W., Walters, F. H., Wilson, T. D., and Stuart, J. D.,** *Anal. Chem.*, 51, 1338, 1979.
11. **Larsen, B. R. and West, F. G.,** *J. Chromatogr. Sci.*, 19, 259, 1981.
12. **Jones, B. and Gilligan, J. P.,** *J. Chromatogr.*, 266, 471, 1983.
13. **Graser, T. A., Godel, H. G., Albers, S., Földi, P., and Fürst, P.,** *Anal. Biochem.*, 151, 142, 1985.
14. **Bhown, A. S. and Bennett, J. C.,** *Anal. Biochem.*, 137, 256, 1984.
15. **Umagat, H., Kucera, P., and Wen, L.-F.,** *J. Chromatogr.*, 239, 463, 1982.
16. **Winspear, M. J. and Oaks, A.,** *J. Chromatogr.*, 270, 378, 1982.
17. **Buck, R. H. and Krummen, K.,** *J. Chromatogr.*, 303, 238, 1984.
18. **Fleury, M. O. and Ashley, D. V.,** *Anal. Biochem.*, 133, 330, 1983.
19. **Hodgin, J. C., Howard, P. Y., Ball, D. M., Cloete, C., and DeJager, L.,** *J. Chromatogr. Sci.*, 21, 503, 1983.

Section II.V

THE CHROMATOGRAPHIC SEPARATION OF PHENYLTHIOHYDANTOIN AMINO ACIDS

Edman degradation is the preferred technique for determining the primary structure of a protein or peptide.[1] During this process, the amino-terminal residue of the protein is sequentially removed and converted to the phenylthiohydantoin (PTH) amino acid. The PTH derivative can then be identified by thin layer chromatography (TLC), gas chromatography (GC), or high pressure liquid chromatography (HPLC) on reversed-phase supports. HPLC is probably the most commonly used method.

CHROMATOGRAPHIC SYSTEMS

Extensive studies have been made to determine the best chromatographic conditions for the separation of PTH amino acids. Under isocratic conditions using LiChrospher® SUPER CH-8, 4 μm, all the common PTH amino acids were baseline separated.[2] On a Bakerbond® cyano column, an Altex® ODS column, and a Bakerbond® C_{18} column, 20 PTH amino acids were not completely resolved, whereas Bakerbond® wide-pore diphenyl columns from different lots all gave complete separation.[3] A step gradient system using an Altex Ultrasphere® ODS column gave excellent resolution, short analysis time, and easy identification of peaks.[4] Tarr[5] developed an analytical system using isocratic buffered acetonitrile, an ODS stationary phase, and a programmed flow that increased during the run. All the common PTH amino acids, including several derivatives of cysteine and lysine and the methyl esters of aspartic and glutamic acids, were separated using a 6-, 8-, or 9-min program. Kolbe et al.[6] used gradient elution with two solvents, the actual gradient being directly measured in a blank run by monitoring the decreasing absorbance of trifluoracetic and acetic acids, constituents of the solvents, at 214 nm. By thus showing the actual gradient during elution as opposed to the programmed gradient, they made guidelines available for fine tuning the separation of PTH amino acids on Ultrasphere® ODS (C_{18}) columns. A reversed-phase C_{18} column eluted with a concave ethanol gradient in ammonium acetate separated PTH amino acids in under 30 min.[7] PTH arginine and PTH histidine could be placed almost anywhere in the chromatogram by small changes of the conductivity; it was therefore necessary to standardize the ionic strength of the buffer by conductivity titrations.

The analytical system of Bhown and Bennett[8] employed two solvents: A made up of 750 to 1000 μℓ of acetic acid and 350 μℓ of acetone to 1 ℓ of water, and B being methanol containing 900 μℓ of triethylamine to 1 ℓ. The computer-controlled precise mixing of solvents A and B achieved accurate pH and avoided the pH adjustment of a buffer. Cunico et al.[9] found that a Micropak® CN column was better for resolving the later eluting peaks methionine, isoleucine, and tyrosine, while a Micropak® C_{18} column best resolved the early eluting components between cystine and methionine. A stationary phase was therefore synthesized that had both CN and alkyl character, adjustment of the relative molar ratios of CN and alkyl carbon providing a column with maximum selectivity.

The absolute retention of PTH amino acids varied only slightly among three columns, dimethyl-*n*-octyl (C_8), dimethylcyanopropyl (CN), and dimethylphenethyl (PE).[10] Lower retentions were found on the CN column than on the C_8 and PE columns for most PTH amino acids under a given set of mobile phase conditions, despite the fact that all the phases had about the same bonded-phase coverage. This was to be anticipated, since a cyano phase is generally less hydrophobic (weaker stationary phase in reversed-phase chromatography) than alkyl phases. Utilizing relatively simple statistical methods, retention data for PTH amino acids on these columns were used to predict optimum separation conditions for combinations of stationary phase/mobile phase gradients. Optimum separation was obtained using a PE column with a ternary gradient containing phosphate buffer, methanol, and tetrahydrofuran.

VARIATIONS BETWEEN COLUMNS

Considerable differences in chromatographic behavior have been observed between often supposedly similar columns used for the separation of PTH amino acids. A Hibar® column with a Merck RP-18 packing employing irregular-shaped 10-μm-diameter silica and a Dupont Zorbax® ODS column employing "spherical" silica 5 to 6 μm in diameter had similar compositions, but different separation characteristics.[11] Treatment of μBondapak® columns with dimethyldichlorosilane eliminated slight variations between batches of support.[12] For the separation of the PTH derivatives, Hawke et al.[13] utilized a Zorbax® ODS column, but the system had to be abandoned when a new batch of Zorbax® ODS was introduced. An Ultrasphere® ODS column and a trifluoroacetate-acetate buffer in the solvent system then gave excellent peak sharpness, separation, and reproducibility.

MICROBORE COLUMNS

One method of obtaining high sensitivity analysis of PTH amino acids is the use of microbore HPLC columns to achieve increased solute detectability by reducing peak dilution. Since peak volume is proportional to column cross-sectional area, reduction of column internal diameter can produce a marked increase in peak concentration for the same sample mass. Several studies have demonstrated that microbore systems can indeed be used for the determination of PTH amino acids at high sensitivity.[2,9,14] Silver et al.[15] have compared a large-volume mobile phase injection and a large-volume "noneluting" solvent injection using microbore columns. A 20-μℓ mobile phase injection containing PTH amino acids showed a lack of resolution and peak tailing. A similar injection using a "noneluting" solvent exhibited greatly increased resolution which was attributed to a concentrating effect, whereby the sample remained at the head of the column and was not eluted before injection was complete.

CONCLUSIONS

Glajch and Kirkland[10] have compared the performance of an optimum gradient-elution separation with that of an isocratic separation. The isocratic analysis offered a shorter analysis time, could be performed using simpler equipment, and offered better reproducibility. On the other hand, the gradient elution system allowed better resolution for many of the early eluting peaks. Of more practical importance, additional "open" space was available in the chromatogram for possible artifacts and impurities that can occur in samples. Finally, the potential for higher sensitivity exists, especially for later eluting compounds, because the peak width in gradient elution is essentially constant during the separation and peak height does not decrease (for equal sample concentration) as a function of retention time, as occurs in isocratic separation.

REFERENCES

1. **Edman, P. and Begg, G.,** *Eur. J. Biochem.,* 1, 80, 1967.
2. **Lottspeich, F.,** *J. Chromatogr.,* 326, 321, 1985.
3. **Kruggel, W. G. and Lewis, R. V.,** *J. Chromatogr.,* 342, 376, 1985.
4. **Black, S. D. and Coon, M. J.,** *Anal. Biochem.,* 121, 281, 1982.
5. **Tarr, G. E.,** *Anal. Biochem.,* 111, 27, 1981.
6. **Kolbe, H. V. J., Lu, R. C., and Wohlrab, H.,** *J. Chromatogr.,* 327, 1, 1985.
7. **Fohlman, J., Rask, L., and Peterson, P. A.,** *Anal. Biochem.,* 106, 22, 1980.
8. **Bhown, A. S. and Bennett, C. J.,** *Anal. Biochem.,* 150, 457, 1985.
9. **Cunico, R. L., Simpson, R., Correia, L., and Wehr, C. T.,** *J. Chromatogr.,* 336, 105, 1984.
10. **Glajch, J. L. and Kirkland, J. J.,** *J. Chromatogr. Sci.,* 25, 4, 1987.
11. **Rose, S. M. and Schwartz, B. D.,** *Anal. Biochem.,* 107, 200, 1980.
12. **Henderson, L. E., Copeland, T. D., and Oroszlan, S.,** *Anal. Biochem.,* 102, 1, 1980.
13. **Hawke, D., Yuan, P.-M., and Shively, J. E.,** *Anal. Biochem.,* 120, 302, 1982.
14. **Godtfredsen, S. E. and Oliver, R. W. A.,** *Carlsberg Res. Commun.,* 45, 35, 1980.
15. **Silver, M. R., Trosper, T. D., Gould, M. R., Dickinson, J. R., and Desotelle, G. A.,** *J. Liquid Chromatogr.,* 7, 559, 1984.

Section II.VI

THE CHROMATOGRAPHIC SEPARATION OF DANSYL AMINO ACIDS

1-*N,N'*-Dimethylaminonaphthalene-5-sulfonyl (dansyl) derivatives have been widely used in qualitative amino acid analysis, the determination of the amino acid terminus of peptides, and for manual microsequence determination. The preparation of dansyl amino acid derivatives is rapid and under carefully controlled conditions is essentially quantitative. Recent studies have demonstrated that these derivatives can be analyzed with good selectivity and detection sensitivity by both normal and reversed-phase HPLC.

THE DANSYLATION OF AMINO ACIDS

Tapuhi et al.[1] devised a dansylation procedure that permits high yields of mono- and didansyl derivatives. The yield is independent of the ratio of dansyl chloride to amino acid, at least over a 1000-fold range (Table 1). The method is convenient and can be used with primary and secondary amino acids as well as amines. A decrease in the dansyl amino acid content of analysis samples contained in glass vials and exposed to daylight for prolonged times was observed. The low recovery of dansyl tyrosine from a chromatogram was avoided when samples were analyzed within 1 day after derivatization while protected from direct light.[2] However, for greatest accuracy the sample should be injected onto the column as soon as possible after derivatization.

THE EFFECT OF pH ON CHROMATOGRAPHIC SEPARATION

Dansyl amino acids show marked changes in retention time and selectivity on reversed phases as the pH of the eluent is altered. Over the pH range 2.0 to 8.0 the changes in retention to alkylsilicas reflect changes in the ionization state of the dimethylamino group and the α- and side chain carboxyl groups.[3] At pH values above the pK of the dimethylamino group (pK = 4.07) fully derivatized dansyl amino acids behave as weak acids, their capacity factors decreasing as the pH is increased up to about pH 8.0. Below about pH 3.5 where the α- and side chain carboxyl groups are essentially unionized, retention decreases due to the increased polarity of the dansyl amino acid protonated dimethylamino group. These opposing pH dependencies produce retention maxima near pH 3.5 to 4.0.

At pH 2.3 dansyl amino acids eluted from μBondapak®-alkylphenyl, C_{18}, and C_{30} columns of equivalent dimensions in the order of their amino acid side chain hydrophobicities essentially in three groups, acidic amino acid derivatives showing the least retention, neutral aliphatic amino acid derivatives second, and finally the didansyl derivatives. At this low pH value dansyl-asparagine is retained less then dansyl-aspartic acid and dansyl-glutamine elutes before dansyl-glutamic acid. ε-Dansyl-lysine and dansyl-arginine exhibit relatively short retention times, but coelute with dansyl-aspartic acid or dansyl-glutamic acid, respectively.

COLUMNS

The resolution of dansyl amino acids on a μBondapak® alkylphenyl support was in several ways superior to that on a μBondapak® C_{18} support, notably in terms of peak shape that permitted lower detection limits and better reproducibility from column to column.[3] At pH 7.8 on a Hypersil® ODS column using a 200-m*M* NH_4HCO_3-acetonitrile gradient, all the common dansyl derivatives of amino acids were well resolved with an analysis time of under

Table 1
THE REACTION OF DANSYL CHLORIDE WITH AMINO ACIDS

Amino acid	Yield at 35 min (%)	Maximum yield (%)	Time required[a]
Alanine	100	100	16
Arginine	97	97	35
Asparagine	78	90	84
Aspartic acid	90	99	50
Cysteic acid	110	110	37
Cystine	109	113	16
Glutamic acid	90	93	55
Glutamine	102	104	38
Histidine	110	125[b]	25
Isoleucine	102	103	32
Leucine	90	95	19
Lysine	98	100	24
Methionine	97	97	26
Phenylalanine	98	98	35
Proline	93	96	6
Serine	95	97	40
Threonine	97	102	55
Tryptophan	92	98	7
Tyrosine	98	99	21
Valine	98	98	35

Note: Reactions conditions are acetonitrile (33.3%), water (66.7%), Li_2CO_3 (26.7 μmol/mℓ), pH 9.5, dansyl chloride/amino acid, 10.

[a] Time required to reach maximum percentage yield.
[b] High value may be due to an impure standard.

REFERENCE

1. **Tapuhi, Y., Schmidt, D. E., Lindner, W., and Karger, B. L.,** *Anal. Biochem.*, 115, 123, 1981.

90 min. Although absolute retention times could vary by up to about 15 sec from run to run, no difficulties were experienced in making peak assignments or identification. Koroleva et al.[4] have separated dansyl amino acids employing gradients of pH and acetonitrile concentration in the eluent, ultrahigh sensitivity being ensured by the use of capillary microcolumns, 0.5-mm internal diameter, packed with Silasorb®-300-C_{18} sorbent and a fluorometer with a cell volume of 1.3 μℓ. Of several C_8 and C_{18} packings tested, the only one to give acceptable separation of the dansyl derivatives of normal protein amino acids was a Brownlee® RP-300, 10-μm resin.[2] A methyl ethyl ketone-2-propanol-based gradient was employed which gave maximum separation between dansyl histidine and dansyl cysteic acid and polar by-products which eluted with the void volume of the column.

THE DETECTION OF DANSYL AMINO ACIDS

Dansyl amino acids are highly fluorescent compounds and are readily detected with high sensitivity by fluorescence techniques. An excitation wavelength at 330 nm can be employed to produce a fluorescence emission above 500 nm. However, the UV absorbance of dansyl amino acids at 248 nm is at least three times greater than the absorption at 330 nm, and these derivatives can, hence, be detected by their absorption at 248 nm.[5] UV detection is satisfactory for analysis at the 100- to 200-pmol level, but for studies with smaller amounts the sensitivity must be increased by using fluorescence methods. deJong et al.[2] separated dansyl amino acids on a Brownlee Aquapore® RP-300 column employing a gradient of methyl ethyl ketone-2-propanol. The absorption properties of this solvent system are such that elution of the dansyl derivatives can be followed between 320 and 340 nm, the variations in peak/area ratios for the different dansyl amino acids being much less pronounced than when comparing fluorescence yields. The lower limits for which the dansyl derivatization techniques are applicable are defined by the amount of contamination present in or on solutions and surfaces that are in contact with the sample during handling; amounts with which it is convenient to work are in the 100- to 300-pmol range.

REFERENCES

1. **Tapuhi, Y., Schmidt, D. E., Lindner, W., and Karger, B. L.,** *Anal. Biochem.,* 115, 123, 1981.
2. **DeJong, C., Hughes, G. J., van Weiringen, E., and Wilson, K. J.,** *J. Chromatogr.,* 241, 345, 1982.
3. **Grego, B. and Hearn, M. T. W.,** *J. Chromatogr.,* 255, 67, 1983.
4. **Koroleva, E. M., Maltseo, V. G., Belenkii, B. G., and Viska, M.,** *J. Chromatogr.,* 242, 145, 1982.
5. **Oray, B., Lu, H. S., and Gracy, R. W.,** *J. Chromatogr.,* 270, 253, 1983.

Section II.VII

THE CHROMATOGRAPHIC SEPARATION OF PHENYLTHIOCARBAMYL AMINO ACIDS

A recently described approach to the precolumn derivatization and analysis of amino acids is based on the formation of the phenylthiocarbamyl derivatives. The technique was first used by Koop et al.[1] for the analysis of free amino acids liberated by carboxypeptidase Y digestion of peptides from cytochrome P-450. The phenylthiocarbamyl derivatives are obtained by reaction of the amino acids with phenylisothiocyanate, the reaction being essentially complete in a few minutes at ambient temperature.[2] Reaction times as short as 5 to 10 min can be used on a standard amino acid mixture with little change in yield.[3] After the reagent is removed from the reaction mixture under vacuum, the derivatized amino acids can be stored dry and frozen for several weeks without appreciable degradation.

Ebert has separated phenylthiocarbamyl amino acids on a 3-μm Spherisorb® ODS-2 column, using gradient elution with solvent A containing sodium acetate plus triethylamine adjusted to pH 6.40 and solvent B consisting of 50% solvent A, 40% acetonitrile, and 10% methanol.[4] Using the same gradient program, variation of the triethylamine concentration within relatively narrow limits affected column selectivity in a predictable manner and provided a rational basis for optimizing chromatographic conditions. Both octyl (C8) and octadecyl (C18) columns were found to effectively separate the derivatives; solvent systems used were based on aqueous ammonium acetate with acetonitrile, methanol, or both as organic solvent.[3] Phenylthiocarbamyl amino acids have been separated on Pico-Tag® columns, the solvent system consisting of two eluents, an aqueous buffer and 60% acetonitrile in water.[2] At the 250-pmol level all the amino acids were well resolved, the technique showing excellent reproducibility.

A Pico-Tag® octadecylsilane column, particle size 5 μm, in the presence of 0.1% sodium dodecyl sulfate (SDS) was used by Shoji et al.[5] A solvent gradient consisting of two eluents, A, pH 6.4 acetate buffer containing 6% acetonitrile and 0.05% triethylamine and B, 60% acetonitrile in water, was employed. This method may be suitable for the microanalysis of the hydrolyzates of peptides and proteins separated and stained on SDS-polyacrylamide slab gels. Scholze found that coupling of amino acids with phenylisothiocyanate is complete within 3 to 5 min, but thorough removal of HCl after hydrolysis by repeated evaporation of the sample with the coupling buffer is absolutely necessary.[6] The amino acids in protein hydrolyzates were separated in 40 min on octadecylsilyl LiChrosphere® and Spherisorb® columns. Scholze suggested that amino acid analysis by reversed-phase HPLC will become an alternative to the more costly ion exchange method.

Quantitation of phenylthiocarbamyl amino acids is simplified by the finding that with the exception of lysine, all the derivatives have nearly identical molar extinction coefficients.[3] Lysine with its two amino groups gives a value about twice as high. Heinrikson and Meredith[3] claim that the use of a UV absorbing reagent overcomes difficulties associated with fluorescent derivatives.

REFERENCES

1. **Koop, D. R., Morgan, E. T., Tarr, G. E., and Coon, M. J.,** *J. Biol. Chem.*, 257, 8472, 1982.
2. **Bidlingmeyer, B. A., Cohen, S. A., and Tarvin, T. L.,** *J. Chromatogr.*, 336, 93, 1984.
3. **Heinrikson, R. L. and Meredith, S. C.,** *Anal. Biochem.*, 136, 65, 1984.
4. **Ebert, R. F.,** *Anal. Biochem.*, 154, 431, 1986.
5. **Shoji, S., Ichikawa, M., Yamaoka, T., Funakoshi, T., and Kubota, Y.,** *J. Chromatogr.*, 354, 463, 1986.
6. **Scholze, H.,** *J. Chromatogr.*, 350, 453, 1985.

Section II.VIII

THE CHROMATOGRAPHIC SEPARATION OF DIMETHYLAMINOAZOBENZENESULFONYL (DABS)-AMINO ACIDS

Chang et al.[1] demonstrated in 1981 that amino acids labeled with dimethylaminoazobenzenesulfonyl chloride (DABS-Cl) could be separated using a Zorbax® ODS column and detected in the visible region. A technique described later which gave a complete baseline separation of DABS-amino acids employed a LiChrosorb® RP-18 column operated at 50°C in conjunction with a complex gradient based on dimethylformamide, acetonitrile, and phosphate or acetate buffer.[2] Although different columns such as Supelcosil® LC-18 and Vydac® C-18 gave complete separation of DABS-amino acids, Knecht and Chang preferred a Merck C-18 column, principally because of its lower cost.[3]

Winkler et al.[4] attempted to optimize the chromatographic conditions so as to obtain complete separation of DABS-amino acids at room temperature. This study involved the analysis of different kinds of organic anion in the eluent (acetate, formate, trifluoroacetate), ion-pairing agents (dodecylamine, dodecyl sulfate), organic solvents (acetonitrile, ethanol, methanol), pH, flow rate, gradient forms, and columns (alkyl phenyl, μBondapak®-phenyl, 5 and 3 μm Ultrasphere® ODS). Best results were obtained using a 5-μm ODS column and pH 4.11 sodium acetate buffer with acetonitrile as organic solvent. This column successfully separated all the amino acids, including leucine and isoleucine, by a single gradient. The most critical parameter was pH. Optimum pH conditions had to be ascertained for each special column, and also for each new column from the same manufacturer.

It is claimed that amino acid analysis obtained with the DABS-Cl method at the picomole level is just as reliable as that obtained with a standard amino acid analyzer at the nanomole level.[1] Additionally, the DABS-Cl method can analyze imino acids together with amino acids at the same level of sensitivity. The DABS-Cl method requires no postcolumn derivatization apparatus and the same instrument can be used for peptide isolation. Moreover, DABS-amino acids are stable compounds and unusual amino acid derivatives can be recovered for further study.

REFERENCES

1. **Chang, J.-Y., Knecht, R., and Braun, D. G.,** *Biochem. J.*, 199, 547, 1981.
2. **Chang, J.-Y., Knecht, R., and Braun, D. G.,** *Biochem. J.*, 203, 803, 1982.
3. **Knecht, R. and Chang, J.-Y.,** *Anal. Chem.*, 58, 2375, 1986.
4. **Winkler, G., Heinz, F. X., and Kunz, C.,** *J. Chromatogr.*, 297, 63, 1984.

Section II.IX

THE CHROMATOGRAPHIC SEPARATION OF 4-*N,N*-DIMETHYLAMINOAZOBENZENE-4′-THIOHYDANTOIN (DABTH)-AMINO ACIDS

The success of the Edman method for the sequential degradation of peptides has led to the synthesis of modified isothiocyanate reagents designed to improve the ease and sensitivity of the procedure. One such reagent, 4-*N,N*-dimethylaminoazobenzene-4′-isothiocyanate (DABITC) is capable of the stepwise degradation of minute quantities of peptides, producing colored 4-*N,N*-dimethylaminoazobenzene-4′-thiohydantoin (DABTH)-amino acids.[1] The use of DABITC as a substitute for phenylisothiocyanate in the Edman degradation has several advantages. DABTH-amino acids have an absorption maximum at 436 nm. This obviates interference from UV-absorbing impurities and baseline rise due to alterations of the solvent gradient during HPLC. Moreover, the extinction coefficients are about twice those of PTH amino acids.

All the DABTH-amino acids, except those of leucine and isoleucine, were satisfactorily separated by TLC on a two-dimensional polyamide sheet. Silica gel plates, however, successfully discriminated between the derivatives of leucine and isoleucine.[2] The sensitive azo group allowed the detection of DABTH-amino acids as red spots down to picomole amounts. Thin layer techniques, however, only give qualitative results. Chang et al.[3] reported the quantitative analysis of DABTH-amino acids by HPLC on an RP-8 column. Resolution of three DABTH-amino acid pairs, namely, the derivatives of threonine and glutamine, methionine and proline, and leucine and isoleucine, however, was not achieved. Yang and Wankel[4] separated all the common DABTH-amino acids on a Spherisorb® ODS column, using a solvent mixture of sodium acetate buffer and 1% ethylene dichloride in acetonitrile. DABTH-Lysine and DABTH-arginine did not leave the column. The quantitative determination of the derivatives at picomole concentrations in the relatively short time of 30 to 40 min was possible.

Foriers et al.[5] demonstrated that an Altex Ultrasphere® ODS column operated at 32°C efficiently separated the common DABTH-amino acids. Solvent A was 10 m*M* phosphate buffer, pH 6.6, containing 23% methanol and 0.5% *n*-butanol, while solvent B was methanol. A gradient of 48 to 67% solvent B in 9 min was established, isocratic elution then being performed for 21 min. Winkler et al.[6] found that the use of ion-pairing agents was particularly effective in separating DABTH-amino acids. Decisive factors in an efficient separation were the addition of 2% triethylamine to the eluent and the use of a 3-μm Ultrasphere® ODS column.

REFERENCES

1. **Chang, J.-Y., Creaser, E. H., and Bentley, K. W.,** *Biochem. J.*, 153, 607, 1976.
2. **Chang, J.-Y.,** *Anal. Biochem.*, 102, 384, 1980.
3. **Chang, J.-Y., Lehmann, A., and Wittmann-Liebold, B.,** *Anal. Biochem.*, 102, 380, 1980.
4. **Yang, C.-Y. and Wakil, S. J.,** *Anal. Biochem.*, 137, 54, 1984.
5. **Foriers, A., Lauwereys, M., and De Neve, R.,** *J. Chromatogr.*, 297, 75, 1984.
6. **Winkler, G., Heinz, F. X., and Kunz, C.,** *J. Chromatogr.*, 297, 63, 1984.

Section II.X

THE DETERMINATION OF PROLINE AND HYDROXYPROLINE BY DERIVATIZATION WITH 4-CHLORO- OR 4-FLUORO-7-NITROBENZOFURAZAN (NBD-Cl AND NBD-F)

The hydroxyprolines are the most characteristic amino acids found in collagens. There are two position isomers, 4-hydroxyproline which is the most abundant, and 3-hydroxyproline which occurs only in minor amounts and only in some forms of collagens.[1] 4-Hydroxyproline is a marker for all types of collagens, while 3-hydroxyproline is a marker for basement membrane collagen. Additionally, some noncollagen proteins such as elastin, acetylcholine esterase, and the lung protein alveolyn contain small amounts of 4-hydroxyproline. The development of sensitive methods of determination of proline and hydroxyprolines is thus of considerable importance. High performance liquid chromatography (HPLC) or ion exchange chromatography (IEC) followed by fluorometric detection has proved a sensitive way of determining amino acids in biological materials. The method, however, has the drawback that secondary amino acids such as proline and hydroxyproline do not react with a typical fluorogenic reagent, *o*-phthalaldehyde. Alternative procedures for the determination of these imino acids have therefore been devised. This section describes one of these techniques, derivatization with 4-chloro- or 4-fluoro-7-nitrobenzofurazan (NBD-Cl or NBD-F).

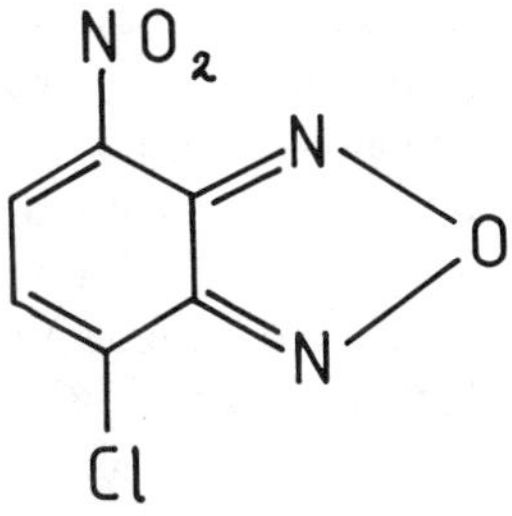

THE PREPARATION OF NBD DERIVATIVES

The reaction rates of proline and hydroxyproline with NBD-Cl are one order of magnitude higher than those of primary amino acids; hydroxyproline reacts quantitatively with NBD-Cl under milder conditions than do primary amines.[2] A shorter reaction time therefore favors the selective derivatization of secondary imino acids. A yield of about 95% is obtained when hydroxyproline reacts with NBD-Cl in methanolic solution; in acetonitrile or acetone lower yields of 65 and 15%, respectively, are obtained. Similar observations have been made with proline. The derivatization reaction of 4-hydroxyproline showed a linear relationship with NBD-Cl concentration, before plateauing at 2.0 m*M* NBD-Cl.[3] This indicates that for hydroxyproline concentrations as high as 1 m*M*, NBD-Cl at 2.0 m*M* is not rate limiting. Reaction for 20 min at 37°C produced a high rate of imino acid derivatization while minimizing the amino acid reaction rate.

Umagat et al.[4] have studied the kinetics of the reaction of proline with NBD-Cl, relating fluorescent peak height to reaction time. On reaction for 5 min the maximum fluorescent response was reached at temperatures of 60 to 80°C. Since the reagent blank peak increased as the temperature increased, NBD-proline reactions were performed at 60°C where interference from the reagent was minimal. The proline derivative response was linear over the 5- to 25-nmol/mℓ range. This derivative is very stable, especially when the reaction is carried out in the dark.

Yoshida et al.[5] have used NBD-Cl derivatization for the determination of both primary amino and secondary imino acids. The detection limit of glycine was found to be 30 pmol, while that of proline was 1 pmol. Imai and Watanabe[6] have claimed that the fluorogenic reaction of proline with NBD-F is superior in terms of reactivity and fluorescence yield, to the reactions with the analogous NBD-Cl and NBD-Br compounds. With NBD-F the reagent blank fluorescence can be suppressed by adjusting the pH of the medium to about 1 with HCl. Many secondary imino acids can be determined by reaction with NBD-F at pH 7.5 and 70°C for 5 min, with subsequent acidification to pH 1.

FLUORESCENCE DETECTION OF NBD-AMINO ACIDS

The high quantum efficiency in medium polar solvents and the high absorptivity means that fluorescence is a very suitable method for the detection of compounds such as NBD-hydroxyproline.[2] The quantum efficiencies are similar to those for dansyl- and bansyl-amino acids. However, the decrease in quantum efficiency when changing from isobutyl methyl ketone to water as solvent is much greater than the decrease observed for DNS-tryptophan when changing from dioxane to water. Thus, for NBD derivatives in aqueous solutions fluorescence detectors may not be more sensitive than UV detectors, although their selectivity is greater.

CHROMATOGRAPHIC TECHNIQUES

Yoshida et al.[5] have described the application of NBD-Cl to the postcolumn detection of picomole quantities of primary and secondary amino acids. Amino acids were eluted from a column of TSK IEX 215 (sulfonated porous polystyrene resin, 5 μm) with 0.07 *M* citrate buffer, pH 3.3 at 52°C. The column eluate was mixed with 0.4 *M* borate buffer, pH 10.5, to adjust the pH to 8.3 and then reacted with ethanolic NBD-Cl reagent (250 mg/ℓ) in Teflon tubing for 1.5 min at 70°C. After reaction the solution was mixed with an equal volume of organic solvent containing 1 *M* HCl before measurement of the fluorescence. Watanabe and Imai[7] used NBD-F as a precolumn fluorescent labeling reagent for high performance liquid chromatography of amino acids, including proline and hydroxyproline. Derivatization was carried out at pH 8.0 and 60°C for 5 min, the NBD derivatives of aspartic acid, glutamic acid, hydroxyproline, serine, glycine, threonine, alanine, and proline being separated on a μBondapak® C_{18} column with 0.1 *M* phosphate buffer, pH 6.0, containing 6.75% methanol and 1.8% tetrahydrofuran. Amino acids could be detected at the 10-fmol level. In a further study, 18 amino acids derivatized with NBD-F were well separated on a μBondapak® C_{18} column in a number of solvent systems.[8]

EXPERIMENTAL PROCEDURES

Reaction with NBD-Cl

A sample of a protein hydrolyzate obtained using methanesulfonic acid is neutralized with a 4 *N* NaOH.[4] An aliquot of the sample is then diluted, when necessary, to obtain a final concentration of about 20 nmol/mℓ of proline. Equal volumes of the sample 0.4 *M* borate buffer (pH adjusted to 9.5 with 4 *N* NaOH) and NBD-Cl solution (concentration 2 mg/mℓ in MeOH) are combined, and the mixture is then heated for 5 min at 60°C in a closed screw-capped vial. After cooling the mixture to 0°C to stop the reaction, aliquots of 100 μℓ are injected into the ODS column.

Reaction with NBD-F

An aqueous solution (0.1 mℓ) containing up to 10 nmol of amino acids and 2.8 mℓ of 0.1 *M* borate buffer, pH 7.5, containing 50% ethanol are well mixed.[6] Then 0.1 mℓ of a

1-m*M* NBD-F solution in ethanol is added, and after heating at 70°C for 5 min the reaction mixture is cooled immediately with ice water and allowed to stand for a few minutes at room temperature; 0.1 mℓ of 3 *M* HCl is then added and the fluorescence intensity is measured at 530 nm with excitation at 470 nm.

The Thin Layer Chromatographic Determination of Hydroxyproline in Protein Hydrolyzates

The NBD technique for the determination of hydroxyproline isomers is far more sensitive than many ion exchange methods, allowing for the accurate measurement of picomole amounts of amino acids. For this to be achieved the reaction must be performed in the absence of mineral ions. Primary amino acids, although their fluorescence is 10 to 100 times lower than that of hydroxyprolines, may interfere with the determination of the latter when they occur in very large amounts or when a solution has to be greatly concentrated to improve the sensitivity. Bellon et al.[9] have described a rapid procedure involving two stages of IEC for completely desalting solutions or urines before derivatization of the proline or hydroxyproline. In addition, the procedure is completely specific in that prior to the formation of the fluorophore by reaction with NBD-Cl any remaining primary amines are destroyed by reaction with *o*-phthalaldehyde.

Procedure (1): Desalting

Collagen, protein, or a peptide mixture (for instance, originating from urine) is hydrolyzed with 6 *M* HCl at 105°C for 18 hr and the hydrolyzate evaporated to dryness. The residue is dissolved in 1 mℓ of 0.20 *M* citrate buffer, pH 2.2, and a 0.20-mℓ aliquot layered on the top of a minicolumn of ion exchange resin. This column is made by drilling a 1-mm-diameter hole through the bottom of a conical plastic Eppendorf 1.5-mℓ microtest tube. The hole is covered with glass wool and 0.8 g of M 72 (Beckman) resin, equilibrated with 0.20 *M* citrate buffer, pH 3.2, is introduced. The Eppendorf microtube is forced into the top of an 8-mℓ glass centrifuge tube of the same internal diameter and spun in a centrifuge fitted with horizontal tubes. The resin is washed four times with the same buffer and centrifuged each time. In the case of urine analysis, all pigments and the basic amino acids remain on the column which can be regenerated by 1 mℓ of 0.2 *M* NaOH followed by 3 mℓ of 0.20 *M* citrate buffer, pH 3.2.

The effluent from the minicolumn is then passed through a 4 × 0.8-cm column of Dowex® W-50-X2 in the H^+ cycle. Proline and hydroxyproline are bound and after several washes with distilled water are eluted with 15 mℓ of 2 *M* NH_4OH. The solution is then evaporated to dryness under nitrogen. The residue is carefully dissolved in a minimum of 2 *M* NH_4OH and is transferred to an Eppendorf conical test tube. The chromatographic tube is washed with a second volume of 2 *M* NH_4OH, which is added to the first one. The contents of the Eppendorf tube are then evaporated to dryness. The final residue can be dissolved in only 20 μℓ of distilled water.

Procedure (2): Derivatization and Thin Layer Chromatography

The *o*-phthalaldehyde reagent is prepared by dissolving 8 mg in 1 mℓ of 30% triethylamine solution; 20 μℓ of the reagent are added to 20 μℓ of the amino acid solution, the mixture being shaken gently and allowed to stand at room temperature for 5 min. The derivatizing solution is prepared by dissolving 600 mg of NBD-Cl in 2 mℓ of chloroform. The chloroform solution is washed with 0.01 *N* HCl until the yellow color of the aqueous phase disappears. The chloroform is evaporated to dryness under a stream of nitrogen, the purified NBD-Cl is dissolved in 100 mℓ of ethanol, and the solution stored at 4°C in the dark, under which conditions it is stable for several weeks. This reagent (20 μℓ) is added to the mixture of amino acids and *o*-phthalaldehyde, and the mixture is allowed to react at 65°C for 30 min in the dark. A 5-μℓ aliquot of the NBD-amino acids is spotted on a silica gel plate, which is developed with chloroform/triethylamine/methanol, 80:10:10. The plate is dried at 65°C for 5 min and the fluorescent spots are scanned in a spectrofluorometer equipped with a thin-layer plate recording device. The surface areas of recorded peaks are measured and compared with standards deposited on the same plate. There is a linear relationship between the area of the peak and the amount of hydroxyproline in the range of 1 to 400 pmol.

REFERENCES

1. **Bisker, A., Pailler, V., Randoux, A., and Borel, J. P.,** *Anal. Biochem.,* 122, 52, 1982.
2. **Ahnoff, M., Grundevik, I., Arfwidsson, A., Fonselius, J., and Persson, B.-A.,** *Anal. Chem.,* 53, 485, 1981.
3. **Lindblad, W. J. and Diegelmann, R. F.,** *Anal. Biochem.,* 138, 390, 1984.
4. **Umagat, H., Kucera, P., and Wen, L.-F.,** *J. Chromatogr.,* 239, 463, 1982.
5. **Yoshida, H., Sumida, T., Masujima, T., and Imae, H.,** *J. High Resolution Chromatogr. Chromatogr. Commun.,* 5, 509, 1982.
6. **Imai, K. and Watanabe, Y.,** *Anal. Chim. Acta,* 130, 377, 1981.
7. **Watanabe, Y. and Imai, K.,** *Anal. Biochem.,* 116, 471, 1981.
8. **Imai, K., Watanabe, Y., and Toyo'oka, T.,** *Chromatographia,* 16, 214, 1982.
9. **Bellon, G., Berg, R., Chastang, F., Malgras, A., and Borel, J. P.,** *J. Chromatogr.,* 278, 167, 1983.

Section III

Detection Reagents

Section III.I

DETECTION METHODS FOR AMINO ACIDS AND AMINES

This section describes general and specific detection and determination methods for amino acids and amines, often those designed for the detection and determination of specific compounds.

1. The determination of D-amino acids
2. The use of amino acid oxidase for the postcolumn determination of amino acids
3. The determination of imino acids in human blood plasma
4. The resolution of amino acid enantiomers by thin layer chromatography
5. Micellar enhanced fluorometric determination of dansyl and *o*-phthalaldehyde derivatives of amino acids
6. The fluorescence detection of amino acids
7. The electrochemical detection of amino acid-*o*-phthalaldehyde derivatives
8. Prediction of the retention of phenylthiohydantoin amino acids
9. The determination of *N*-acetylcysteine in human plasma
10. The determination of cysteine sulfinic acid and cysteic acid in rat brain
11. The determination of *S*-sulfocysteine in urine by HPLC
12. The fluorescence detection of cystine by *o*-phthalaldehyde derivatization
13. Cysteine — a potential source of error in amino acid analysis
14. The determination of free trimethyllysine
15. The determination of D- and L-aspartic acid by HPLC
16. The determination of methionine in crude plant materials by gas chromatography
17. The gas chromatographic determination of methylhistidine isomers
18. The chromatographic separation of (3R)- and (3S)-β-leucine as diastereomeric derivatives
19. The determination of 5-pyrrolidone-2-carboxylic acid in tissue homogenates
20. The determination of dimethylglycine in biological fluids
21. The rapid microscale determination of tryptophan
22. Microelectrodes for the quantitative determination of methylamines
23. The detection of tertiary aliphatic amines
24. The reaction of histamine with *o*-phthalaldehyde
25. The gas chromatographic determination of histamine and its basic metabolites
26. The chromatographic determination of histamine and N^{τ}-methylhistamine using electrochemical detection
27. The fluorometric estimation of histamine
28. The determination of cystamine by HPLC

1. THE DETERMINATION OF D-AMINO ACIDS

1-Fluoro-2,4-dinitrophenyl-5-L-alanine amide (FDNP-L-Ala-NH_2) contains a reactive fluorine atom. Its reaction with a mixture of L- and D-amino acids gives a quantitative yield of diastereomers that can be separated and estimated by HPLC. This enables the relative amounts of D- and L-isomers in the amino acid mixture to be determined.

Procedure

An aqueous solution (50 m*M*) of D- and L-amino acids is used; 50 μℓ (2.5 μmol) of the solution is placed in a 2-mℓ plastic microcentrifuge tube. To it is added 100 μℓ of a 1% acetone solution of FDNP-L-Ala-NH_2 (1 mg, 3.6 μmol) followed by 20 μℓ of 1 *M* $NaHCO_3$

(20 μmol). The contents are mixed and heated on a hotplate at 30 to 40°C for 1 hr with frequent mixing. After cooling to room temperature, 10 μℓ (20 μmol) of 2 *M* HCl is added and the contents are mixed and dried in a vacuum desiccator over NaOH pellets. The residue is dissolved in 0.5 mℓ of dimethylsulfoxide and an aliquot used for HPLC.

REFERENCE

1. **Marfey, P.,** *Carlsberg Res. Commun.*, 49, 591, 1984.

2. THE USE OF AMINO ACID OXIDASE FOR THE POSTCOLUMN DETERMINATION OF AMINO ACIDS

Kiba and Kaneko[1] have used immobilized amino acid oxidase as a postcolumn reactor in the HPLC of amino acids. An L-amino acid oxidase catalyzes the deamination of L-amino acids.

$$\text{L-amino acid} + O_2 + H_2O = \text{2-keto acid} + NH_3 + H_2O_2$$

Coimmobilized amino acid oxidase-peroxidase was prepared by covalent coupling to an aminoaryl derivative of controlled pore glass (CPG) beads and packing the enzymes into a column. The column reactor was placed in a continuous-flow system for HPLC and used for the simultaneous determination of tyrosine, phenylalanine, tryptophan, and methionine. The eluent from the chromatographic column passed through the reaction column thermostatted at 40°C and then to a spectrofluorometer, the excitation and emission wavelengths being 315 and 425 nm, respectively. Tyrosine, phenylalanine, and methionine in the range 0.25 to 5 nmol and tryptophan in the range 0.5 to 5 nmol can be determined with a coefficient of variation not exceeding 10%.

REFERENCE

1. **Kiba, N. and Kaneko, M.,** *J. Chromatogr.*, 303, 396, 1984.

3. THE DETERMINATION OF IMINO ACIDS IN HUMAN BLOOD PLASMA

Tsuchiya et al.[1] have described the specific determination of free proline and hydroxyproline in human blood plasma by HPLC. The plasma sample is reacted with formaldehyde, and then the imino acids are converted to their dansyl derivatives and subjected to chromatography.

Procedure

A 500-μℓ vol of aqueous ethanol is added to 30 μℓ of plasma and the mixture is shaken vigorously. After centrifuging at 7000 × *g* for 15 min the supernatant is evaporated to dryness. The residue is dissolved in 100 μℓ of 0.1 *M* phosphate buffer, pH 6.8, containing 10% formaldehyde by sonication for 2 min. Reaction of compounds containing primary amino groups with formaldehyde then occurs simultaneously with dissolution of the residue. Imino acids are then derivatized with dansyl chloride. A 50-μℓ vol of 0.5 *M* $NaHCO_3$ and 100 μℓ of a solution of dansyl chloride in acetone (2 mg/mℓ) are added to the solution and the mixture is kept in the dark at 37°C for 20 min. A 50-μℓ aliquot of the resulting solution is injected onto the chromatograph column. Any insoluble material is removed by centrifugation before injection. Chromatography is performed on a 250 × 4.0-mm stainless steel column packed with LiChrosorb® RP-18, particle size 5 μm, operated at 50°C. Gradient elution is carried out with two solvents, the first being 10% acetonitrile in 50 m*M* acetate buffer, pH 4.0, and the second being 70% acetonitrile in water. Dansyl amino acids are detected spectrofluorometrically. Percentage recoveries of proline and hydroxyproline added to plasma are 108 and 95%, respectively.

REFERENCE

1. **Tsuchiya, H., Hayashi, T., Tatsumi, M., Fukita, T., and Takagi, N.,** *J. Chromatogr.*, 339, 59, 1985.

4. THE RESOLUTION OF AMINO ACID ENANTIOMERS BY THIN LAYER CHROMATOGRAPHY

Weinstein[1] has described the separation of enantiomers of the dansyl derivatives of all the protein amino acids, except proline, by thin layer chromatography (TLC).

Procedure

Reversed-phase TLC plates, 10 × 20 cm (Merck) or 5 × 20 cm (Whatman), are developed prior to application of the dansyl amino acids, in 0.3 *M* sodium acetate in acetonitrile/water, 40:60, adjusted to pH 7 with acetic acid (buffer A). After fan drying, the plates are immersed in a solution of 8 m*M* *N,N*-di-*n*-propyl-L-alanine and 4 m*M* Cu(II) acetate in acetonitrile/water, 97.5:2.5, for 1 hr and up to overnight, or sprayed with the solution and left to dry in the air. The plates are stable and can be stored for future use. Dansyl amino acids in aqueous solution are applied to the plate and developed using buffer A with or without *N,N*-di-*n*-propyl-L-alanine (4 m*M*) and Cu(II) acetate dissolved in it. The enantiomers are detected by irradiating with UV light (360 nm), which causes them to be revealed as yellow-green fluorescent spots. Dansylic acid produced during derivatization of the amino acids has a blue fluorescence. The acetonitrile concentration of the solvent may be varied according to the amino acids to be studied: 97.5% acetonitrile is preferred for glutamic and aspartic acids, serine, and threonine, and 40% acetonitrile for other amino acids. Quantitative results may be obtained by measuring the fluorescence or UV absorption of the extracted spots.

REFERENCE

1. **Weinstein, S.,** *Tetrahedron Lett.*, 25, 985, 1984.

5. MICELLAR-ENHANCED FLUOROMETRIC DETERMINATION OF DANSYL AND *o*-PHTHALALDEHYDE DERIVATIVES OF AMINO ACIDS

Dansyl chloride and *o*-phthalaldehyde are two of the most widely used fluorometric reagents for the determination of amino acids. The fluorescence yields of the resulting amino acid derivatives are greatly dependent on the composition of the solvent in which they are dissolved. As the dielectric constant of the solvent is decreased, fluorescence enhancements of 2 to 10 are observed for dansyl amino acids.[1,2] The intensity of fluorescence of *o*-phthalaldehyde derivatives is similarly increased by 1.7 to 17 times in the presence of aqueous dimethylsulfoxide mixtures.[3] This solvent-induced enhanced fluorescence does not usually imply a more sensitive determination procedure, because the increased fluorescence is more than offset by the dilution produced by addition of the cosolvent to the aqueous amino acid solution.

The use of aqueous micellar systems as an alternative to mixed solvent systems in fluorometric amino acid assays has been explored.[4] On reaching a certain minimum concentration, the critical micelle concentration, amphiphilic surfactant molecules tend to associate in aqueous solution to form molecular aggregates (micelles). The microenvironment of a solute associated with a micellar system may be greatly different from that in a homogeneous solvent system. For many solutes in micelles significant increases in fluorescence have been observed. Specifically, the fluorescence intensity of dansyl glycine was remarkably enhanced in the presence of hexadecyltrimethylammonium chloride, dodecyltrimethylammonium chloride, and *N*-dodecyl-*N,N*-dimethylammonium-3-propane-1-sulfonic acid micellar systems. Similarly, the lysine derivative of *o*-phthalaldehyde-2-mercaptoethanol showed intensified fluorescence in the presence of Brij-35, Triton® X-100, and sodium dodecyl sulfate micelles. Fluorescence enhancements of from 8 to 20 were observed in comparison with that of water alone, and the sensitivity of fluorometric methods for the determination of these derivatives was correspondingly increased.

Procedure (1): The Determination of Dansyl Glycine

To a measured aliquot of dansyl glycine solution is added sufficient hexadecyltrimethylammonium chloride to ensure the formation of micelles, i.e., about 1.3×10^{-3} *M*. After mixing, the fluorescence intensity is measured, using an excitation wavelength of 350 nm and an emission wavelength of 510 nm. A blank containing only the surfactant solution is run simultaneously. Dansyl glycine can be determined in the 1.0×10^{-9} to 3.2×10^{-5} *M* concentration range by comparison with known standards or from calibration graphs obtained under the same experimental conditions.

Procedure (2): The Determination of Lysine

A 2-mℓ aliquot of the aqueous lysine solution is buffered to pH 10.10 using sodium tetraborate. To this solution are added 30 μℓ of a solution of *o*-phthalaldehyde in ethanol (10 mg/mℓ), 15 μℓ of a solution of 2-mercaptoethanol in ethanol (10 μℓ/mℓ), and 100 μℓ of a concentrated solution of surfactant. Standard calibration graphs are prepared by varying the lysine concentration by serial dilution of a concentrated stock solution. Lysine can be determined in the 1×10^{-8} to 4×10^{-5} *M* concentration range.

REFERENCES

1. **Chen, R. F.,** *Arch. Biochem. Biophys.,* 120, 609, 1967.
2. **Froehlich, P. M. and Murphy, L. D.,** *Anal. Chem.,* 49, 1606, 1977.
3. **Chen, R. F., Scott, C., and Trepman, E.,** *Biochim. Biophys. Acta,* 576, 440, 1979.
4. **Singh, H. N. and Hinze, W. L.,** *Analyst,* 107, 1073, 1982.

6. THE FLUORESCENCE DETECTION OF AMINO ACIDS

When using fluorescence for the detection of amino acids a light source is generally used for the excitation of fluorophores. As a way to excite the fluorophore instead of irradiation, reaction of oxalic esters with hydrogen peroxide has been proposed. Kobayashi and Imai[1] studied the applicability of this technique to the determination of dansyl amino acids following their separation by HPLC. Bis(2,4,6-trichlorophenyl)oxalate (TCPO) was used as the oxalic ester because it is stable and easy to prepare, and high quantum yields are obtainable with its use. The concentration of TCPO and hydrogen peroxide and the constituents of the reaction medium all affected the intensity and duration of the chemiluminescence of dansyl amino acids. Optimal conditions for the detection of these compounds eluted from a column of μBondapak® C_{18} were determined. The sensitivity of the method, in which 10 fmol of dansyl amino acids is detectable, is much greater than that of the conventional fluorescence method. Miyaguchi et al.[2] then separated dansyl amino acids on a reversed-phase column, TSK, ODS-120A eluted with imidazole-nitrate buffer and acetonitrile. Using the chemiluminescence reaction the detection limit for each amino acid was 2 to 5 fmol. In a later development the use of a microbore column, 250 × 1 mm I.D. of ODS-2, 10 μm, lowered the detection limits for the dansyl amino acids to 2×10^{-16} mol.[3]

REFERENCES

1. **Kobayashi, S.-I. and Imai, K.,** *Anal. Chem.*, 52, 424, 1980.
2. **Miyaguchi, K., Honda, K., and Imai, K.,** *J. Chromatogr.*, 303, 173, 1984.
3. **Miyaguchi, K., Honda, K., and Imai, K.,** *J. Chromatogr.*, 316, 501, 1984.

7. THE ELECTROCHEMICAL DETECTION OF AMINO ACID-*o*-PHTHALALDEHYDE DERIVATIVES

The sensitivity of ion exchange methods for amino acid determination has been increased by the use of fluorogenic reagents, of which *o*-phthalaldehyde-mercaptoethanol is outstanding because of its low cost, lack of intrinsic fluorescence, and stability in water. *o*-Phthalaldehyde derivatives of amino acids can be resolved on reversed-phase columns, and the technique of precolumn derivatization with *o*-phthalaldehyde and reversed-phase HPLC with fluorescence detection has been used to determine amino acids in biological materials.

Electrochemical detection is a sensitive and specific method for compound with intrinsic electrochemical activity, but most amino acids with the exception of tryptophan, tyrosine, and cysteine are not electrochemically active within the useful potential range. *o*-Phthalaldehyde derivatives of amino acids contain an isoindole grouping, and in view of the electrochemical activity of isoindoles such as tryptophan, the possibility exists of detecting these derivatives electrochemically. Joseph and Davies[1] showed that this was the case, and demonstrated the HPLC analysis of amino acids using fluorometric and electrochemical detectors in series.

A 15 × 4.6-cm column packed with Hypersil® ODS, 5 μm, a guard column, and phosphate buffers with methanol as organic modifier were used. Under isocratic conditions a group of slower running (i.e., more hydrophobic) *o*-phthalaldehyde-amino acid derivatives was separated in about 20 min. Each peak on the fluorescence detector had an accompanying electrochemical peak with the working electrode set at 1.0 V. For resolution of the full range of amino acids, gradient elution was employed. Each fluorescence peak was again accompanied by an electrochemical peak. The *o*-phthalaldehyde derivatives were active at quite low operating potentials of the electrochemical detector. The ratio of fluorescent-to-electrochemical activity at a given potential can be used to characterize different amino acids. *o*-Phthalaldehyde derivatives of some amino acids, such as cysteine, kynurenine, and 3-hydroxykynurenine, show little or no fluorescence. These amino acids, however, do form derivatives that can be detected electrochemically. Electrochemical detection at settings that give approximately equal sensitivity to fluorometric detection for tryptophan gives 3 to 8 times more sensitivity than fluorometric detection for dipeptides and 8 to 12 times more sensitivity for tetrapeptides. Electrochemical detection thus offers promise for the determination of the *o*-phthalaldehyde derivatives of small peptides following HPLC separation.

REFERENCE

1. **Joseph, M. H. and Davies, P.,** *J. Chromatogr.*, 277, 125, 1983.

8. PREDICTION OF THE RETENTION OF PHENYLTHIOHYDANTOIN AMINO ACIDS

The chromatographic separation of PTH amino acids is of the greatest importance in protein and peptide structure studies. Jinno[1] has indicated that the retention behavior of PTH

Table 1
COMPARISON OF MEASURED AND PREDICTED CAPACITY FACTORS OF PHENYLTHIOHYDANTOIN AMINO ACIDS

	k′	
Amino acids	**Predicted**	**Measured**
Alanine	1.86	2.01
Arginine	1.16	1.19
Asparagine	0.54	0.51
Aspartic acid	0.84	0.81
Glutamic acid	1.06	1.11
Glutamine	0.96	0.83
Glycine	1.34	1.33
Histidine	0.87	0.81
Isoleucine	8.72	9.11
Leucine	9.82	9.56
Lysine	10.7	10.5
Methionine	4.95	4.70
Norvaline	7.91	7.05
Phenylalanine	8.77	8.56
Proline	4.38	3.40
Serine	0.76	0.72
Threonine	1.28	1.23
Tryptophan	7.70	7.41
Tyrosine	2.61	2.65
Valine	4.99	4.70

Conditions:	Packing:	Jasco Fine® SIL C-18, 10 μm
	Column:	25 cm × 0.5 mm I.D. fused silica capillary
	Solvent:	Acetonitrile/2% tetrahydrofuran in sodium acetate buffer, pH 5.4, 35:65
	Temperature:	25°C
	Detection:	UV at 254 nm

amino acids can be predicted on the premise that relationships exist between physical parameters representing molecular properties such as structure and/or shape and retention. A novel retention-solubility parameter, R, was defined and the performance of the R value evaluated by a comparison of measured and predicted retention data. Excellent agreement was found between the two values (Table 1). The R value has a high potential for describing the retention of PTH amino acids using C-18, C-8, and phenethyl stationary phases. However, it did not prove useful for describing retention on a CN column. This means that there may be some differences in retention mechanisms between hydrocarbonaceous stationary phases and a CN phase. The system data from the concept put forward by Jinno can be stored in a microcomputer-assisted retention prediction system, which should facilitate optimization of the separation parameters of PTH amino acids.

REFERENCE

1. **Jinno, K.,** *Chromatographia,* 20, 743, 1985.

9. THE DETERMINATION OF *N*-ACETYLCYSTEINE IN HUMAN PLASMA

N-Acetylcysteine is effective as an antidote for the treatment of acetaminophen poisoning and an improved method for its determination has been developed. After incubation with dithiothreitol to release bound *N*-acetylcysteine, free *N*-acetylcysteine and the internal standard *N*-acetylpenicillamine are derivatized with 2,4-dinitro-1-fluorobenzene (DNFB). The samples are deproteinized by ultrafiltration, the dinitrophenyl derivatives are extracted from the acidified ultrafiltrate and purified by back extraction, and they are then separated from naturally occurring plasma components and reagent impurities by HPLC.

Procedure: Extraction

Aliquots (1.5 mℓ) of samples are pipetted into 15.0-mℓ glass extraction tubes provided with PTFE liners and 100 μℓ of *N*-acetylpenicillamine solution added to give a concentration of 40 μg/mℓ in plasma. A vol of 125 μℓ of dithiothreitol solution (10 mg/mℓ) is added to each tube which is vortex mixed, capped, and heated in a water bath at 37°C for 10 min. To each tube is added 1.0 mℓ of 1% $NaHCO_3$ and 180 μℓ of 10% DNFB in methanol; the tubes are vortex mixed, capped, and heated in a water bath at 60°C for 30 min. The tube contents are decanted into ultrafiltration cones that are placed in 30-mℓ glass vials and centrifuged at 1000 × *g* for 20 min at 20°C. A vol of 400 μℓ of the ultrafiltrate is pipetted in 15-mℓ screw-capped extraction tubes containing 3.0 mℓ of 0.2 *M* HCl. Diethylether (8.1 mℓ) is added to each tube which is shaken mechanically for 5 min, followed by centrifuging at 750 × *g* for 5 min at 10°C.

The ether layers are then transferred using Pasteur pipettes into 15-mℓ screw-capped extraction tubes containing 2.0 mℓ of 1% $NaHCO_3$. The tubes are shaken and centrifuged and the ether layers aspirated to waste. A volume of 100 μℓ of 6 *M* HCl and 8.0 mℓ of diethyl ether is added to each tube, which is shaken and centrifuged. The ether layers are transferred with Pasteur pipettes to glass vials and evaporated to dryness under a gentle stream of nitrogen at room temperature. The residues are stored at −20°C until they are dissolved in 100 μℓ of the mobile phase just before chromatography, when 20 μℓ is injected onto the column.

Chromatography

Chromatography is performed on a 25-cm × 4.6-mm I.D. stainless steel column prepacked with Ultrasphere® ODS, 5 μm, material. The mobile phase consists of methanol/0.067 *M* phosphate buffer, pH 8.0, containing 0.005 *M* tetrabutyl ammonium hydrogen sulfate, 50:50. It is filtered through a 0.5-μm PTFE membrane filter, degassed by sonication, purged with pure helium, and delivered to the column at a flow rate of 1.0 mℓ/min. The eluate from the chromatogram is monitored at 360 nm. The retention times for *N*-acetylcysteine and *N*-acetylpenicillamine are 7.8 and 13.3 min, respectively.

REFERENCE

1. **Lewis, P. A., Woodward, A. J., and Maddock, J.,** *J. Chromatogr.*, 327, 261, 1985.

10. THE DETERMINATION OF CYSTEINE SULFINIC ACID AND CYSTEIC ACID IN RAT BRAIN

Male Wistar rats weighing 180 to 200 g are killed by microwave irradiation, the brains quickly removed, chilled on ice, and dissected. Tissue samples are immediately homogenized in 10 vol of ice-cold 10% trichloroacetic acid and the homogenates centrifuged at 8000 × *g* for 30 min at 4°C. The supernatant is extracted four times with 3 vol of water-saturated ether to remove trichloroacetic acid; aliquots are lyophilized and dissolved in distilled water. Of each extract, 100 μℓ is applied to a column, 6 × 120 mm, of cation exchange resin, Dowex® 50W-X8, 200-400 mesh in the hydrogen form, and the column is eluted with distilled water. The first 1.0 mℓ of the eluate is discarded and the following 1.0 mℓ containing sulfinic acid, cysteic acid, and taurine is collected, relyophilized, and redissolved in distilled water. Aliquots of this solution are then studied by HPLC. Chromatography is conducted using a column 25 cm × 4 mm I.D. packed with anion exchange resin ISA-07/S 2504. The effluent from the column is mixed with *o*-phthalaldehyde and fluorescence measured using excitation and emission wavelengths of 348 and 450 nm, respectively. The peak for each amino acid is identified by comparison with the retention time of authentic compounds. The retention times of cysteine sulfinic acid, cysteic acid, phosphoserine, and taurine in the system are, respectively, 22.5 ± 0.3, 34.2 ± 0.5, 40.2 ± 0.6, and 7.2 ± 0.1 min. The fluorescence peak area for cysteine sulfinic acid is linear between 20 pmol and 5 nmol and that for cysteic acid between 10 pmol and 5 nmol.

REFERENCE

1. **Ida, S. and Kuriyama, K.,** *Anal. Biochem.,* 130, 95, 1983.

11. THE DETERMINATION OF *S*-SULFOCYSTEINE IN URINE BY HPLC

Existing TLC or GC techniques proved to be insufficiently sensitive and specific for the determination of the fairly low concentrations of *S*-sulfocysteine (cysteine-*S*-sulfonic acid) present in the urine of normal human beings. A method for the determination of this compound in urine was therefore developed. A preliminary separation of sulfocysteine from interfering compounds by ion exchange chromatography (IEC) is followed by its conversion to the dansyl derivative. The latter is then determined by HPLC with fluorometric detection.

Procedure

A 1-mℓ aliquot of human urine collected for 24 hr with thymol-isopropanol as a preservative is applied to a column of AG 50W-X8 resin, 3.2 × 1.0 cm, and the column is washed with 6 mℓ of water. The combined effluent and washings are neutralized with 1 *M* NaOH and applied to a column of AG3-X4A resin, 3.2 × 1.0 cm. The resins AG 50W-X8 (H^+, 100-200 mesh) and AG3-X4A (Cl^-, 200-400 mesh) from Bio-Rad Laboratories are used as received. The AG3-X4A column is washed with 8 mℓ of water, the effluent and washings being discarded. Sulfocysteine is then eluted with 8 mℓ of 1 *M* NaCl. A 1-mℓ aliquot of the eluate is transferred to a 10-mℓ test tube with a PTFE-lined screw cap and 0.5 mℓ of 0.12 *M* Li_2CO_3, adjusted to pH 9.5 with HCl, and 1 mℓ of a solution of dansyl chloride in acetonitrile (1.5 g/ℓ) are added. After vortex mixing the mixture is left in the dark at room temperature overnight and is then evaporated to dryness at 60°C using a vortex evaporator.

The residue is suspended in 0.5 mℓ of the chromatographic mobile phase (toluene/pyridine/acetic acid/ethanol, 8:0.2:2:1) by vortex mixing for two periods of 3 min separated by 3 min of sonification. After centrifuging, 100 μℓ of the clear supernatant is taken for chromatography. Chromatography is conducted on a column, 250 × 4.5 mm, of Apex® silica, 5-μm particle size. The fluorometric detector is operated with a mercury lamp, a 360-nm excitation filter, and a 418- to 700-nm emission filter. The mobile phase is delivered to the column at a flow rate of 1 mℓ/min at room temperature. Peak heights are measured for the quantitative determination of sulfocysteine and a standard curve is prepared using sulfocysteine solutions of known concentration. The standard curve is linear up to concentrations of sulfocysteine corresponding to 15 μmol/ℓ of urine. The recovery of sulfocysteine added to urine is 91 ± 11% and the detection limit is 0.1 μmol/ℓ.

REFERENCE

1. **Kagedal, B., Kallberg, M., and Sorbo, B.,** *J. Chromatogr.*, 276, 418, 1983.

12. THE FLUORESCENCE DETECTION OF CYSTINE BY *o*-PHTHALALDEHYDE DERIVATIZATION

The *o*-phthalaldehyde derivatization of amino acids is ideally suited to their detection in physiological fluids using liquid chromatography. Since the reaction only proceeds in the presence of a thiol, cystine is reduced to cysteine. However, cysteine, unlike other amino acids, yields a derivative with minimal fluorescence.[1] This problem may be overcome by oxidizing cystine or cysteine to cysteic acid, which on reaction with *o*-phthalaldehyde yields a derivative with fluorescent properties comparable to those of other amino acid derivatives. Because one oxidizing and one reducing reaction are involved in the derivatization, it is difficult to obtain between-sample reproducibility in the determination. A simple procedure to overcome this problem by pretreating cysteine with iodoacetic acid before *o*-phthalaldehyde-2-mercaptoethanol derivatization has been described.[2] The method produces a cysteine derivative with a fluorescence intensity similar in magnitude to that of other amino acid-*o*-phthalaldehyde derivatives, and allows rapid separation of the cysteine derivative by HPLC and its quantification in biological samples.

Procedure

The following volumes of reagents were used to optimize conditions. To 100 μℓ of cystine standards were added 200 μℓ of 20 μ*M* homocysteic acid to act as internal standard, 500 μℓ of 70 m*M* perchloric acid containing concentrations of 2-mercaptoethanol ranging from 2.5 to 125 m*M*, and 200 μℓ of 3 *M* NaOH. After mixing, 200 μℓ of iodoacetic acid of concentrations ranging from 10 to 100 m*M* was added. To 200 μℓ of the solution were added 10 μℓ of *o*-phthalaldehyde-2-mercaptoethanol reagent (Section II.IV) and 20 μℓ immediately injected onto the chromatographic column.

When assaying serum or urine, to 100 μℓ of sample were added 200 μℓ of 200 μ*M* homocysteic acid to act as internal standard and 500 μℓ of protein precipitant (25 m*M* 2-mercaptoethanol in 70 m*M* perchloric acid). After centrifuging for 2 min, 400 μℓ of the supernatant was added to 100 μℓ of 3 *M* NaOH. To 200 μℓ of this solution were added 100 μℓ of *o*-phthalaldehyde-2-mercaptoethanol reagent and 200 μℓ immediately injected onto the chromatographic column. The optimal conditions for the reduction of cystine were 25 m*M* 2-mercaptoethanol in 70 m*M* perchloric acid. Maximum fluorescence was obtained using 50 m*M* iodoacetic acid. The fluorescence response of the cysteine *o*-phthalaldehyde derivative is linear with different cystine concentrations, allowing quantitative, precise, and rapid estimations.

REFERENCES

1. **Roth, M.,** *Anal. Chem.*, 43, 880, 1971.
2. **Cooper, J. D. H. and Turnell, D. A.,** *J. Chromatogr.*, 227, 158, 1982.

13. CYSTEINE — A POTENTIAL SOURCE OF ERROR IN AMINO ACID ANALYSIS

Mercaptoethane sulfonic acid has been used as an agent for hydrolysis of proteins and peptides prior to amino acid analysis, since tryptophan is not destroyed and hydrolyzates after partial neutralization and dilution can be applied directly to ion exchange columns without needing to be dried down. Moore and Stein's original procedure for HCl hydrolysis of proteins included reconstitution of the hydrolysate at pH 6.5 in air before reducing the pH to about 2 for loading on the column.[1] This step is intended to allow the oxidation of cysteine to cystine, a reaction that proceeds rapidly in air at alkaline or neutral pH, but which is omitted by many analysts who reconstitute hydrolysates directly in citrate buffer at pH 2.2. Mercaptoethanol or thioglycolic acid is often added to the HCl to minimize losses of methionine.

Gardner[2] conducted experiments to evaluate possible sources of error in these procedures, reduced glutathione (1.63 μmol) being hydrolyzed with mercaptoethane sulfonic acid (1 mℓ/3 mol/ℓ) in sealed tubes or in vacuo with HCl (1 mℓ, 6 mol/ℓ) at 110°C for 24 hr.[2] In some cases the HCl contained mercaptoethanol (1:2000) or thioglycolic acid (1:1000). The hydrolyzates were analyzed on an automatic analyzer with ninhydrin detection. The results showed that a peak that could not be distinguished from proline in elution time and was closely similar to proline in the ratio of its absorbance at 570 to 440 nm occurred on analysis of the mercaptoethane sulfonic acid hydrolyzates. It also occurred to a lesser extent in HCl hydrolyzates if these were reconstituted directly at pH 2.2 without aerobic oxidation of cysteine at neutral or alkaline pH being allowed to occur. This was particularly the case if the hydrolysis had been carried out in the presence of mercaptoethanol or thioglycolic acid. Gardner emphasizes that although the difficulties inherent in the estimation of cystine in hydrolyzates are well known, the possibility that proline estimations may be erroneously high requires consideration, especially if mercaptoethane sulfonic acid is used for hydrolysis. With HCl hydrolysis reconstitution at neutral or alkaline pH should avoid this potential error.

REFERENCES

1. **Hirs, C. H. W., Stein, W. H., and Moore, S.,** *J. Biol. Chem.*, 211, 941, 1954.
2. **Gardner, M. L. G.,** *Anal. Biochem.*, 141, 429, 1984.

14. THE DETERMINATION OF FREE TRIMETHYLLYSINE

6-*N*-Trimethyllysine forms a minor constituent of proteins in a variety of organisms. It has been identified in cytochrome *c* of yeast and wheat germ, myosin, and histone and is also present in the plasma and urine of human subjects. The determination of free trimethyllysine is important in understanding trimethyllysine metabolism and in delineating the regulation of carnitine biosynthesis.

A method for the determination of trimethyllysine in urine was described by Hoppel et al.[1] Trimethyllysine and the internal standard triethyllysine were isolated from aqueous samples by ion exclusion chromatography on a microcolumn. The specimens were then derivatized with 1-fluoro-2,4-dinitrobenzene and by-products extracted with organic solvents. The trimethyllysine and internal standard derivatives were separated by reversed-phase paired-ion HPLC and were determined spectrophotometrically. The procedure was used to determine trimethyllysine in human urine.

This method did not afford sensitivity adequate for measurement of free trimethyllysine in plasma or tissue specimens. An alternative procedure was therefore devised for the measurement.[2] Using this technique trimethyllysine and an internal standard are isolated from plasma and tissue homogenates by protein precipitation and combined ion exchange-ion exclusion chromatography of the protein-free supernatant. Reversed-phase ion-pair HPLC is followed by postcolumn reaction with *o*-phthalaldehyde and 2-mercaptoethanol and fluorometric detection. This procedure is described.

Sample Preparation

Plasma

Whole blood is collected in heparinized tubes, chilled on ice, and centrifuged. A 2-mℓ volume of chilled plasma or standard solution is combined with 0.5 mℓ of 12% perchloric acid and 350 μℓ of triethyllysine internal standard working solution in a disposable polypropylene test tube, 16.8 × 95 mm. The tubes are allowed to stand on ice for 1 hr and are then centrifuged. The supernatant is transferred to another polypropylene test tube containing 0.5 mℓ of cold 2 *M* $KHCO_3$, vortexed, placed on ice for 1 hr, and centrifuged. Columns made from 5-mℓ disposable polypropylene pipette tips, 10 × 116 mm, are filled with 2.5 mℓ of Dowex® 1-X8 (OH^-) resin followed by 2.5 mℓ of Dowex® 50W-X8 (NH_4^+) resin; the layer of the first resin is allowed to settle completely before addition of the second resin. The entire sample supernatant is added to the column, excluded and weakly retained species are eluted with 6 mℓ of 1 *M* NH_4OH, and trimethyllysine and the internal standard are eluted with 3 mℓ of 1 *M* NH_4OH and 5 mℓ of deionized water. The combined 8 mℓ of effluent are evaporated to dryness under vacuum at 38°C and are redissolved in 250 μℓ of 10^{-3} *M* HCl before injection into the chromatograph.

Tissue Specimens

To 1 g of rat liver, kidney, heart, or mixed skeletal muscle is added 3 mℓ of ice cold 6% perchloric acid. The mixture is homogenized and 2 mℓ of the suspension combined with 350 μℓ of the internal standard working solution in a disposable polypropylene test tube. The tubes are placed on ice for 1 hr, centrifuged, and the supernatant transferred to another polypropylene test tube containing 0.5 mℓ of 2 *M* $KHCO_3$. This, in turn, is vortexed, placed on ice for 1 hr, and centrifuged. The supernatant is decanted and subjected to ion exchange-ion exclusion chromatography as described for plasma. The combined effluent fractions are evaporated to dryness under vacuum at 38°C and the sample is redissolved in 250 μℓ of 10^{-3} *M* HCl for injection into the chromatograph.

Chromatography

Chromatographic separation is accomplished on a 10 × 0.5-cm cartridge of 10 μm nominal particle diameter Radial-Pak® C_{18} (Waters). A precolumn packed with a Co:Pell® ODS reversed-phase pellicular chromatographic medium (Whatman) precedes the column compression unit in the eluent stream. The chromatographic eluent is 30% acetonitrile in water containing 5 × 10^{-2} *M* sodium dodecyl sulfate (SDS) and 5 × 10^{-2} *M* NaH_2PO_4. To 700 mℓ of water are added 6.9 g of NaH_2PO_4 and 14.4 g of SDS. The solution is adjusted to pH 2.75 with concentrated H_2SO_4, filtered through a cellulose acetate membrane, and 300 mℓ of acetonitrile is added with magnetic stirring. The derivatization reagent is prepared by dissolving 300 mg of *o*-phthalaldehyde and 500 μℓ of 2-mercaptoethanol in 10 mℓ of a 3% solution of Brij-35 in 95% ethanol and adding this to 500 mℓ of a 0.5 *M* sodium borate solution of pH 10.4. The postcolumn derivatization reagent is introduced into the eluent stream. A coil of steel tubing provides a short reaction time prior to detection. The fluorometric detector is operated at an excitation wavelength of 229 nm with an emission filter cutoff wavelength of 418 nm. Standard curves of trimethyllysine/triethyllysine peak height or area ratios vs. concentration were obtained daily over a trimethyllysine standard solution concentration range of 0.5 to 4.0 nmol/mℓ.

REFERENCES

1. **Hoppel, C. L., Weir, D. E., Gibbons, A. P., Ingalls, S. T., Brittain, A. T., and Brown, F. M.,** *J. Chromatogr.*, 272, 43, 1983.
2. **Davis, A. T., Ingalls, S. T., and Hoppel, C. L.,** *J. Chromatogr.*, 306, 79, 1984.

15. THE DETERMINATION OF D- AND L-ASPARTIC ACID BY HPLC

A simple technique has been described for the determination of D- and L-aspartic acid in the low picomolar range. The method is a modification of the *o*-phthalaldehyde precolumn derivatization procedure for amino acid analysis, which is modified by the substitution of the chiral sulfhydryl reagent, *N*-acetyl-L-cysteine, in place of 2-mercaptoethanol or ethanethiol to form the thiol adduct that reacts with primary amines.

Procedure: Preparation of the Reagent

o-Phthalaldehyde, 4.0 mg, is dissolved in 300 μℓ of methanol. The following are then added in the indicated order: 250 μℓ of 0.4 *M* sodium borate, pH 9.4, 390 μℓ of distilled water, and 60 μℓ of 1.0 *M N*-acetyl-L-cysteine, brought to pH 5.0 to 6.0 with NaOH. The reagent remains useful for at least 3 weeks when stored at 4°C.

Derivatization of Amino Acids

Twenty microliters of amino acid solution, containing between 10 and 1000 pmol of each amino acid, is thoroughly mixed with 5 μℓ of the reagent in a polyethylene "microfuge" tube. After 2 to 3 min 475 μℓ of 50 m*M* sodium acetate, pH 5.2, is added. The solution is then centrifuged at 10,000 × *g* for 30 sec to sediment any extraneous particulate material and 250 μℓ of the solution is injected directly into the HPLC system.

The reversed-phase column, 12 cm × 4.6 mm, contains Hypersil®-ODS, 3 μm diameter. Solvent A is 50 m*M* sodium acetate, pH 5.9; solvent B is 80% methanol and 20% solvent A. Isocratic elution with 90% solvent A, 10% solvent B is carried out for 5 min, the content of solvent B being then increased linearly to 100% in 5 min. At a flow rate of 1.0 mℓ/min the fluorescent derivatives of D- and L-aspartic acid elute reproducibly at 3.3 and 3.7 min, respectively, followed by L-glutamic acid and L-serine. The elution order is the same as when 2-mercaptoethanol or ethanethiol is used in the derivatization reaction.

The fluorescence responses of both D- and L-aspartic acid derivatives are linear over the range of 5 to 500 pmol per injection. The specific fluorescence of D-aspartic acid is 10 to 15% lower than that of L-aspartic acid. The reaction of the reagent with D- and L-aspartic acid is maximal at 2 min and is constant for at least 10 min. D-Aspartic acid (5 pmol) can be detected in the presence of a 100-fold excess of L-aspartic acid in a total analysis time, including derivatization of 10 min. In principle, the technique should be applicable to the separation of enantiomers of other amino acids and primary amines.

REFERENCE

1. **Aswad, D. W.,** *Anal. Biochem.*, 137, 405, 1984.

16. THE DETERMINATION OF METHIONINE IN CRUDE PLANT MATERIALS BY GAS CHROMATOGRAPHY

The reaction of cyanogen bromide with peptide-bound methionine releases methylthiocyanate which may be quantitatively determined by gas chromatography. This procedure allows the determination of methionine in proteins and crude plant materials.

Procedure

The following solutions, which are stable for at least 4 months at 4°C, are prepared: solution a, 1% cyanogen bromide in 70% aqueous formic acid; solution b, 1000 ppm 5-nonanone in 97% formic acid; solution c, 2000 ppm methyl thiocyanate in 97% formic acid. The reaction is carried out in 1.5 mℓ Silli Vials® (Applied Science Laboratories) with Teflonlaminated disk stoppers. To dry samples of 20 to 30 mg of finely ground seed meal or 2 to 10 mg of protein are added 750 μℓ of solution a and 50 μℓ of solution b. A standard test mixture is prepared daily from 700 μℓ of solution a and 50 μℓ each of solutions b and c. The mixture is incubated at 95°C for 1 hr with seed meal and for 20 min with proteins. After centrifugation at 3000 × g for 5 min, an aliquot (1 to 2 μℓ) of the clear supernatant is injected into the gas chromatograph, which is equipped with a flame ionization detector and automatic integrator. The glass column, 100 cm × 2 mm I.D., is packed with 10% Carbowax®-1000 and 2% Ucon® LB 550X on Chromosorb® W-AW, 80-100 mesh, the oven temperature is 105°C, and the nitrogen carrier gas flow rate rate is 40 mℓ/min.

The methionine content of the sample is calculated from the formula

$$\text{Methionine, \%(w/w)} = \frac{A_{IS}/A_{M})_{STD} \times A_M/A_{IS})_{sample}}{W_{sample}} \times 19.16$$

where A_{IS} and A_M represent peak areas for 5-nonanone and methyl thiocyanate, respectively, W is the weight of the sample in milligrams, and 19.16 is the product of four factors: the molecular ratio (1.794) of anhydromethionine to methylthiocyanate, the amount of methylthiocyanate in the standard mixture (0.100 mg), the density of methylthiocyanate (1.068 mg/mℓ) at 25°C, and 100 (%).

One percent of cyanogen bromide is sufficient to ensure complete reaction with bean seed meal. Lower concentrations (e.g., 0.5%) give variable results while higher concentrations give no additional recovery of methylthiocyanate. Optimal conversion of methionine is obtained with formic acid concentrations greater than 70%. The internal standard, 5-nonanone, is stable and separates well from methyl thiocyanate. The sulfoxide and sulfone of methionine do not react with cyanogen bromide. *S*-Methylcysteine and the dipeptide glutamyl-*S*-methylcysteine, both of which are present in legumes, make negligible contributions to the analysis of methionine. There is close agreement between the value for the methionine content of seed meals obtained by the procedure and the result of determinations of methionine as methionine sulfone after performic acid oxidation and subsequent amino acid analysis. The method is rapid, economical, and precise and allows more than 50 samples to be analyzed per man-day.

REFERENCE

1. **Apostolatos, G. and Hoff, J. E.,** *Anal. Biochem.*, 118, 126, 1981.

17. THE GAS CHROMATOGRAPHIC DETERMINATION OF METHYLHISTIDINE ISOMERS

Interest in the determination of methylhistidine isomers, 1-methylhistidine, and 3-methylhistidine has recently increased. The latter is present in the peptide chains of actin and myosin, and serves as a urinary marker of muscular turnover, while the former has been isolated in a peptide of the neuromuscular junction. Mussini et al.[1] described a method for the isolation of the methylhistidine isomers from biological specimens and their determination by glass capillary gas chromatography. This method, however, required sophisticated equipment and experienced maintenance staff and was thus not adapted for routine clinical analysis. Marcucci et al.,[2] therefore developed a simpler technique using a normal gas chromatograph apparatus and a packed column.

Isolation of Methylhistidine Isomers

A chromatographic column is prepared by packing a column, 1.5 × 1.0 cm I.D., with charcoal-Celite®, 1:1 (w/w), and buffering it at pH 5 with 0.33 *M* acetate buffer. Dried hydrolyzates of biological samples are dissolved in 0.33 *M* acetate buffer, pH 5. An amount corresponding to 0.25 mℓ of urine or 15 mg of wet muscle is applied to the column, which is then washed with 20 mℓ of water and 5 mℓ of 80% acetone. The methylhistidine isomers are eluted with 30 mℓ of dichloromethane/methanol/33% ammonium hydroxide, 70:25:5, and the eluates are evaporated to dryness under vacuum.

Derivatization for Gas Chromatography

The carboxy group of 1- and 3-methylhistidine is esterified with 5 mℓ of a mixture of 5% dry acetyl chloride in *n*-propanol. The tube is sealed and the contents mixed and allowed to react overnight at 90°C in a Reacti-Therm® heating module (Pierce). The sample is evaporated to dryness under vacuum and *N*-acylated with 150 μℓ of trifluoroacetic anhydride and 200 μℓ of dichloromethane for 30 min at room temperature. The solution is evaporated to dryness under vacuum and the residue dissolved in ethyl acetate containing as reference standard for the gas chromatography 2-*N*-methylamino-5-chlorobenzophenone (MACB), 25 μg/mℓ.

Gas Chromatography

For gas chromatographic analysis, a glass column, 2 m × 4 mm I.D., is silanized and packed with 1.5% SP-2250 and 1.95% SP-2401 on Supelcoport®, 100-120 mesh. The column is maintained at 210°C, and carrier gas nitrogen at a flow rate of 33 mℓ/min. A flame ionization detector is employed.

Comments

The sensitivity limit of the technique is about 4 μg/mℓ, compared with 2 μg/mℓ using the glass capillary method. The use of an electron capture detector increases the sensitivity about 100-fold, but this detector is not routinely used because human urine contains large amounts of 1- and 3-methylhistidines. This detector should be useful for measuring low concentrations of the isomers as, for example, in muscle proteins. Quantitation is by the internal standard method using MACB. The calibration curves for the two isomers show a linear response over the range 12.5 to 100 ng/μℓ. Average recovery is 97 ± 2%. The authors conclude that the method is satisfactory with respect to sensitivity, accuracy, precision, and economy. The procedure is clinically applicable to the routine analysis of methylhistidine isomers in biological samples.[2]

REFERENCES

1. **Mussini, E., Cotellessa, L., Colombo, L., Cani, D., Sfondrini, P., Marcucci, F., and Poy, F.,** *J. Chromatogr.*, 224, 94, 1981.
2. **Marcucci, F., Colombo, L., and Mussini, E.,** *J. Chromatogr.*, 336, 356, 1984.

18. THE CHROMATOGRAPHIC SEPARATION OF (3R)- AND (3S)-β-LEUCINE AS DIASTEREOMERIC DERIVATIVES

In connection with studies on the stereochemistry of leucine 2,3-aminomutase, Aberhart et al.[1] devised a method for the separation of (3R)- and (3S)-β-leucine as diastereomeric derivatives.

Derivatization

(3R,S)-β-Leucine, 0.7 mg or 0.5 μmol, is treated with 20 μℓ of a 1% solution of Marfey's reagent [N^2-(5-fluoro-2,4-dinitrophenyl)-L-alaninamide] in acetone, plus 40 μℓ of 1 *M* $NaHCO_3$ at 40°C for 1 hr. After cooling to 25°C and the addition of 20 μℓ of 2 *M* HCl, the solution is filtered through a Gelman Acro® LC-13 filter and injected directly into the chromatograph.

Chromatography

A 3.9-mm × 15-cm Nova-Pak® C_{18} column is used with an eluting solvent of 30% acetonitrile/0.05 *M* triethylamine phosphate, pH 3.0, and a flow rate of 2.0 mℓ/min. Detection is made at 340 nm. The (3R) diastereomer is eluted before the (3S) diastereomer.

REFERENCE

1. **Aberhart, D. J., Cotting, J.-A., and Lin, H.-J.,** *Anal. Biochem.*, 151, 88, 1985.

19. THE DETERMINATION OF 5-PYRROLIDONE-2-CARBOXYLIC ACID IN TISSUE HOMOGENATES

5-Pyrrolidone-2-carboxylic acid, a cyclic lactam of glutamic acid, occurs in cells and body fluids both in the free state and as the N-terminal residue of peptides and proteins. Bousquet et al.[1] have described a method for the determination of 5-pyrrolidone-2-carboxylic acid in which this compound is derivatized with 4-nitrophenylacyl bromide. The 4-nitrophenylacyl ester has a relatively high stability, can be readily separated by reversed-phase HPLC, and allows the estimation of nanomolar concentrations of 5-pyrrolidone-2-carboxylic acid.

Procedure: Derivatization in Tissue Homogenates

Male Swiss albino mice, about 30 g, are maintained on a standard laboratory cube diet for 1 week before sacrifice. The animals are decapitated, and brain, liver, and kidneys are rapidly removed and homogenized in a glass homogenizer fitted with a motor-driven Teflon pestle at 0 to 4°C in the presence of 5% trichloroacetic acid. After standing for 30 min at 0°C the homogenate is centrifuged at 50,000 rpm for 10 min. The pellet is resuspended in an equal volume of 5% trichloroacetic acid and the supernatants from both centrifugations are combined. The combined supernatants are extracted six times with ether and 1-mℓ aliquots of the ether-free aqueous solution are dried at 40°C under nitrogen. The dried residue is suspended in 10 mℓ of acetonitrile with 2 to 3 μℓ of triethylamine and 2 mg of 4-nitrophenylacyl bromide, and made up at 25 mℓ in a volumetric flask with acetonitrile. After standing for 2 hr at room temperature the solution is ready for analysis by HPLC. Mouse skin is treated with a depilatory agent and homogenized at 0 to 4°C in the presence of 5% trichloroacetic acid. Further processing of the homogenate is carried out as described above.

Chromatography

Chromatographic analysis is carried out on a Micropak® MCH-10 (RP-18) column, 30 × 0.4 cm, at room temperature. The developing solvent consists of 60% acetonitrile and 40% 0.01 *M* monobasic sodium phosphate with a flow rate of 1 mℓ/min. Absorbance measurements for the 4-nitrophenylacyl ester of 5-pyrrolidone-2-carboxylic acid are made at 263 nm. Between each analysis the column is rinsed with 1 mℓ of acetonitrile. The recovery of 5-pyrrolidone-2-carboxylic acid exceeds 90%.

REFERENCE

1. **Bousquet, E., Guarcello, V., Morale, M. C., and Rizza, V.,** *Anal. Biochem.*, 131, 135, 1983.

20. THE DETERMINATION OF DIMETHYLGLYCINE IN BIOLOGICAL FLUIDS

N,N-Dimethylglycine in biological fluids can be quantitatively determined by cation exchange HPLC, the amino acid being assayed in the eluate using dimethylglycine dehydrogenase.

Sample Preparation and HPLC

Whole blood is centrifuged to remove cells and the plasma centrifuged through a filter at 1000 to 2000 × *g*. Filtered material is used immediately or stored at −20°C. Urine is treated in the same way except that the initial centrifugation is omitted. An aliquot of the material is mixed with an equal volume of 0.325 *M* sodium citrate, pH 3.28, and 80-μℓ samples eluted isocratically from a Bio-Rad Aminex® A-9 column, 150 × 4 mm, in the Na^+ form with 0.325 *M* sodium citrate, pH 3.28, at 0.2 mℓ/min at room temperature. Seven-drop (182-μℓ) fractions are collected and assayed for dimethylglycine. The column is regenerated by a 20-min wash 0.2 *N* NaOH, and 2 m*M* EDTA which has been passed through a 0.22-μm filter, followed by a 45-min wash with the elution buffer.

Enzymatic Assay

To each 182-μℓ fraction is added 392 μℓ of 1 *M* K_2HPO_4, 283 μℓ of H_2O, 33 μℓ of 3 m*M* 2,6-dichlorophenol indophenol, and 100 μℓ of 22 m*M* phenazine methosulfate. The blank rate of decrease of absorbance at 600 nm is determined and the reaction initiated by adding dimethylglycine dehydrogenase. The initial rate of decrease in absorbance at 600 nm is determined. A standard curve showing the rate of change of absorbance against the concentration of dimethylglycine is prepared. The amount of dimethylglycine in each fraction is determined from the initial and blank absorbance measurements, the standard curve, and the volume of the enzymatic assay. Solutions of 2,6-dichlorophenol indophenol and phenazine methosulfate used in the assay are freshly prepared every day.

Recoveries of added dimethylglycine are 99 to 104%, with no endogenous dimethylglycine being found in rat plasma or normal human urine. Human plasma contained a small quantity of endogenous dimethylglycine. The cation-exchange chromatography separates dimethylglycine from other compounds able to serve as substrates for dimethylglycine dehydrogenase.

REFERENCE

1. **Porter, D. H., Lin, M., and Wagner, C.,** *J. Chromatogr.*, 151, 299, 1985.

21. THE RAPID MICROSCALE DETERMINATION OF TRYPTOPHAN

A rapid micromethod using acid hydrolysis for the determination of the tryptophan content of proteins has been described.

Procedure

For hydrolysis, a glass tube with an outer diameter of 4.0 mm and an inner diameter of 2.5 mm is flamed to form a bottle of 100 μℓ. A solution of the protein (5 μg) is poured into the open end, is sedimented to the bottom of the bottle, and is dried in vacuo; 30 μℓ of 3 *M* mercaptoethane sulfonic acid is added, and the bottle is evacuated and sealed. After hydrolysis at 166°C for 25 min in an oil bath, the tube is cooled, opened, and 50 μℓ of 1.38 *M* NaOH added to the hydrolysate. The hydrolysate is analyzed on a Durrum D500 automatic analyzer using a single column system, 500 × 1.75 mm, at a temperature of 50°C and a flow rate of 10 mℓ/hr. A sodium citrate buffer step gradient is applied: pH 3.25 (66 m*M*) for Asp, Thr, Ser, Glu, Pro, Gly, Ala, Cys, and Val; pH 4.25 (66 m*M*) for Met, Ile, Leu, Tyr, Phe, and His; pH 5.28 (366 m*M*) for Trp, Lys, and Arg. Elution times for tryptophan and arginine are 62 and 128 min, respectively. The amount of tryptophan is calculated from a color constant obtained using a known standard.

REFERENCE

1. **Maeda, K., Scheffler, J. J., and Tsugita, A.,** *Hoppe Seylers Z. Physiol. Chem.*, 365, 1183, 1984.

22. MICROELECTRODES FOR THE QUANTITATIVE DETERMINATION OF METHYLAMINES

A method for measuring methylamino compounds using a glass microelectrode has been described by Bagnasco.[1] The assay is suitable for use with column chromatographic techniques of separation of methylamines.

Technique

Microelectrodes are fabricated from borosilicate glass capillaries (Kwik-Fill® glass capillaries 1 B 120 F, WPI Instruments Inc.) using a pipette puller. The tips of the pipettes are broken after pulling. The pipettes are then silanized by dipping their tips into a mixture containing 3% Prosil®-28(R) (SCM Specialty Chemicals) and 1-chloronaphthalene, subsequently heated at 150°C for 1 hr. They are filled with a mixture containing 5% potassium tetrakis (*p*-chlorophenyl) borate and 3-nitro-*o*-xylene as ion exchanger in which a silver-silver chloride wire is enclosed. A calomel electrode is used as reference. Voltage is recorded by a dual electrometer.

Chromatographic elution of methylamines in amounts of 10 to 50 μmol is performed using the cation exchange resin AG 50 W-X8, 200-400 mesh, hydrogen form, and a 0.7 × 20-cm column. Water (30 mℓ) and 0.4 *N* HCl, 250 mℓ, are used as eluants at a flow rate of 0.5 mℓ/min. Individual 3-mℓ fractions are directly assayed with the electrode, the difference in voltage compared to 0.4 *N* HCl alone being measured. The voltage is converted to concentration using standard curves. Standard solutions containing 0.1, 0.5, 1.0, 5.0, or 10.0 m*M* concentrations of amine are made by dissolving the appropriate amount in NaCl or HCl solutions. Standard curves are obtained from the linear regression of voltage and logarithm of concentration. The sensitivity is highest for choline and trimethylamine, followed by trimethylamine oxide and betaine, all of which are readily measurable from 0.1 to 10.0 m*M*. Linear and reproducible standard curves are also obtained for dimethylamine and L-carnitine from 0.5 to 10.0 m*M*. The sensitivity of the electrode is maintained even in solutions of high ionic strength and low pH. The electrode is virtually insensitive to sodium from 0.1 to 10.0 m*M*. A typical electrode has a lifetime of at least 48 hr.

REFERENCE

1. **Bagnasco, S.,** *Anal. Biochem.*, 149, 572, 1985.

23. THE DETECTION OF TERTIARY ALIPHATIC AMINES

A 1.0% solution of citric acid in acetic anhydride forms the color reagent. HPLC separation of tertiary aliphatic amines is performed on a 250 × 4-mm I.D. stainless steel column packed with Nucleosil® 5 $N(CH_3)_2$, a high-pressure pump being used for delivery of the color reagent. PTFE tubing, 10 × 0.25 mm, is used as the reaction coil in a temperature-controlled oil bath. Hexane/chloroform/ethyl acetate, 80:10:10, is used as the mobile phase at a flow rate of 2.0 mℓ/min, and the color reagent is delivered at a flow rate of 1.0 mℓ/min. The reaction temperature is 120°C and the detection wavelength 550 nm. The reagent gives a wine-red color with tertiary aliphatic amines and does not react with primary and secondary amines. Water in the reaction system interferes with the color formation and anhydrous citric acid should therefore be used. The concentration of citric acid affects the sensitivity of detection; optimal results are obtained at a concentration of 1%. Calibration graphs are linear over the range 0.05 to 0.5% amine, the lower limit of detection being 0.01%.

REFERENCE

1. **Kudoh, M., Matoh, I., and Fudano, S.,** *J. Chromatogr.*, 261, 293, 1983.

24. THE REACTION OF HISTAMINE WITH *o*-PHTHALALDEHYDE

The reaction of *o*-phthalaldehyde with primary amines and amino acids has been used under a variety of conditions for the formation of fluorescent products that can be detected in very small amounts. With amino acids the reaction is conducted in the presence of a thiol-like mercaptoethanol; histamine and many other amines do not require mercaptoethanol to produce highly fluorescent products.[1] The reaction of amino acids with *o*-phthalaldehyde and a thiol compound gives 2-alkylthio-substituted isoindole derivatives as the main product. Ronnberg et al.[2] have produced evidence that the stable fluorescent main product formed from histamine and *o*-phthalaldehyde is a substituted dihyrophenantroline derivative, and the reaction, therefore, requires at least 2 mol of *o*-phthalaldehyde per mole of histamine. The product on reversed-phase HPLC gave three chromatographic peaks, suggesting the existence of three structurally different isomers that are easily equilibrated.[3]

Some conclusions regarding the reaction of *o*-phthalaldehyde with histamine can be drawn. The reaction in alkaline solution is likely to produce a very unstable intermediate that is the precursor of the stable fluorescent species obtained on acidification; to obtain maximum fluorescence this intermediate must be "trapped" by acidification. The reaction of histamine with *o*-phthalaldehyde in alkaline solution reaches an optimum fluorescence yield after a few minutes at room temperature; the fluorescence intensity is increased by a factor of about 18 on acidification. The optimal reaction time is decreased by an increase in temperature with a simultaneous decrease in the overall fluorescence yield. The fluorescence yield is almost constant over the pH range 12.1 to 12.8; the pH, however, influences the fluorescence yield obtained on acidification, producing an optimum at pH 12.6.

Under the conditions used for histamine, the reaction rate for the formation of the fluorescent product from histidine and spermidine is much lower, yielding no optimum within 15 min. On acidification the reaction product of histidine affords a product of lower fluorescence intensity, whereas spermidine gives about a ninefold fluorescence intensity compared with the product in alkaline solution.

REFERENCES

1. **Allenmark, S., Bergstrom, S., and Enerback, L.,** *Anal. Biochem.*, 144, 98, 1985.
2. **Ronnberg, A. L., Hansson, C., Drakenberg, T., and Hakansson, R.,** *Anal. Biochem.*, 139, 329, 1984.
3. **Ronnberg, A. L., Hansson, C., and Hakansson, R.,** *Anal. Biochem.*, 139, 338, 1984.

25. THE GAS CHROMATOGRAPHIC DETERMINATION OF HISTAMINE AND ITS BASIC METABOLITES

Wollin and Navert[1] have described a procedure consisting of ion exchange chromatography and chromatography on silicic acid for the isolation and purification of histamine and its basic metabolites. The isolated compounds are then prepared for gas chromatography by derivatization with heptafluorobutyric anhydride and acetic anhydride.

Ion Exchange Chromatography

Cell fractions or tissue samples are homogenized in 0.4 *N* perchloric acid and centrifuged. Human urine is collected in the presence of 0.5 g of sodium metabisulfite as preservative. If the sample cannot be analyzed the next day, the pH is brought to less than 2 with HCl and the sample stored at 5°C. Before analysis the sample is adjusted to pH 6.5 and aliquots are diluted with 25 mℓ of sodium phosphate buffer, pH 6.5 (EDTA, 10^{-6} *M*) and 25 mℓ of distilled water. After addition of 10 nmol of 2-methylhistamine as an internal standard, the diluted sample is passed through a column of Amberlite® CG-50 (Na^+), 0.7 × 7 cm, previously washed with 15 mℓ of the buffer. The column is then washed with 25 mℓ of 0.1 *N* sodium phosphate-citric acid buffer, pH 5.0 and 25 mℓ of water and eluted with 6 mℓ of 0.8 *N* HCl. The eluate is collected in 20-mℓ culture tubes with Teflon-lined screwcaps in which derivatization is performed.

Derivatization with Heptafluorobutyric Anhydride

The sample is lyophilized, and 3 mℓ of distilled ethyl acetate, 10 μℓ of 0.1 *M* triethylamine, and 300 μℓ of heptafluorobutyric anhydride are added. After being kept at 50°C with occasional shaking for 1 hr the reaction vessels are centrifuged and the supernatant is transferred to a smaller tube and evaporated under vacuum nearly to dryness in a Roto-Vac® (Savant).

Silicic Acid Chromatography

A silicic acid column is prepared by suspending 0.3 g of silicic acid in 2 mℓ of ethyl acetate and allowing it to settle in a glass column, 0.7 × 10 cm. The column is washed with 1 mℓ of ethyl acetate/isopropanol, 3:7, followed by 1 mℓ of *n*-hexane, and the sample is transferred to the column with 2 × 250 μℓ of methylene chloride/ethyl acetate, 5:1. After washing with 2 mℓ of ethyl acetate followed by 1 mℓ of ethyl acetate/isopropanol, 9:1, the heptafluorobutyryl derivatives of histamine and its metabolites are eluted with 4 mℓ of acetone/isopropanol, 3:7, and the eluate is evaporated to dryness under vacuum.

Acylation of Histamine

The sample is suspended in 150 to 200 μℓ of ethyl acetate and centrifuged. An aliquot of 100 to 200 μℓ of the supernatant is transferred to a 200-μℓ reaction vial and one or two crystals of sodium acetate and 2 μℓ of acetic anhydride are added. The vial is immediately capped, the contents mixed, and after 45 min at room temperature with occasional mixing, 2 μℓ of the sample is chromatographed.

Gas Chromatography

Chromatography is performed on a 2-mm × 2-m glass coil packed with 3% OV-275 on Chromosorb® W with helium as carrier gas at a flow rate of 40 mℓ/min. The column temperature is maintained at 170°C and a nitrogen-phosphorus detector is used. The detection limit for N^1-methylhistamine and histamine is 32 pmol and recovery ranges between 77 and 91%.

REFERENCE

1. **Wollin, A. and Navert, H.,** *Anal. Biochem.*, 145, 73, 1985.

26. THE CHROMATOGRAPHIC DETERMINATION OF HISTAMINE AND N^τ-METHYLHISTAMINE USING ELECTROCHEMICAL DETECTION

A procedure has been described for the determination of histamine and N^τ-methylhistamine in rat brain. Perchloric acid extracts of brain are chromatographed on a strong cation exchange resin and the amines in the eluate are derivatized with the water-soluble Bolton-Hunter reagent (sulfo B-H, sulfosuccinimidyl-3-[4-hydroxyphenyl] propionate). The derivatives are purified using a cellulose phosphate fibrous cation exchanger and are quantified with an electrochemical detector.

Tissue Preparation

Rat brains are quickly removed after decapitation of the animals, and the hypothalami are dissected and immediately frozen and weighed. A hypothalamus or a part of one is homogenized with 0.5 mℓ of ice-cold 3% perchloric acid in a glass homogenizer. The homogenizer is rinsed three times with 0.5 mℓ of 3% perchloric acid, the homogenates and rinses are combined, a suitable amount of N^{τ}-methylhistamine is added as an internal standard, and the mixture is centrifuged at 10,000 × *g* for 30 min at 4°C. A chromatographic column is prepared by sequentially washing Dowex® 50 W cation exchange resin, dry mesh 100-200, with excess 2 *M* HCl, 2 *M* NaOH, and water. The washed resin is equilibrated with 0.2 *M* sodium phosphate buffer, pH 6.5, and is packed into a glass column 16 × 5 mm I.D. The supernatant of the tissue sample is poured on the column which is then washed with 5 mℓ of water and 4 mℓ of 2 *M* HCl. The amines are eluted with 2.5 mℓ of 3.5 *M* HCl, and the eluate is collected in a silanized glass tube and evaporated to dryness.

Derivatization

The dry sample is dissolved in 0.8 mℓ of water, and 150 μℓ of reaction buffer and 100 μℓ of a 0.02 *M* sulfo B-H solution are added. The reaction buffer is prepared by mixing 0.1 *M* Na_2CO_3 and 0.1 *M* $NaHCO_3$ in the ratio of 10:1 by volume. Sulfo B-H solution is prepared fresh each day. To prepare standards, 150 μℓ of reaction buffer and 100 μℓ of 0.02 *M* sulfo B-H are added to 1 mℓ of a standard solution containing known amounts of N^{τ}- and N^{α}-methylhistamine. The pH of the mixture is between 9.8 and 10.0 when the HCl is completely evaporated. The samples are shaken vigorously for 30 sec using a vortex mixer. The derivatives are extracted using a cellulose phosphate fibrous cation exchanger packed into a glass column, 17 mm I.D. × 13 mm, which is successively washed with 0.1 *M* NaOH, water, 0.1 *M* HCl, and finally with enough water to obtain a pH between 5 and 6. The reaction mixture is brought within the range pH 5.5 to 6.0 with 0.1 *M* HCl and applied to the column, which is then washed with 5 mℓ of water and 6 mℓ of 1.0 *M* HCl. The derivatives are eluted with 2.5 mℓ of 0.1 *M* HCl, the first 0.5 mℓ being discarded and the next 2.0 mℓ being evaporated and stored at 4°C in the dark prior to chromatographic analysis.

Chromatography

Methanol is added to the dry sample followed by water to a final volume range of 110 μℓ to 1 mℓ of 10% methanol solution. One hundred microliters of the solution is injected into the chromatographic system. The column used is Beckman Ultrasphere® ODS, 250 × 4.6 mm I.D., with a Whatman® CSKI guard column packed with Waters μBondapak® C_{18}/Corasil®. The mobile phase consists of 0.14 *M* sodium acetate/methanol, 17:73, containing 3.89 *M* 1-octanesulfonic acid and 56 mg of EDTA, the solution being adjusted to pH 3.48 by adding glacial acetic acid and then degassed under vacuum. Chromatography is performed at room temperature with a flow rate of 1.0 mℓ/min. The electrochemical detector is operated in the oxidative screen mode.

The sulfo B-H derivatives of histamine and N^{τ}- and N^{α}-methylhistamine are completely separated. The pH is critical for separation, since the retention times of the derivatives are sensitive to even small changes, increasing linearly with increasing pH. The addition of 1-octanesulfonic acid to the mobile phase is important, because it selectively affects the retention of the imidazole amine derivatives and enables their separation from reaction products such as *p*-hydroxyphenylpropionic acid. The B-H reagent has a highly electroactive phenolic ring, and its hydroxy group is excellent for quantifying low concentrations of the histamine derivatives with electrochemical detection. Additionally, the high stability of the derivatives permits a high degree of precision in the determination. The detection limits are 0.1 pmol of histamine and 0.2 pmol of N^{τ}-methylhistamine.

REFERENCE

1. **Mine, K., Jacobson, K. A., Kirk, K. L., Kitajima, Y., and Linnoila, M.,** *Anal. Biochem.*, 152, 127, 1986.

27. THE FLUOROMETRIC ESTIMATION OF HISTAMINE

After extraction from biological samples, histamine is converted to the *o*-phthalaldehyde derivative. A sample of 1 mℓ is made alkaline (pH 12.1) by adding 0.2 mℓ of 2 *N* NaOH. Derivatization is carried out by mixing with 0.1 mℓ of 0.1% *o*-phthalaldehyde in methanol. The fluorometric reaction is allowed to proceed for 40 min at 0°C in the dark and is then stopped by acidifying with 0.2 mℓ of 1 *N* H_2SO_4. The fluorescence is stable for 2 hr.

REFERENCE

1. **Robert, J. C., Vatier, C., Nguyen Phuoc, B. K., and Bonfils, S.,** *J. Chromatogr.*, 273, 275, 1983.

28. THE DETERMINATION OF CYSTAMINE BY HPLC

A sensitive and specific assay for cystamine using HPLC has been described. The method is based on postcolumn derivatization of cystamine with *o*-phthalaldehyde in the presence of 2-mercaptoethanol and sodium hypochlorite.

Sample Preparation

Male rats, 160 to 180 g, are decapitated and tissues (brain, heart, kidney, liver, stomach, spleen, and testis) removed and placed on ice. The tissue is quickly homogenized in 4 vol of ice-cold 5% trichloroacetic acid and the homogenate centrifuged at 8000 × *g* for 30 min at 4°C. The supernatant is extracted four times with 3 vol of water-saturated ether, aliquots of the supernatant are lyophilized, redissolved in 0.1 *N* sodium citrate buffer, pH 2.2., and used as the "trichloroacetic acid extract".

Pretreatment of Trichloroacetic Acid Extract

The trichloroacetic acid extract is applied to a column of cation exchange resin, Dowex® 50W X8, 200-400 mesh in the hydrogen form. The column, 6 × 40 mm, is eluted with 4 mℓ of 3 *N* HCl, followed by 1.2 mℓ of 6 *N* HCl, and the eluates are discarded. The next 2 mℓ eluated with 6 *N* HCl is collected as the "cystamine fraction". This fraction is lyophilized and the residue redissolved in 0.1 *M* sodium citrate buffer, pH 2.2. An aliquot is then applied to the HPLC column. The column, 5 cm × 4 mm I.D. packed with a cation exchange resin, ISC-05(SO 504), is used at 70°C. The mobile phase consists of 0.1 *M* sodium borate, pH 11.10, used at a flow rate of 0.7 mℓ/min. For postcolumn derivatization the column effluent is mixed with 0.08% sodium hypochlorite, followed by the addition of 0.08% *o*-phthalaldehyde in the presence of 0.2% 2-mercaptoethanol. The fluorescence of the eluate is measured spectrofluorometrically at excitation and emission wavelengths of 348 and 450 nm, respectively.

Cystamine is identified by its retention time, 7.5 ± 0.1 min, and is quantified by peak height measurement. The concentration of added cystamine and the fluorescence intensity exhibit a linear relationship over the concentration range of 2 to 200 pmol. Since cysteamine in biological materials is quantitatively converted to cystamine during the sample preparation, the method is suitable for assaying the cysteamine plus cystamine content in organs and tissues.

REFERENCE

1. **Ida, S., Tanaka, Y., Ohkuma, S., and Kuriyama, K.,** *Anal. Biochem.*, 136, 352, 1984.

Section III.II

SUMMARY TABLES FOR DETECTION REAGENTS

The following tables represent summary information on detection reagents and the sensitivity of detection of amino acid derivatives.

Table 1
THE SENSITIVITY OF DETECTION OF AMINO ACIDS USING A RUBIDIUM SILICATE BEAD DETECTOR (AFID) AND A FLAME IONIZATION DETECTOR (FID)

	Signal-to-noise ratio		
Amino acid	**FID**	**AFID**	**Enhancement**
Alanine	12.4	1320	106
Arginine	3.7	780	210
Cystine	4.5	610	135
Glutamic acid	10.0	910	91
Leucine	14.1	1370	97
Lysine	7.3	1360	186
Methionine	7.6	690	90
Proline	10.0	2000	200
Serine	7.6	1430	188
Threonine	11.3	1320	116

Note: *N(O,S)*-Pentafluoropropionyl amino acid *n*-propyl esters were separated by capillary gas chromatography on Chirasil-Val®. The sensitivity of the two detectors was compared using identical chromatographic conditions.

REFERENCE

1. **Frank, H., Vujtovic-Ochenga, N., and Rettenmeier, A.,** *J. Chromatogr.*, 279, 507, 1982.

Table 2
DETECTION LIMITS OF *o*-PHTHALALDEHYDE DERIVATIVES OF AMINO ACIDS

Amino acid	Detection limit (pmol)[a]	Sensitivity (relative to histidine)
Alanine	0.25	4.3
Arginine	0.24	4.5
Asparagine	0.21	5.2
Aspartic acid	0.17	6.3
Carboxymethyl cysteine	0.24	4.5
Cysteic acid	0.17	6.5
Glutamic acid	0.20	5.5
Glutamine	0.51	2.1
Glycine	0.42	2.5
Histidine	1.07	1.0
Hydroxylysine	0.46	2.3
Isoleucine	0.18	5.9
Leucine	0.23	4.7
Lysine	1.07	1.0
Methionine	0.28	3.8
Methionine sulfoxide	0.19	5.6
Phenylalanine	0.34	3.1
Proline	5.95	
Serine	0.28	3.9
Threonine	0.29	3.7
Tryptophan	0.46	2.3
Tyrosine	0.25	4.3
Valine	0.25	4.3

Note: The *o*-phthalaldehyde derivatives of amino acid were separated on an Ultrasphere® ODS column and detected fluorometrically.

[a] Detection limit is expressed as the minimum amino acid mass placed on the column corresponding to twice signal-to-noise ratio.

REFERENCE

1. **Umagat, H., Kucera, P., and Wen, L.-F.,** *J. Chromatogr.*, 239, 463, 1982.

Table 3
THE RELATIVE FLUORESCENCE OF *o*-PHTHALALDEHYDE-3-MERCAPTOPROPIONIC ACID AMINO ACID DERIVATIVES

Amino acid	FK′	Relative fluorescence
Alanine	15.5	0.86
α-Aminoadipic acid	16.0	0.89
α-Aminobutyric acid	15.0	0.83
Arginine	16.5	0.92
Asparagine	20.3	1.13
Aspartic acid	13.5	0.75
Carnosine	22.0	1.22
Citrulline	17.0	19.4
Glutamic acid	14.0	0.78
Glutamine	18.0	1.00
Glycine	17.0	0.94
Histidine	8.5	0.47
Isoleucine	18.3	1.02
Leucine	18.3	1.02
Lysine	2.9	0.16
Methionine	16.5	0.92
1-Methylhistidine	17.8	0.99
3-Methylhistidine	19.5	1.08
Norvaline	21.0	1.17
Ornithine	3.3	0.18
Phenylalanine	18.0	1.00
Serine	16.5	0.92
Taurine	8.5	0.47
Threonine	15.0	0.83
Tyrosine	15.0	0.83
Valine	16.3	0.91

Note: The calculated apparent fluorescence constants, FK′, were calculated by considering the respective area (μV output unit) as an expression of the fluorescence yield per 1 pmol of the individual amino acid derivative. The relative fluorescence was calculated by taking phenylalanine as the reference amino acid (Phe = 100).

REFERENCE

1. **Godel, H., Graser, T., Földi, P., Pfaender, P., and Fürst, P.,** *J. Chromatogr.*, 297, 49, 1984.

Table 4
THE FLUORESCENCE INTENSITY OF AMINO ACIDS AFTER DERIVATIZATION WITH 4-CHLORO-7-NITROBENZOFURAZAN (NBD-Cl)

Amino acid	Fluorescence intensity
Alanine	7.0
Arginine	12.0
Aspartic acid	10.0
Cystine	0.9
Glutamic acid	12.0
Glycine	1.5
Histidine	1.2
Hydroxylysine	0
4-Hydroxyproline	98
3-Hydroxyproline	101
Isoleucine	0
Leucine	0
Lysine	1.2
Methionine	0.8
Phenylalanine	10.0
Proline	100
Serine	0.7
Threonine	0.9
Tryptophan	3.6
Tyrosine	1.4
Valine	0

Note: The NBD derivatives of amino acids were separated on silica gel plates, 0.2 mm thickness, using methanol/toluene/acetone/tributylamine, 15:40:40:5, as solvent. The fluorescence of known amounts of the NBD derivatives was measured and calculated for an equal concentration with the fluorescence of the NBD derivative of proline taken as 100%.

REFERENCE

1. **Bisker, A., Pailler, V., Randoux, A., and Borel, J. P.,** *Anal. Biochem.*, 122, 52, 1982.

Section IV

Methods of Sample Preparation Including Derivatization

Section IV

METHODS OF SAMPLE PREPARATION INCLUDING DERIVATIZATION

This section describes procedures for the preparation and derivatization of samples prior to chromatography. It is hoped that the methods can, in many cases, be used directly by researchers for the preparation of samples for chromatography, but they should at least serve as a guide illustrative of the kind of steps which are required for workers designing their own procedures.

1. The hydrolysis of peptides
2. Gas phase acid hydrolysis of proteins
3. The preparation of free amino acids from strong acid salts
4. The alkaline hydrolysis of proteins for tryptophan determination
5. The removal of glucose interference during amino acid analysis of plasma by gas chromatography (GC)
6. A cleanup procedure for serum amino acids
7. The preparation of amino acid-*o*-phthalaldehyde derivatives
8. The preparation of *o*-phthalaldehyde derivatives of phosphoamino acids for high performance liquid chromatography (HPLC)
9. The preparation of the phenylthiocarbamyl derivatives of amino acids for HPLC
10. The conversion of anilinothiazolinones to phenylthiohydantoins
11. *p-N,N*-Dimethylaminophenylisothiocyanate (DMAPI) as an electrochemical label for HPLC determination of amino acids
12. The preparation of dimethylaminoazobenzenethiohydantoin (DABTH) derivatives of iodoamino acids for thin layer chromatography (TLC)
13. The dansylation of amino acids for HPLC analysis
14. Derivatization of amino acids with 4-dimethylaminoazobenzene-4′-sulfonyl chloride (DABS-Cl) for quantitative determination
15. Improved preparation of DABS-amino acids for HPLC
16. The derivatization of amino acids with 9-fluorenylmethyl chloroformate (FMOC-Cl) for HPLC
17. The preparation of heptafluorobutyryl and trifluoroacetyl amino acid *n*-butyl esters for GC
18. The preparation of *N(O,S)*-pentafluoropropionyl amino acid *n*-propyl esters for GC
19. The preparation of t-butyldimetnylsilyl derivatives of amino acids for GC
20. Isocyanates as reagents for the separation of enantiomers of amino acids and *N*-methylamino acids
21. The derivatization of enantiomeric amino acids to give diastereomeric thiourea derivatives
22. The preparation of the chiral reagent L-aspartylcyclohexamide (AspcHex) for the chromatographic resolution of enantiomeric amino acids
23. The derivatization of low molecular weight aliphatic amines in air for liquid chromatographic determination
24. Benoxaprofen as a fluorescence marker for TLC and HPLC assay of amines
25. The preparation of a bonded *p*-acetylphenoxy stationary phase for the liquid chromatography of amines
26. Derivatization of aliphatic amines for HPLC and TLC
27. The derivatization of amines with phenylisocyanate for HPLC
28. The preparation of dansyl amines

1. THE HYDROLYSIS OF PEPTIDES

For the quantitative determination of amino acids, peptide samples are divided into two aliquots. One aliquot is hydrolyzed under reducing conditions for the analysis of all amino acids except proline and cystine. The other is treated with performic acid before hydrolysis and is used for the determination of proline and cystine as cysteic acid.

Hydrolysis

Heavy-wall Pyrex® ignition tubes, 10 × 70 mm, are used for hydrolysis. The tubes are freed from amino acid contamination by baking at 500°C overnight in a muffle oven, the tubes then being protected from dust until use. For hydrolysis the peptide sample in 10- to 500-$\mu\ell$ solution is divided into two aliquots which are added to hydrolysis tubes and dried in vacuo. To one of the tubes 50 $\mu\ell$ of constant boiling HCl containing 2% (v/v) thioglycolic acid is added. The middle part of the tube is heated in a torch flame and pulled to form a tube neck suitable for flame sealing. The open end of the tube is attached by heavy-wall Tygon® tubing to a vacuum system and the tube is evacuated using a high vacuum after freezing the sample in liquid nitrogen. The sample is then allowed to melt while under vacuum and the tube is flame sealed at the construction.

The peptide aliquot in the second tube is treated with performic acid, 90 $\mu\ell$ of formic acid, and 10 $\mu\ell$ of 30% hydrogen peroxide and the reaction is allowed to proceed at room temperature for 15 min. After drying in a vacuum desiccator, 50 $\mu\ell$ of constant boiling HCl is added and the tube sealed as described previously. Hydrolysis is conducted at 110°C for 20 hr or longer as required. After hydrolysis the tubes are opened, dried in a vacuum, and the residue dissolved in 130 $\mu\ell$ of buffer for loading on the chromatography column.

REFERENCE

1. **Böhlen, P. and Schroeder, R.,** *Anal. Biochem.,* 126, 144, 1982.

2. GAS PHASE ACID HYDROLYSIS OF PROTEINS

The sensitivity limit of amino acid analysis depends on the background contaminants. When nanogram quantities of proteins were hydrolyzed for amino acid composition analysis, the most likely source of amino acid contaminants was attributed to aqueous 6 *N* HCl. Gas phase acid hydrolysis reduced the background observed during liquid phase hydrolysis to about 25%. Gas phase hydrolysis is also technically easier to perform than liquid phase hydrolysis. It does not require vacuum sealing of each sample tube, because more than 12 sample tubes can be sealed under vacuum in the same hydrolysis vessel.

Procedure

Approximately 0.05 to 2 μg of protein samples, which is preferably free from salt, is placed in hydrolysis tubes, 50 × 4 mm I.D., and dried in a vacuum centrifuge. Twelve samples can be placed simultaneously in a Waters® gas phase hydrolysis vessel. At least one standard is included in each batch for hydrolysis; one or two blanks are included when less than 0.1 μg of protein is hydrolyzed. About 400 μℓ of 6 *M* HCl is then placed in the bottom of the vessel, which is purged with argon for 10 sec and is finally sealed through the control of the Teflon® valve with a vacuum of lower than 0.1 mbar. Hydrolysis is carried out at 110°C for 24 hr. After hydrolysis the tubes are dried under vacuum.

REFERENCE

1. **Knecht, R. and Chang, J.-Y.,** *Anal. Chem.*, 58, 2375, 1986.

3. THE PREPARATION OF FREE AMINO ACIDS FROM STRONG ACID SALTS

Acidic and neutral amino acids may be recovered from their strong acid salts or from large excesses of strong acids by passing them through a column of cross-linked poly-4-vinylpyridine. Basic amino acids are eluted as mono acid salts. High recoveries are obtained with submicromolar samples.

Procedure

Four hundred milligram (3.4 mEq) of poly-4-vinylpyridine are packed in a Pasteur pipette (6 × 30-mm bed) above a glass wool plug. The column is saturated with water and allowed to equilibrate for several hours. The pipette is then immersed in a boiling water bath and 6 mℓ of water is passed through the resin using gentle air pressure. The column is placed in a 10-mℓ flask with a neck long enough to surround the resin bed. The flask is immersed to the top of the neck in boiling water. In a typical procedure for hydrophobic amino acids, leucine hydrochloride, 1 mEq (167.6 mg), dissolved in 4 mℓ water is preheated in the bath and applied to the column. Two 1-mℓ washes of hot water are added, and gentle nitrogen pressure is applied to give a flow rate of about 1 mℓ/min. The column effluent is evaporated to dryness using a rotary evaporator to give a yield of 131.0 mg (99.9%) of leucine. In a typical procedure for water-soluble amino acids, a column of poly-4-vinylpyridine is packed, equilibrated, and washed as already described. Subsequent steps are carried out at room temperature. Cysteine hydrobromide, 1 m*M* (202.1 mg), dissolved in 1 mℓ of water is applied to the column followed by two 1-mℓ washes of water. The product is dried to give L-cysteine in a yield of 120.5 mg (99.4%).

REFERENCE

1. **Jewett, D. M. and Ehrenkaufer, R. L.,** *Anal. Biochem.*, 122, 319, 1982.

4. THE ALKALINE HYDROLYSIS OF PROTEINS FOR TRYPTOPHAN DETERMINATION

Tryptophan analysis of proteins requires a hydrolysis step that can cause variable losses of this amino acid. Acid hydrolysis techniques can cause significant losses of tryptophan, particularly in the presence of carbohydrate, and alkaline hydrolysis is often the method of choice when precise determination of tryptophan is required. Levine[1] has described a method of alkaline hydrolysis using a simple bench-top block heater. Following hydrolysis, tryptophan may be quantitated using the sensitive fluorometric method of Nakamura and Pisano[2] or by other methods.

Procedure

The hydrolysis solution of 5 *M* NaOH with 3% (v/v) thiodiglycol added to minimize loss of tryptophan is made fresh daily. Hydrolysis vials (Wheaton® 1-dram borosilicate glass vials with a PTFE on silicone cap liner) are used without additional treatment or cleaning. Unlined caps are used, and if rubber-lined caps are supplied the liner must be removed before inserting the PTFE/silicone liner. The caps are boiled in water to remove loosely adhering particles. The vials are fitted into a heating block (Pierce Reacti-Block® C) seated in a standard, small, bench-top heater. The temperature is set at 155°C; with the cover in place temperature variations are no greater than 1°C.

In the hydrolysis vial are placed 25 μℓ of protein solution, 2 to 200 nmol in tryptophan, followed by 500 μℓ of hydrolysis solution. The vial is flushed with a stream of argon or nitrogen for 20 sec, and the cap with the PTFE liner in place is screwed on. The vial is placed in the heating block for the desired time; 30 to 45 min is adequate for most proteins. Extraction of silicate from the glass increases considerably when hydrolysis is conducted for more than 45 min. Heating for 90 min or longer decreases tryptophan recovery. After cooling, 500 μℓ of 25% acetic acid and then 2.5 mℓ of 200 m*M* borate, pH 9.0, are added. The borate solution is prepared by titrating boric acid with NaOH. The vial is recapped and centrifuged for 10 min to remove insoluble silicates. A 100-μℓ aliquot of the supernatant may be used for tryptophan determination. The ease of hydrolysis makes it possible to obtain results at several times of hydrolysis by using different vials. It is then possible to extrapolate to infinite hydrolysis time.

REFERENCES

1. **Levine, R. L.,** *J. Chromatogr.*, 236, 499, 1982.
2. **Nakamura, H. and Pisano, J. J.,** *Arch. Biochem. Biophys.*, 172, 98, 1976.

5. THE REMOVAL OF GLUCOSE INTERFERENCE DURING AMINO ACID ANALYSIS OF PLASMA BY GAS CHROMATOGRAPHY

Studies on the GC analysis of amino acids in deproteinized plasma demonstrated that, although most amino acids were successfully resolved, there was a major component responsible for several chromatographic peaks eluting over a broad area, and preventing the resolution of hydroxyproline, methionine, aspartic acid, and phenylalanine. This component was shown to be glucose. The interference may be due to the hydroxyl groups of glucose reacting with the acylating agent, heptafluorobutyric anhydride, forming derivatives that coelute with some of the amino acids. Alternatively, glucose may react with free amino acids forming the so-called Maillard compounds, which may directly contribute to the spurious peaks. However, it was possible to overcome this interference by the enzymic conversion of glucose to glucose-6-phosphate.

Procedure

To 500 μℓ of plasma are added aqueous hexokinase solution (200 U in 200 μℓ) and 40 μℓ of 0.08 *M* ATP solution, and the pH is adjusted to 7.4 with 0.5 *M* NH_4OH. The mixture is incubated at 30°C for 15 min, deproteinized by addition of 2 mℓ of methanol/12 *M* HCl (4:1), and centrifuged at 1600 × *g* for 10 min. The supernatant is taken to dryness for derivatization and chromatography.

REFERENCE

1. **Labadarios, D., Shephard, G. S., Moodie, I. M., Jardine, L., and Botha, E.,** *J. Chromatogr.*, 339, 365, 1985.

6. A CLEANUP PROCEDURE FOR SERUM AMINO ACIDS

The GC determination of amino acids in biological samples generally requires the preliminary removal of proteinaceous material. Physiological fluids are usually treated with picric acid, sulfosalicylic acid, ethanol, or chloroform; precipitated proteins are removed by centrifuging; and amino acids are isolated by cation exchange cleanup. A simple technique involving dilution of the sample with acetic acid eliminates the protein precipitation step and in combination with cation exchange cleanup has been used as a preliminary step in the determination of free amino acids in physiological fluids. The yields of some amino acids, however, proved to be higher with the picrate method, rendering the acetic acid procedure less attractive.

Husek et al. made a comprehensive study of the acetic acid method and found that the degree of cross-linking of the ion exchange polystyrene matrix with divinylbenzene markedly influenced the yields of some amino acids.

Procedure

Thin-wall polyethylene tubing, 3 to 3.2 mm I.D., is cut into pieces about 25 mm long, and a plug of glass wool 3 to 4 mm is placed at one end. A slurry of Dowex® 50W-X2 (H^+) resin, 100-200 mesh, in water is sucked in the tubing with the help of a syringe to form a 15-mm-high column of wet resin. A peristaltic pump is used to wash the resin bed with 1 mℓ of *N* HCl and 2 mℓ of distilled water and for the subsequent uptake of amino acids by suction.

Serum or urine, 50 to 100 μℓ, is placed in a 1.5-mℓ Silli-Vial® (Pierce) and mixed with ten times the volume of 25% acetic acid containing internal standard. The solution is sucked slowly through the plastic column, which is immersed in the liquid by the open end and connected by a piece of glass tubing, 3.2 mm O.D. and 2 mm I.D., to the silicone tubing of the peristaltic pump. After the fluid has been completely sucked off, 250 to 500 mℓ of water is added and sucked in the resin at an increased flow rate. To elute the amino acids a polypropylene tip filled with glass wool is placed in the lower "suck-in" end of the column. The "suck-off" end of the column is connected to a 1-mℓ syringe that is filled with 0.7 mℓ of 2 *N* NH_4OH and the amino acids are eluted by pushing the syringe plunger down for 15 to 20 sec.

Husek et al.[1] found that there was a continuous decrease in uptake of some amino acids as the percentage of cross-linkage of the resin employed in the method increased. When X8 resin was used in place of X2 resin the recoveries of phenylalanine, tryosine, and α-aminocaproic acid, the internal standard for GC, decreased to 70 to 80%, while the recoveries of arginine and tryptophan were only 40 to 50%. Yields were even lower when X12 resin was used. The X2 resin increased recoveries for arginine and tryptophan to 70 to 80%, recoveries for other protein amino acids being close to 100%. The use of reversed flow for the uptake made recoveries of all protein amino acids more reproducible.

REFERENCE

1. **Husek, F., Herzogova, G., and Felt, V.,** *J. Chromatogr.*, 236, 493, 1982.

7. THE PREPARATION OF AMINO ACID-*o*-PHTHALALDEHYDE DERIVATIVES

o-Phthalaldehyde has found acceptance as a postcolumn derivatization reagent for the determination of amino acids. Additionally, however, the chromatographic separation of these amino acid derivatives has been studied as a method of amino acid determination. A considerable number of procedures for the preparation of amino acid-*o*-phthalaldehyde derivatives have been suggested, several of which are now described.

Technique I

Graser et al.[1] have used a two-step procedure for the *o*-phthalaldehyde derivatization of amino acids, which minimized the occurrence of *o*-phthalaldehyde adducts that interfere with the baseline during chromatography at high sensitivities. The derivatization reagent is prepared by dissolving 50 mg of *o*-phthalaldehyde in 4.5 mℓ of methanol, and 50 μℓ of 3-mercaptopropionic acid together with 0.5 mℓ of 1.0 *M* sodium borate, pH 9.5. The reagent is kept in the dark at 4°C and is freshly prepared every week. In step 1 of the derivatization, 5 μℓ of a standard solution of amino acids or 20 μℓ of a sample to be analyzed is mixed with 200 μℓ of methanol, 200 μℓ of 1.0 *M* sodium borate, pH 9.5, and 50 μℓ of derivatizing reagent. In step 2 after a 2.5-min incubation, 200 μℓ of 12.5 m*M* sodium phosphate buffer, pH 7.2, is added to 5 μℓ of the standard mixture or 20 μℓ of the sample mixture. After mixing thoroughly, 20 μℓ of the resulting solution is injected into the liquid chromatograph.

Technique II

The derivatization reagent is prepared by dissolving 60 mg of *o*-phthalaldehyde in 10 mℓ of methanol, and the solution is prepared fresh daily and is kept in the dark on ice.[2] Immediately before use, 250 μℓ of this reagent is added to 250 μℓ of 0.4 *M* boric acid buffer, pH 9.4, and 25 μℓ of 2-mercaptoethanol; 250 μℓ of this mixture is vortex-mixed with 25 μℓ of amino acid standard (125 nmol/ℓ per amino acid, 3 mmol/ℓ of total amino acid) for 2 min. A 10-μℓ aliquot is removed and injected onto the column of the liquid chromatograph.

Technique III

A derivatizing reagent is prepared by adding to 50 μℓ of a 0.745 *M* ethanolic solution of *o*-phthalaldehyde (100 mg/mℓ) 6 μℓ of ethanethiol, 25 μℓ of Brij (10 mg/mℓ in water), and 1.25 mℓ of a 0.5 *M* Na_2HPO_4, pH 10.3 buffer.[3] The buffer is prepared by dissolving 70.98 g of Na_2HPO_4 in 900 mℓ of water, titrating to pH 10.3 with 1.0 *M* NaOH and bringing up to a volume of 1 ℓ with water. A fresh solution of the reagent is prepared daily. The freshly prepared reagent is opaque, presumably due to excess ethanethiol. Derivatization of an amino acid solution, containing about 1 mg/mℓ of amino acid, is performed by placing a 10- to 20-μℓ sample in a 500-μℓ conical reaction vial. An aliquot of the *o*-phthalaldehyde reagent equal to twice the volume of the amino acid solution is added to the mixture and allowed to react for 1.0 min. An aliquot, 2 to 10 μℓ, of the reaction mixture is removed and injected into the HPLC column.

Technique IV

The derivatizing solution is prepared by dissolving 50 mg of *o*-phthalaldehyde in 1.25 mℓ of absolute methanol and adding 50 μℓ of 2-mercaptoethanol and 11.2 mℓ of 0.4 *M* sodium borate, pH 9.5.[4] The mixture is flushed with nitrogen, stored in the dark, and used after 24 hr, 10 μℓ of 2-mercaptoethanol being added every 2 days to maintain the reagent strength. New solution is prepared every 2 weeks. For derivatization, 10-μℓ aliquots of standards or unknown plasma samples are mixed with 10 μℓ of a 2% sodium dodecyl sulfate solution in 0.4 *M* sodium borate, pH 9.5. Ten microliters of the derivatizing solution are added and the sample is mixed thoroughly. After 1 min 20 μℓ of 0.1 *M* potassium phosphate, pH 4.0, is added and the sample again mixed. An aliquot of 20 μℓ is then injected into the reversed-phase chromatographic column.

Technique V

A reagent solution is prepared by dissolving 50 mg of *o*-phthalaldehyde in 1.25 mℓ of absolute methanol followed by the addition of 50 μℓ of 2-mercaptoethanol and 11.2 mℓ of 0.4 *M* sodium borate, pH 9.5.[5] The solution is stable for 1 to 2 weeks. For derivatization, 5- to 10-μℓ aliquots of amino acid standards at a concentration of 25 or 50 pmol/mℓ for each amino acid or unknown samples are mixed with 5 μℓ of the reagent. After 1 min, 20 to 100 μℓ of 0.1 *M* sodium acetate, pH 7.0, is added, the solution is mixed, and a 20-μℓ sample subjected to analysis.

Technique VI

o-Phthalaldehyde solution is prepared by dissolving 270 mg of the compound in 5 mℓ of 99.5% ethanol.[6] Borate buffer is prepared by adjusting 0.4 *M* boric acid solution to pH 9.5 with 1 *M* NaOH. The reagent solution is prepared by adding 200 μℓ of 2-mercaptoethanol and 5 mℓ of *o*-phthalaldehyde solution to the borate buffer. The volume is then adjusted to 50 mℓ with buffer. The reagent is allowed to "age" for at least 24 hr prior to use, its strength being maintained by addition of 20 μℓ of 2-mercaptoethanol every 3 to 4 days. For derivatization of mixed solutions of amino acid standards containing approximately 40 pmol of each amino acid, 20 μℓ is reacted with 100 μℓ of *o*-phthalaldehyde reagent at room temperature. Then 5 to 10 μℓ is injected into the chromatographic column after 2 min. For physiological fluids, varying amounts of phosphate buffer are added for dilution purposes before derivatization. Injection into the LC column is made after a precise 2-min incubation period.

Technique VII

The derivatizing solution is prepared by dissolving 50 mg of *o*-phthalaldehyde in 4.5 mℓ of ethanol.[7] Ethanethiol, 50 μℓ, and borate buffer, 0.5 mℓ, are then added and the solution is mixed. The reagent solution should be protected from light and used for only 1 day. The borate buffer is prepared by adding enough boric acid with heat to 1 ℓ of water to form a saturated solution. After cooling the supernatant phase is filtered and adjusted to pH 9.5 with NaOH. For the derivatization of amino acid standards, to a 5.0-mℓ volumetric flask are added 1.00 mℓ of a 5.0-μ*M* amino acid solution, 100 mℓ of the borate buffer, and 0.50 mℓ of the derivatizing solution. The mixture is diluted to volume with methanol and allowed to react at room temperature for 1 min. For the derivatization of serum samples, 100 μℓ of serum is added to 2.00 mℓ of methanol, and the solution is mixed and then centrifuged. A 1.00-mℓ aliquot of the clear supernatant phase is then derivatized using the same procedure as used for amino acid standards. A 5.0-μℓ aliquot of the derivatized amino acid standard or serum extract is analyzed by HPLC.

Technique VIII

Derivatization conditions have been described that are designed to provide the best overall analysis for the primary amino acids present in plasma.[8] Whole blood is collected in tubes containing heparin anticoagulant, the plasma being separated by centrifuging and the plasma supernatant being deproteinized by ultrafiltration. The derivatizing reagent is prepared by dissolving 25 mg of *o*-phthalaldehyde in 5 mℓ of methanol and adding 45 mℓ of 0.1 *M* sodium borate, pH 9.5, followed by 200 μℓ of 2-mercaptoethanol. The solution is stored in an amber bottle and is freshly prepared every 2 weeks. If the reagent is exposed to light during storage it forms a yellow-colored product, which is no longer effective in derivatization. For derivatization, 10 μℓ of filtered plasma is taken, 100 μℓ of the reagent solution added, and the mixture is vortexed for 30 sec and incubated at room temperature for 2 min. Immediately preceding injection into the chromatographic column, 100 μℓ of methanol is added to the solution, which is vortexed, and 50 μℓ injected for LC. Standards for calibration curves are dissolved in physiological saline (0.9% NaCl), are derivatized, and 50 μℓ of solution injected into the LC column.

REFERENCES

1. **Graser, T. A., Godel, H. G., Albers, S., Foldi, P., and Furst, P.,** *Anal. Biochem.*, 151, 142, 1985.
2. **Price, S. J., Palmer, T., and Griffin, M.,** *Chromatographia*, 18, 62, 1984.
3. **Larsen, B. R. and West, F. G.,** *J. Chromatogr. Sci.*, 19, 259, 1981.
4. **Hogan, D. L., Kraemer, K. L., and Isenberg, J. I.,** *Anal. Biochem.*, 127, 17, 1982.
5. **Jones, B. N. and Gilligan, J. P.,** *J. Chromatogr.*, 266, 471, 1983.
6. **Lindroth, P. and Mopper, K.,** *Anal. Chem.*, 51, 1667, 1979.
7. **Hill, D. W., Walters, F. H., Wilson, T. D., and Stuart, J. D.,** *Anal. Chem.*, 51, 1338, 1979.
8. **Halfpenny, A. P. and Brown, Ph. R.,** *J. High Resolution Chromatogr. Chromatogr. Commun.*, 8, 243, 1985.

8. THE PREPARATION OF *o*-PHTHALALDEHYDE DERIVATIVES OF PHOSPHOAMINO ACIDS FOR HPLC

The derivatizing solution contains 10 mg (75 μmol) of *o*-phthalaldehyde, 14.9 mℓ of methanol, 5.3 mℓ of saturated boric acid solution, pH 9.5, and 25 μℓ of ethanethiol. It is prepared fresh daily. Amino acid samples containing 30 pmol to 30 nmol of each amino acid are mixed with 1 volume of derivatizing solution and incubated for 3 min before being immediately injected onto the column.

REFERENCE

1. **Carlomagno, L., Huebner, V. D., and Matthews, H. R.,** *Anal. Biochem.*, 149, 344, 1985.

9. THE PREPARATION OF THE PHENYLTHIOCARBAMYL DERIVATIVES OF AMINO ACIDS FOR HPLC

Technique I

Bidlingmeyer et al.[1] have described a new approach to the precolumn derivatization and analysis of amino acids, based on the use of a sample preparation module, the Pico-Tag® Workstation (Waters Assoc.). The unit incorporates two separate functions for the batchwise preparation of protein or peptide hydrolyzates and for the derivatization of amino acids. A vacuum pump equipped with cold and acid traps is connected to a manifold that allows attachment of a vacuum vial containing 12 sample tubes. The vials are equipped with a resealable closure that is drilled to allow application of a vacuum. The use of HCl/water vapor for hydrolysis, rather than placing 6 *M* HCl in the sample tubes, results in cleaner hydrolysis and quicker sample drying after hydrolysis. Removal of traces of oxygen from hydrolysis vials is accomplished by alternate evacuating and nitrogen flushing. Twelve samples can be sealed under vacuum in less than 5 min.

Hydrolysis of Proteins and Peptides

Proteins or peptides are dissolved in water or 0.1 *M* HCl at a concentration of 1.0 mg/mℓ, and for high sensitivity work at a concentration of 20 μg/mℓ. A volume corresponding to 0.1 to 5.0 μg is pipetted into a 50 × 6-mm tube; up to 12 tubes may be placed in the vacuum vial. The solvent is removed under vacuum, the vacuum released, and 200 μℓ of constant boiling HCl containing 1% (v/v) phenol pipetted into the bottom of the vacuum vial (not the hydrolysis tubes). The vacuum vial is reattached to the manifold, evacuated, and sealed under vacuum. The samples are hydrolyzed at 150°C for 1 hr or at 108°C for 24 hr and the residual HCl inside the vacuum vial is then removed under vacuum.

Derivatization of Amino Acids with Phenylisothiocyanate

Samples containing up to 25 nmol of each amino acid are placed in 50 × 6-mm tubes and dried under vacuum. They are then dried down again after adding 10 to 20 μℓ of ethanol/water/triethylamine, 2:2:1, to each tube. The derivatization reagent consisting of ethanol/triethylamine/water/phenylisothiocyanate, 7:1:1:1, is prepared fresh daily. Phenyl isothiocyanate is stored at −20°C under nitrogen to prevent the formation of breakdown products. Phenylthiocarbamyl amino acids are formed by adding 20 μℓ of reagent to the dried samples and sealing them in the vacuum vials for 20 min at room temperature. The reagents are then removed under vacuum. Longer reaction times and higher temperatures do not increase the yields for any amino acids, but decreased yields of glutamic and aspartic acids are observed at elevated temperatures. Liquid chromatography of the derivatives allows analysis of primary and secondary amino acids in protein hydrolyzates with a sensitivity of 1 pmol and an analysis time of 12 min.

Similar methods of precolumn derivatization have been used by other workers.

Technique II[2]

Protein and peptide samples are hydrolyzed in vacuo for 24 hr at 110°C in 6*N* HCl and the acid is removed by rotary evaporation. A sample containing about 25 nmol of each amino acid is dissolved in 100 μℓ of coupling buffer (acetonitrile/pyridine/triethylamine/water, 10:5:2:3) and dried by rotary evaporation. The amino acids are then dissolved once more in 100 μℓ of coupling buffer. Phenylisothiocyanate (5 μℓ) is added, and after the reaction is allowed to proceed at room temperature for 5 min, the solution is rotary evaporated to dryness under high vacuum. The resulting phenylthiocarbamyl derivatives are analyzed by reversed-phase HPLC.

Technique III[3]

To the solution of amino acids is added 3 μℓ of 50% ethanol followed by 7 μℓ of a mixture of 90% ethanol/triethylamine/phenylisothiocyanate, 7:2:1. After the reaction has proceeded for 10 min at room temperature, the mixture is dried under vacuum and dissolved in buffer for HPLC analysis.

REFERENCES

1. **Bidlingmeyer, B. A., Cohen, S. A., and Tarvin, T. L.,** *J. Chromatogr.*, 336, 93, 1984.
2. **Heinrikson, R. L. and Meredith, S. C.,** *Anal. Biochem.*, 136, 65, 1984.
3. **Koop, D. R., Morgan, E. T., Tarr, G. E., and Coon, M. J.,** *J. Biol Chem.*, 257, 8472, 1982.

10. THE CONVERSION OF ANILINOTHIAZOLINONES TO PHENYLTHIOHYDANTOINS

In the well-known Edman degradation[1] of proteins or peptides, the end product is a very unstable anilinothiazolinone. This thiazolinone is converted to a more stable phenylthiohydantoin derivative. This has been brought about by heating the thiazolinone under nitrogen in (1) 1.0 *N* HCl, (2) 25% trifluoroacetic acid, (3) 1.0 *N* HCl in methanol, or (4) 25% trifluoroacetic acid in methanol. In commercial sequencers this procedure has been automated. The methods for conversion and for identification or the final PTH-amino acids presented a number of problems. Conversion with aqueous acid produced deamidation of glutamine and asparagine, while methanolic acid caused esterification of acidic residues. Bhown and Bennett[2] overcame these problems by using trifluoroacetic acid in ethyl acetate as a conversion reagent and by modifying the chromatographic technique.

Procedure

Trifluoroacetic acid (10%) in sequencer-grade ethyl acetate containing 15 mg/ℓ of dithiothreitol is used as the conversion reagent. For manual conversion, 100 μℓ is added to a test tube containing 200 to 250 pmol of each amino acid derivative, the tube is flushed with nitrogen, incubated at 80°C for 10 min, and the solvents evaporated under nitrogen. For autoconversion 25 mℓ of conversion reagent is used and the reaction is performed at 50°C.

REFERENCES

1. **Edman, P.,** *Protein Sequence Determination,* Needleman, S. B., Ed., Springer-Verlag, Berlin, 1970, 215.
2. **Bhown, A. S. and Bennett, J. C.,** *Anal. Biochem.,* 150, 457, 1985.

11. DMAPI AS AN ELECTROCHEMICAL LABEL FOR HPLC DETERMINATION OF AMINO ACIDS

Amino acids react with DMAPI to form the corresponding substituted phenylthiohydantoins. The derivatized amino acid can be separated on a bonded C_8 column in 0.1 *M* phosphate buffer/25% acetonitrile (pH 2 or 6) and detected at a glassy carbon electrode.

Procedure

To a solution of amino acids in a 15 × 100-mm screw-cap tube containing 10 to 1000 ng of each amino acid is added 2 mℓ of freshly prepared pH 9.5 buffer prepared by mixing 50 mℓ water, 50 mℓ acetone, 5 mℓ of 0.2 *M* acetic acid, 1 mℓ triethylamine, and sufficient pyridine to bring the pH to 9.5. Five milliliters of a 10-ppm solution of DMAPI in acetone, or a minimum of a twofold excess of DMAPI by weight is added and the tube purged with nitrogen. The test tube is covered and heated at 50°C in a water bath for 20 min, and the solution is acidified with 5 mℓ of a solution of one part of glacial acetic acid saturated with HC1 and one part of water. After purging with nitrogen the mixture is again heated at 50°C for 20 min and then allowed to stand for 20 min. The solvent is removed in a vacuum and the resulting solid washed several times with isooctane to remove excess isothiocyanate. The purified product is dissolved in a measured volume of 0.1 *M* phosphate buffer, pH 2, and 10 μℓ is injected into the chromatographic system.

REFERENCE

1. **Mahachi, T. J., Carlson, R. M., and Poe, D. P.,** *J. Chromatogr.,* 298, 279, 1984.

12. THE PREPARATION OF DABTH DERIVATIVES OF IODOAMINO ACIDS FOR TLC

Iodotyrosines at 0.4 m*M* concentration are dissolved in water; iodothyronines are first dissolved in a small volume of concentrated NH_4OH and diluted with water to 0.4 m*M*. These solutions are stored at −20°C in the dark, and an aliquot is dried under vacuum in the dark just before use. An iodotyrosine or iodothyronine sample (159 nmol) dissolved in 40 μℓ of water/acetone/triethylamine/0.2 *M* acetic acid, 50:50:0.2:0.5, pH 9.5, is reacted with 4-*N,N*-dimethylaminoazobenzene-4′-isothiocyanate (DABITC) (40 nmol) in 20 μℓ of acetone for 75 min at 52°C with gentle shaking. After drying in the dark under vacuum over P_2O_5 the residue is dissolved in 50% trifluoroacetic acid, the tube is filled with nitrogen, stoppered, and incubated at 52°C for 45 min. The mixture is desiccated and the residue dissolved in 40 μℓ of ethanol. An aliquot of 0.05 to 0.1 μℓ of the solution is used to identify the DABTH-iodoamino acids by chromatography on polyamide sheets.

REFERENCE

1. **Marriq, C., Rolland, M., and Lissitzky, S.,** *Anal. Biochem.* 116, 89, 1981.

13. THE DANSYLATION OF AMINO ACIDS FOR HPLC ANALYSIS

Dansyl chloride (Dns-Cl) is a derivatizing agent for the trace detection of amino acids. Dns-amino acids have been separated by TLC and by HPLC using silica gel or chemically bonded reversed-phase packings. The analysis of Dns-amino acids by HPLC is an attractive proposition because of the speed, automation, and low detection limits possible. A high yield on derivatization, however, is required for the successful quantitative analysis of amino acids. The percentage conversion of amino acids to the dansyl derivatives was reported to depend on the ratio of dansyl chloride to amino acid.[1,2] The formation of dansyl/alanine was 89% when the ratio was 1:1, but at a ratio of 22:1 the yield was only 61%. The difference in yield was explained in terms of three competing reactions:

$$\text{AA-NH}_2 + \text{Dns-Cl} \rightarrow \text{AA-NH-Dns} + \text{HCl} \quad (1)$$

$$\text{H}_2\text{O} + \text{Dns-Cl} \rightarrow \text{Dns-OH} + \text{HCl} \quad (2)$$

$$\text{AA-NH-Dns} + \text{Dns-Cl} \rightarrow \text{Dns-NH}_2 + \text{other products} \quad (3)$$

where AA represents the amino acid, Dns-OH is dansyl sulfonic acid, and Dns-NH_2 is dansyl amide.

Reaction 1 is accelerated by high pH values, which also favor reaction 2. Conversion of the appropriate amino acids to didansyl derivatives is also pH dependent. An optimum pH must therefore be employed. Reaction conditions must also be chosen that minimize decomposition of Dns-amino acids (reaction 3). Tapuhi et al. found reaction conditions that afforded high yields of the mono- and didansyl amino acid derivatives.[3] Importantly, the yield was independent of the ratio of dansyl chloride to amino acid, at least over a 1000-fold range. The procedure provides a means of quantitative analysis for amino acids.

Procedure

The reagent solution consists of dansyl chloride dissolved in acetonitrile (1.5 mg/mℓ, 5.56 m*M*). In order to avoid possible interference, HPLC-grade acetonitrile is distilled over dansyl chloride. Dansyl chloride containing none of its hydrolysis products is used; partially hydrolyzed dansyl chloride does not dissolve completely in acetonitrile. Amino acids are dissolved in 40 m*M* lithium carbonate buffer, brought to pH 9.5 with HCl. For reaction at room temperature 1 mℓ of dansyl chloride solution is rapidly added to 2 mℓ of amino acid solution (0.001 to 1 m*M* in the buffer solution), the mixture is gently shaken for 2 min and is then allowed to stand at room temperature. The precipitate that initially appears on mixing dissolves immediately. The reaction is terminated by the addition of 100 μℓ of 2% ethylamine or methylamine hydrochloride solution. All amino acids are derivatized in high yields after reaction for 35 min with dansyl chloride.

For reaction at higher temperatures, 2 mℓ of the amino acid solution are preheated to the required temperature in a Pierce Reacti-Therm® heating module, and then 1 mℓ of dansyl chloride solution is added. After shaking for a short time, the reaction vial is maintained at the desired temperature in the heating block. The reaction is terminated by adding 100 μℓ of 2% ethylamine solution. Reaction vials in all cases are wrapped in aluminum foil to exclude light. Derivatization at 60°C can give reasonable results even in 5 min. Lower yields are obtained at 95°C with 5-min reaction times, but the reaction can be conveniently carried out at temperatures of up to 60°C with short reaction times. A small amount of dansyl amide is produced in any dansylation reaction of amino acids. This arises partly from the small amount of ammonia in both the buffer solution and amino acid samples and partly from reaction 3.

REFERENCES

1. **Seiler, N.,** *Methods of Biochemical Analysis* Vol. 18, Glick, D., Ed. Interscience, New York, 1970, 259.
2. **Neadle, D. J. and Pollitt, R. S.,** *Biochem. J.*, 97, 607, 1965.
3. **Tapuhi, Y., Schmidt, D. E., Lindner, W., and Karger, B. L.,** *Anal. Biochem.*, 115, 123, 1981.

14. DERIVATIZATION OF AMINO ACIDS WITH DABS-Cl FOR QUANTITATIVE DETERMINATION

Samples of a lyophilized protein or peptide are hydrolyzed with constant-boiling 6 *N* HCl under reduced pressure at 110°C, the reaction being stopped at different time intervals by freezing. Derivatization of the free amino acids is performed with the colored reagent DABS-Cl that has been recrystallized from boiling acetone. The hydrolysate or 0.1 to 1 μg polypeptide is transferred to a silicone-treated glass tube with a conical bottom and a screw cap, dried, and redissolved in 0.1 *M* $NaHCO_3$ buffer pH 9.0. The pH is controlled and readjusted (if necessary) with 0.1 *M* NaOH before adding 20 μℓ of a solution of DABS-Cl (4 nmol/μℓ in acetone). The reaction is allowed to proceed at 70°C for 15 min, with brief periods of shaking after the first minute. The sample is then diluted with 100 μℓ of 70% (v/v) ethanol and analyzed be reversed-phase chromatography using a 5-μm ODS column and 50 m*M* sodium acetate buffer at pH 4.11 and acetonitrile as mobile phase.

REFERENCE

1. **Winkler, G., Heinz, F. X., and Kunz, C.,** *J. Chromatogr.*, 297, 63, 1984.

15. IMPROVED PREPARATION OF DABS-AMINO ACIDS FOR HPLC

The previously described conditions for the preparation of DABS-amino acids[1] were improved.[2] Acetone was replaced by acetonitrile for the DABS-Cl solution in the derivatization. Dabsylation of aspartic acid, glutamic acid, serine, threonine, and cystine then gives yields comparable to those of other amino acids. Acetonitrile is also less volatile than acetone.

Procedure

Hydrolyzed samples, standards, and blank samples are dissolved in 20 μℓ of 50 m*M* $NaHCO_3$, pH 8.1. A DABS-Cl solution, 4 nmol/μℓ in acetonitrile, is freshly prepared shortly before dabsylation and 40 μℓ is added to each sample. The samples are sealed with silicon-rubber stoppers and are heated at 70°C for 10 min, the mixture becoming completely soluble after heating. After dabsylation the samples are diluted with 50 m*M* sodium phosphate, pH 7.0/ethanol, 1:1, to suitable volumes and aliquots taken for HPLC analysis. The final volume of the diluted sample is roughly dependent on the amount of protein. When 0.5 μg of protein is hydrolyzed and dabsylated, the sample is usually diluted to 1 mℓ and 20 μℓ taken for HPLC.

REFERENCES

1. **Chang, J.-Y., Knecht, R., and Braun, D. G.,** *Methods Enzymol.*, 91, 41, 1983.
2. **Knecht, R. and Chang, J.-Y.,** *Anal. Chem.*, 58, 2375, 1986.

16. THE DERIVATIZATION OF AMINO ACIDS WITH FMOC-Cl FOR HPLC

The derivatization reagent is prepared by dissolving FMOC-Cl in acetone/acetonitrile, 1:1, to give a concentration of 15 m*M*. Samples (400 μℓ) and buffer (100 μℓ) are mixed in a 3-mℓ reaction vial equipped with a Teflon-lined screw cap (Supelco). The 1 *M* borate buffer is prepared from boric acid and NaOH to give a pH in the range 7.7 to 8.0 in the aqueous reaction solution. The derivatization reagent (500 μℓ) is added and after 40 sec the vial is almost filled with pentane and the mixture is shaken to remove excess reagent. The extraction is repeated twice, the pentane phases being discarded. The sample is acidified with 10 μℓ of glacial acetic acid and is then ready for injection onto the column.

REFERENCE

1. **Einarsson, S., Folestad, S., Joseffson, B., and Lagerkvist, S.,** *Anal. Chem.*, 58, 1638, 1986.

17. THE PREPARATION OF HEPTAFLUOROBUTYRYL- AND TRIFLUOROACETYL AMINO ACID *n*-BUTYL ESTERS FOR GC

The freeze-dried sample containing amino acids, prepared, for example, by the cleanup of urine, is dissolved in 1 mℓ of 2 *M* methanolic HCl, which is obtained by mixing equal amounts of methanol and acetyl chloride. The tightly closed Reacti® vial is heated for 20 min to 120°C in a Reacti-Therm® module (Pierce). After cooling to room temperature, excess methanolic HCl is evaporated to dryness under a stream of nitrogen at 60°C. The residue is transesterified with 3 *M* butanolic HCl prepared in the same way as the methanolic HCl, and heated to 120°C for 40 min. Excess reagents are evaporated under a stream of nitrogen, a few drops of methylene chloride are added, and the solvent is removed under nitrogen. Dichloromethane (50 μℓ) and 150 μℓ of heptafluorobutyric anhydride or tetrafluoroacetic anhydride are added. After ultrasonication for a few minutes, the closed vial is placed in the Reacti-Therm® module at 150°C for 12 min if acylation is with heptafluorobutyric anhydride and for 5 min if acylation is with tetrafluoroacetic anhydride. After cooling to room temperature, the mixture is evaporated under a stream of nitrogen, 150 μℓ of methylene chloride are added, and the solvent is evaporated under nitrogen to remove traces of water. The residue is dissolved in 20 μℓ of dichloromethane and 0.5 μℓ of the solution is used for gas chromatographic analysis.

REFERENCE

1. **Schneider, K., Neupert, M., Spiteller, G., Henning, H. V., Matthaei, D., and Scheler, F.,** *J. Chromatogr.*, 345, 19, 1985.

18. THE PREPARATION OF *N(O,S)*-PENTAFLUOROPROPIONYL AMINO ACID *n*-PROPYL ESTERS FOR GC

Two hundred microliters of *n*-propanol containing 4 mol/ℓ HCl prepared by mixing the appropriate volumes of *n*-propanol and acetyl chloride, and 10 μℓ of ethanethiol are added to dry amino acids (1 to 10 ng of each amino acid). The closed screw-capped vial is heated to 110°C for 30 min, and after cooling to about 50°C, is opened. The reagent is evaporated using a gentle stream of nitrogen at moderate temperature and the residue is dissolved in 50 μℓ of dichloromethane and 50 μℓ of pentafluoropropionic anhydride. The vial is tightly closed and heated to 150°C for 10 min. After cooling to room temperature, the solvent and reagent are evaporated under a gentle stream of nitrogen. For derivatization of histidine, the residue is dissolved in 50 μℓ of dichloromethane, 10 μℓ of ethyl chloroformate are added, and the mixture is heated at 100°C for 10 min. After cooling the solvent and reagent are evaporated under a stream of nitrogen. At this stage excessive drying should be avoided or the more volatile amino acids — alanine, valine, and threonine — may suffer losses. The residue is dissolved in an appropriate volume of toluene for GC.

REFERENCE

1. **Frank, H., Vijtovic-Ockenga, N., and Rettenmeier, A.,** *J. Chromatogr.*, 279, 507, 1983.

19. THE PREPARATION OF t-BUTYLDIMETHYLSILYL DERIVATIVES OF AMINO ACIDS FOR GC

For analysis of labeled or unlabeled amino acids, 4 μmol of each compound as an aqueous solution is transferred to a 4-mℓ derivatization vial (Kimble Products, Toledo, Ohio) and dried by lyophilization. *N*-Methyl-*N*-(t-butyldimethylsilyl) trifluoroacetamide (MTBSTFA) and acetonitrile (200 μℓ of each) are added to the vial, which is then closed with a Teflon-lined screw cap and allowed to incubate at room temperature overnight. The derivatized sample, 2 to 10 μℓ, is injected onto the gas chromatograph.

REFERENCE

1. **Schwenk, W. F., Berg, P. J., Beaufrere, B., Miles, J. M., and Haymond, M. W.,** *Anal. Biochem.* 141, 101, 1984.

20. ISOCYANATES AS REAGENTS FOR THE SEPARATION OF ENANTIOMERS OF AMINO ACIDS AND *N*-METHYLAMINO ACIDS

The formation of *N*-alkylureido derivatives of amino acid esters increases the selectivity of a chiral stationary phase toward enantiomers, resulting in complete separation, even for proline and isovaline enantiomers. In the reaction of isocyanates with *N*-methylamino acids, *N*-alkylureido/*N*-alkylamide derivatives are formed in one step. By this procedure the enantiomers of *N*-methylamino acids were separated.

Derivative Formation

Amounts of 100 μg of racemic amino acids are esterified in 1 mℓ of 1.5 *N* HCl/isopropanol at 100°C in a screw-capped vial. Excess reagent is removed by a stream of nitrogen; 100 μℓ of dichloromethane, 5 μℓ of dry pyridine, and 100 μℓ of *tert*-butyl isocyanate are added to the sample which is kept at room temperature for 1 hr. Heating the reaction mixture for 15 min at 100°C improves the yield of derivatives in some cases. *N*-Methylamino acids are heated in a screw-capped vial with 150 μℓ of dichloromethane and 150 μℓ of isopropyl- or *tert*-butylisocyanate for 15 min at 100°C. Excess reagents are removed in a stream of nitrogen and the sample dissolved in dichloromethane for GC.

REFERENCE

1. **König, W. A., Benecke, I., Lucht, N., Schmidt, E., Schulze, J., and Sievers, S.,** *J. Chromatogr.*, 279, 555, 1983.

21. THE DERIVATIZATION OF ENANTIOMERIC AMINO ACIDS TO GIVE DIASTEREOMERIC THIOUREA DERIVATIVES

The reagent 2,3,4,6-tetra-*O*-acetyl-β-D-glucopyranosyl isothiocyanate (GITC) reacts with enantiomeric amino acids to give derivatives which are efficiently separated on an octadecylsilyl silica gel column by gradient elution and are detected by their absorbance at 250 nm.

Procedure

To 10 μℓ of 0.1 *M* HCl containing 100 μg/mℓ of each amino acid are added 10 μℓ of acetonitrile containing 50 mg/mℓ of triethylamine and 20 μℓ of GITC reagent. This reagent is prepared by dissolving 10 mg of GITC per milliliter of acetonitrile. The mixture is allowed to stand at room temperature for about 20 min and an aliquot of 3 to 5 μℓ is injected into the chromatograph.

REFERENCE

1. **Nimura, N., Toyama, A., and Kinoshito, T.,** *J. Chromatogr.*, 316, 547, 1984.

22. THE PREPARATION OF THE CHIRAL REAGENT L-ASPARTYLYCYCLOHEXAMIDE (AspcHex) FOR THE CHROMATOGRAPHIC RESOLUTION OF ENANTIOMERIC AMINO ACIDS

The chromatographic resolution of enantiomeric amino acids can be accomplished on reversed-phase columns using aqueous mobile phases containing the copper complex of AspcHex. This reagent has an advantage in that the copper ions allow the detection of nonaromatic amino acids at 230 nm.

Procedure

ZAsp(OBzl)OH (3.7 g, 10.2 mmol) is dissolved in 25 mℓ of dimethylformamide and 1.13 mℓ of *N*-methylmorpholine is added. The mixture is cooled to −15°C and isobutyl chloroformate (1.26 mℓ, 9 mmol) is added, followed after 2 min by freshly distilled cyclohexylamine (1.15 mℓ, 10 mmol). The mixture is kept at −15°C for 1 hr, and then at 0°C 15 mℓ of 2 *M* $KHCO_3$ is added with vigorous stirring. After 30 min 100 mℓ of 25% NaCl is added, and the white precipitate is filtered off and washed with water until pH 7 is reached and no chloride ion can be detected in the washings. The product is then dried; it has a melting point of 139 to 141°C and is obtained in a yield of 85%. To a solution of 3 g of Z(Asp)(OBzl)-cHex in 30 mℓ of 90% acetic acid is added 300 mg of 10% palladium/carbon and the mixture is hydrogenated at room temperature (5 atm). The catalyst is filtered off, the solvent evaporated, and the residue recrystallized from water. AspcHex is obtained with a melting point of 235°C in a yield of 86%.

REFERENCE

1. **Gilon, C., Leshem, R., and Grushka, E.,** *Anal. Chem.* 52, 1206, 1980.

23. THE DERIVATIZATION OF LOW MOLECULAR WEIGHT ALIPHATIC AMINES IN AIR FOR LIQUID CHROMATOGRAPHIC DETERMINATION

A 5- to 100-ℓ volume of air is passed at 0.8 to 1.2 ℓ/min through a Sep-Pak® C_{18}(SP) cartridge (Waters Associates) coated with phosphoric acid. For cigarette smoke, 30 mℓ of smoke is puffed for 2 sec every 30 sec until the cigarette is exhausted. Particulates are removed by filtering the smoke through a glass fiber filter. A 120-mℓ volume from four puffs, corresponding to about one fifth of the cigarette, is sampled through the cartridge. The adsorbed substances on the cartridge are removed by elution with 5 mℓ of methanol. Aliphatic amines in the eluate are derivatized with 7-chloro-4-nitro-2,1,3-benzoxadiazole

(NBD-Cl). The NBD-Cl solution is 10 to 20 μmol/mℓ of NBD-Cl in 36 μmol/mℓ KOH in methanol. The eluate is mixed with 1 mℓ of the NBD-Cl solution and allowed to stand at room temperature for 24 hr. A 2- to 4-μℓ aliquot of the solution is analyzed by HPLC, amines being identified by their retention times and quantified by peak height.

REFERENCE

1. **Nishikawa, Y. and Kuwata, K.,** *Anal. Chem.*, 56, 1790, 1984.

24. BENOXAPROFEN AS A FLUORESCENCE MARKER FOR TLC AND HPLC ASSAY OF AMINES

Benoxaprofen ([R,S]-2-[p-chlorophenyl]-α-methyl-5-benzoxazoleacetic acid) is an antiinflammatory drug which shows an intense fluorescence.

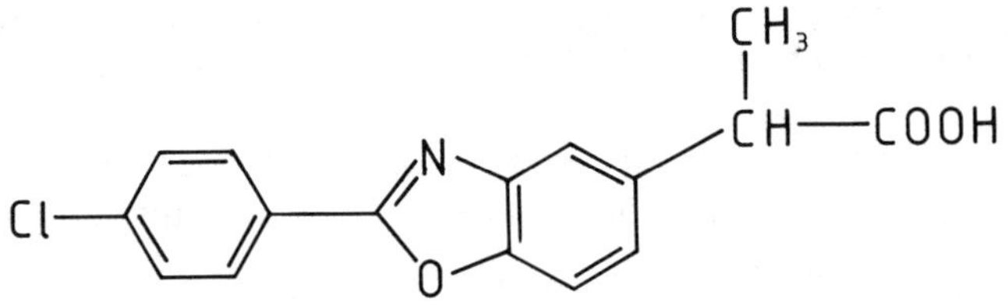

Numerous primary and secondary amines can easily be converted to readily detectable compounds using benoxaprofen chloride and benoxaprofen imidazolide. The derivatives can be prepared at room temperature in dry solvents such as dichloromethane, chloroform, acetonitrile, and toluene. Studies were made to optimize the reaction conditions. In the absence of bases the yield of reaction product using α-methylbenzylamine was about 91% after 1 hr. If Na_2CO_3, K_2CO_3, of freshly distilled triethylamine was added the yield was nearly 100%. After 12 hr identical yields of reaction product, about 100%, were obtained in all the cases. When triethylamine was used as proton acceptor in the reaction mixture even trace amounts of ethylamine and diethylamine present as impurities were detected and interfered in the chromatography. Carbonates sometimes had an unfavorable influence on the TLC of the reaction mixture. In the case of α-methylbenzylamine and tranylcypromine the derivatization is therefore preferably performed without the addition of bases.

Formation of Benoxaprofen Chloride

Benoxaprofen, 600 mg, is dissolved in 50 mℓ of dry toluene, and after the slow addition of 5 mℓ of thionyl chloride freshly distilled over linseed oil, the mixture is refluxed for 30 min. After evaporation to dryness the crystalline residue is recrystallized from dichloromethane if necessary.

Formation of Benoxaprofen Imidazolide

A solution of 500 mg of carbonyldiimidazole in 50 mℓ of dry dichloromethane is added to 100 mg of benoxaprofen. The solution is allowed to stand at room temperature for 30 min, with occasional shaking. Five milliters of glacial acetic acid is added to destroy excess carbonyldiimidazole and the reagent solution is used at once.

Derivatization with Benoxaprofen Chloride

Up to 0.5 μmol of amine and 500 μℓ of benoxaprofen chloride in dry dichloromethane (1 mg/mℓ) are allowed to stand at room temperature for 60 min or heated to 50°C for 30 min.

Derivatization with Benoxaprofen Imidazolide

Up to 0.5 μmol of amine and 500 μℓ of a freshly prepared solution of benoxaprofen imidazolide are allowed to stand at room temperature for 0.5 hr.

REFERENCE

1. **Spahn, H., Weber, H., Mutschler, E., and Möhrke, W.,** *J. Chromatogr.*, 310, 167, 1984.

25. THE PREPARATION OF A BONDED *p*-ACETYLPHENOXY STATIONARY PHASE FOR THE LIQUID CHROMATOGRAPHY OF AMINES

p-Hydroxyacetophenone (27.2 g, 0.20 mol) is reacted with sodium metal (4.6 g, 0.20 mol) in a round-bottom flask containing dry xylene (250 mℓ). Allyl bromide (25.4 g, 0.21 mol) is added dropwise to the heated solution, and the reaction is allowed to proceed at 80 to 90°C for 3 hr. Sodium bromide is filtered off under vacuum and the solution is reacted with chlorodimethylsilane (20.8 g, 0.22 mol) in the presence of H_2PtCl_6 (0.0041 g, 10^{-3} mol) in the dry xylene. The chlorodimethylsilane is added drop by drop to the solution which is heated to 90°C for 3 hr. The 3-(*p*-acetylphenoxy)propylchlorodimethylsilane is isolated by vacuum distillation and added to 100 mℓ of a xylene/pyridine mixture (90:10) containing 5.0 g of microparticle silica (10 μm RSil®, Alltech Associates Inc.). The mixture is allowed to react at 110°C for 12 hr and the resulting bonded phase substrate is filtered, washed thoroughly with dry xylene, isopropanol, and acetone, and dried under vacuum at 80°C overnight. Microanalysis indicates 10.65% carbon by weight, equivalent to 15.78% loading of the bonded phase. The dry phase is end capped by suspending it in dry toluene (100 mℓ) and treating it with trimethylchlorosilane. This mixture is refluxed for 1 hr, filtered, washed successively with dry xylene, isopropanol, and acetone, and dried under vacuum before use.

REFERENCE

1. **Pharr, D. Y., Uden, P. C., and Siggia, S.,** *J. Chromatogr. Sci.*, 23, 391, 1985.

26. DERIVATIZATION OF ALIPHATIC AMINES FOR HPLC AND TLC

1,2-Naphthoylenebenzimidazole-6-sulfochloride may be used as a fluorescent reagent for the derivatization of aliphatic amines prior to their separation and quantitation using HPLC or TLC.

Procedure HPLC

Ten milliliters of a sample containing amines at a concentration of 10^{-4} mol/ℓ or less is mixed with 10 mℓ of a 10^{-3}-mol/ℓ solution of the reagent in acetonitrile, 20 mℓ of acetonitrile, and 10 mℓ of a concentrated $NaHCO_3$ solution. The reaction mixture is shaken vigorously and then kept at 40 to 45°C for 10 min. After adding five drops of concentrated NH_4OH to remove any unreacted reagent, the mixture is shaken and again heated to 40 to 45°C for 10 min. The volume is brought to 100 mℓ with acetonitrile and 10 μℓ of the solution is injected into the chromatograph.

TLC

One milliliter of a saturated solution of the reagent in benzene is mixed with 0.05 to 1 mℓ of the solution of amines in acetonitrile and the mixture is heated for 15 to 20 min at 40 to 45°C. The volume of the mixture is then brought to 10 mℓ with acetonitrile and 20 μℓ spotted on the TLC plate.

REFERENCE

1. **Jandera, P., Pechova, H., Tocksteinova, D., Churacek, J., and Kralovsky, J.,** *Chromatographia,* 16, 275, 1982.

27. THE DERIVATIZATION OF AMINES WITH PHENYL ISOCYANATE FOR HPLC

Aliphatic and aromatic primary and secondary amines react with phenyl isocyanate to give *N,N'*-disubstituted ureas, which are very stable and exhibit high absorption in the UV. The reaction is quantitative and is complete in a few minutes. The derivatives are easily chromatographed in a reverse-phase HPLC system.

Procedure

It is preferable to dissolve samples in *N,N*-dimethylformamide which catalyzes the reaction and is a good solvent for the disubstituted ureas. A 0.5-mℓ volume of the amine (usually less than 10 mg) dissolved in *N,N*-dimethylformamide is pipetted into a glass-stoppered test tube containing a short magnetic rod. Then 0.5 mℓ of phenyl isocyanate is added and the mixture stirred magnetically for 5 min. In order to destroy excess phenyl isocyanate, 0.5 mℓ of an aliphatic alcohol, e.g., methanol, is added. The alcohol must be chosen so that it does not interfere with the chromatographic peaks to be determined. Before injecting the sample into the column, an internal standard, usually another amine derivative or an *O*-alkyl-*N*-phenylurethane, may be added. It is advisable to prepare a blank consisting of "derivatized" reagents. Alcohols, water, phenols, and carboxylic acids, as well as amines, react with phenyl isocyanate. A large excess of the derivatizating agent should, therefore, be used. With very dilute samples water should be evaporated or the amines extracted before derivatization.

REFERENCE

1. **Bjorkqvist, B.,** *J. Chromatogr.,* 204, 109, 1981.

28. THE PREPARATION OF DANSYL AMINES

Aqueous solutions of amines, each at about 50 nmol/mℓ, 200 μℓ, are "whirlimixed" with 400 μℓ of dansyl chloride (5 mg/mℓ in acetone) and then saturated with solid Na_2CO_3. After leaving in the dark at 20 ± 5°C for 15 hr, 400 μℓ of 15% aqueous proline is added to react with residual dansyl chloride. The mixture is "whirlimixed" and left in the dark for at least 1 hr. Dansylated amines are extracted with 0.5 mℓ of ethyl acetate, leaving dansyl proline behind in the aqueous phase.

REFERENCE

1. **Hayman, A. R., Gray, D. A., and Evans, S. V.,** *J. Chromatogr.,* 325, 462, 1985.

29. DERIVATIZATION OF AMINES FOR HPLC WITH ELECTROCHEMICAL DETECTION

N-(4-Anilinophenyl) isomaleimide (APIM) and *N*-(4-anilinophenyl) isophthalimide (APIP) have been prepared and evaluated for precolumn derivatization of amines for HPLC with electrochemical detection.

Preparation of Reagents

APIM

N-(4-Anilinophenyl) maleamic acid (500 mg) and sodium acetate (250 mg) are added to acetic anhydride (4 mℓ). The mixture is heated at 90 to 100°C for 5 to 10 min until the solution is transparent and is then poured into ice water. The brown precipitate is filtered off and recrystallized from ethyl acetate to give reddish-brown needles, mp 158 to 159°C.

APIP

A solution of phthalic anhydride (1.92 g) in dichloromethane (5 mℓ) is added drop by drop to 4-aminodiphenylamine (2.0 g) in dichloromethane (10 mℓ) with ice cooling. After standing overnight at room temperature the precipitate is filtered off, washed with dichloromethane, and dried to give *N*-(4-anilinophenyl) phthalamic acid (2.0 g) as a gray powder. The crude product is used without further purification. A solution of *N,N'*-dicyclohexylcarbodiimide (310 mg) in dichloromethane (2 mℓ) is added drop by drop to a stirred suspension of *N*-(4-anilinophenyl) phthalamic acid (500 mg) in dichloromethane and the mixture is allowed to stand at room temperature for 2 hr. After the removal of dicyclohexylurea by filtration the filtrate is evaporated. The residue is purified by chromatography on a column of silica gel using ethyl acetate-*n*-hexane as eluent. Recrystallization from ethyl acetate-*n*-hexane gives APIP, 200 mg as yellow needles.

Derivatization of Amines

Derivatization of amines (1.0 μg) is performed at room temperature in acetonitrile/0.05 *M* borate buffer, pH 9.0, 1:1 (0.2 mℓ) for 1 hr and is terminated by adding 0.1 mℓ of 2% methylamine hydrochloride solution. An aliquot of the solution is used for HPLC. The derivatization reagents can be stored unchanged in a refrigerator for at least 1 month. APIM was the more suitable derivatization reagent with respect to sensitivity and selectivity when used in connection with electrochemical detection after HPLC.

REFERENCE

1. **Shimada, K., Tanaka, M., and Nambara, T.,** *J. Chromatogr.*, 280, 271, 1983.

30. DERIVATIZATION OF TERTIARY AMINES WITH 2-NAPHTHYL CHLOROFORMATE (NCF)

NCF has been used as a derivatizing agent for drugs containing a tertiary amino group. When heated with NCF the tertiary amines undergo dealkylation forming fluorescent carbamates. The reaction has been applied to the precolumn derivatization of antihistamines.

Procedure

About 10 pmol to 5 nmol of the tertiary amine as the free base is heated with a tenfold molar excess of NCF and 10 mg of K_2CO_3 in 100 μℓ of dry benzene at 100°C for 1 hr in a sealed conical vial. After cooling, excess reagent is removed by shaking the mixture with 300 μℓ of saturated methanolic KOH for 1 min. After phase separation by adding 1 mℓ of water, a sample of the benzene layer is examined by HPLC.

REFERENCE

1. **Göbitz, G., Wintersteiger, R., and Hartinger, A.,** *J. Chromatogr.*, 218, 51, 1981.

31. DERIVATIZATION OF AMINES FOR GC

Acetyl Amines

Amines (10 μℓ or 100 mg of each) are dissolved or suspended in 250 μℓ of ethyl acetate in a flat-bottom glass tube. Acetic anhydride (100 μℓ) and pyridine (50 μℓ) are added, and the tube is capped and heated to 60°C in a heating block for 30 min. Excess reagents are removed in a vacuum overnight and the sample is resuspended in ethyl acetate.

Trifluoroacetyl Amines

Amines (10 μℓ or 10 mg of each) are dissolved or suspended in 250 μℓ of dichloromethane and are treated with trifluoroacetic anhydride/pyridine, 14:1, for 30 min at 60°C or for 18 hr at room temperature. The trifluoroacetyl amines thus produced are washed with 500 μℓ of 1 *N* HCl for 10 min and then with 500 μℓ of 0.1 *N* NaOH for 10 min. The cloudy organic phase containing the amine derivative is then separated and dried over 1 to 2 g of anhydrous sodium sulfate.

Heptafluorobutyryl Amines

Amines to be derivatized (5 μℓ or 5 mg of each) are dissolved or suspended in 250 μℓ of ehtyl acetate in a glass reaction vessel, and 20 μℓ of pyridine and 40 μℓ of heptafluorobutyric anhydride are added. The reaction tube is tightly sealed and after thorough mixing is heated for 30 min at 60°C. When cool the sample is evaporated to dryness in a vacuum desiccator and redissolved in 100 μℓ of ethyl acetate.

REFERENCE

1. **Tavakkol, A. and Drucker, D. B.,** *J. Chromatogr. Sci.*, 22, 12, 1984.

32. THE DERIVATIZATION OF AMINES WITH CHLOROFORMATE ESTERS FOR GC ANALYSIS

The GC analysis of amines after conversion to electron capture-sensitive carbamates in two-phase systems has been studied.

Primary and Secondary Amines

A carbonate buffer, pH 10.3, containing the amines, 10^{-5} to 10^{-7} *M*, is shaken for 20 min with an equal volume of methylene chloride containing 2,2,2-trichloro-*tert*-butyl chloroformate, 5×10^{-3} *M*. A few microliters of the organic phase are used for analysis by GC when employing thermionic detection. If an electron capture detector is used, excess reagent is reacted with *N,N*-dimethylethylenediamine and the resulting compound extracted into an aqueous acidic phase. Toluene is added and the methylene chloride is evaporated. Rapid and quantitative reaction is obtained even at a low reagent concentration. The hydrophobic dodecylamine is derivatized rapidly even below pH 8. The reagent and its byproducts give some peaks with the electron capture detector that are not found using the thermionic detector. With the electron capture detector the reagent must therefore be removed before chromatography.

A Tertiary Amine

N,N-Dimethyl-*n*-octylamine, 10^{-4} *M*, is dissolved in phosphate buffer, ionic strength 0.2 containing iodide ion. 2,4,6-Tribromophenyl chloroformate, 2.5×10^{-3} *M* is dissolved in methylene chloride containing an internal marker. Equal volumes of aqueous and organic phases are shaken together at 20°C for 5 min, and the reaction is then quenched with sulfuric acid. The choice of the chloroformate ester is of vital importance for carbamate formation. *N,N*-dimethyl-*n*-octylamine cannot be derivatized with an alkyl chloroformate ester at ambient temperature and an aryl chloroformate ester must be used to derivatize trialkyl tertiary amines.

REFERENCE

1. **Ahnfelt, N. O., Hartvig, P., and Karlson, K. E.,** *Chromatographia,* 16, 60, 1982.

33. THE DERIVATIZATION OF CHIRAL AMINES WITH ISOPROPYL ISOCYANATE FOR GC

Chiral amines are reacted with isopropyl isocyanate to give the *N*-isopropyl urea derivatives. The derivatives are easily prepared on a microgram scale at room temperature in 10 min and in excellent yields. If the reaction is performed at 100°C the urea derivatives may react further with isopropyl isocyanate, forming derivatives of higher molecular weight.

Procedure

Samples consisting of 0.5 mg of racemic mixtures of chiral amines are dissolved in 200 μℓ of dichloromethane and 100 μℓ of isopropyl isocyanate in a screw-capped vial. After standing for 10 min at room temperature excess reagent is removed with a steady stream of dry nitrogen. The residue is dissolved in 0.5 mℓ of dichloromethane for GC.

REFERENCE

1. **König, W. A., Benecke, I., and Sievers, S.,** *J. Chromatogr.,* 238, 427, 1982.

34. THE PREPARATION OF CHIRAL STATIONARY PHASES FOR THE GC SEPARATION OF ENANTIOMERS OF AMINES AND AMINO ALCOHOLS

Chiral constituents are introduced into silicone OV-225 by reducing its cyano groups to amino groups with lithium aluminum hydride ($LiAlH_4$), and coupling the amino groups with benzyloxycarbonyl-L-valine and benzyloxycarbonyl-L-leucine.[1]

Reduction of OV-225

To a suspension of 50 mg of $LiAlH_4$ in 5 mℓ of dry diethyl ether is slowly added a solution of 100 mg of OV-225 in 3 mℓ of diethyl ether, and the mixture is then refluxed for 5 hr. After decomposition of excess $LiAlH_4$ with water, concentrated NaOH solution is added and the phases separated. The organic phase is removed, the aqueous phase extracted three times with diethyl ether, and the combined extracts dried over Na_2SO_4. Completion of the reaction is ascertained by infrared spectroscopic examination after removal of the solvent.

Coupling of Z-L-Val and Z-L-Leu to OV-225 Amines

Benzyloxycarbonyl amino acids are prepared and are added in equimolar ratio to the OV-225 amine in chloroform solution.[2] After adding a small excess of dicyclohexyl carbodiimide the reaction mixture is stirred at room temperature for 12 hr. The solution is washed with 1 *N* HCl, saturated $NaHCO_3$ solution, and water and dried over Na_2SO_4. Dicyclohexylurea is removed from the product by chromatography on a silica column using ethyl acetate as eluent. Pyrex glass capillary columns are coated with the phase for GC.[1]

REFERENCES

1. **Künig, W. A. and Benecke, I.,** *J. Chromatogr.,* 209, 91, 1981.
2. **Grassman, W. and Wünsch, E.,** *Chem. Ber.,* 91, 462, 1958.

Section V

Products and Sources of Chromatographic Materials

Section V

PRODUCTS AND SOURCES OF CHROMATOGRAPHIC MATERIALS

There is available today an almost bewildering range of different chromatographic materials, such as column packings, resins, ready prepared reagents and buffer solutions, and sophisticated, often expensive instrumentation. The following section of this book does not attempt to provide an exhaustive list of products; it should, however, illustrate the variety available. Many of the products are described because of their relevance to compounds and separations considered in the present volume. Statements as to the performance of particular materials are derived from the manufacturers' literature. The mention of a particular item does not imply the author's recommendation since other products may serve a similar purpose equally well, or even better. Up-to-date information on new materials is provided in the appropriate pages of journals such as *The Journal of Chromatographic Science, Analytical Chemistry Lab Guide,* and the *American Association for the Advancement of Science Guide to Scientific Instruments.*

One difficulty encountered by the research worker using HPLC techniques has been the greatly different behavior of packings that from their technical descriptions might be expected to behave in similar ways. Instances have also occurred where different batches of nominally the same material from a single manufacturer have shown markedly different chromatographic patterns. Strenuous efforts are now being made to overcome these and other probelms. Packings are being prepared to rigid specifications and are being produced in relatively large amounts to avoid batch variations. Certain column packings are designed specifically to mimic the behavior of others. Moreover, packings are being developed for specific purposes and separations, e.g., affinity columns and columns for amino acid separations. Specifications of materials in the manufacturers' literature are frequently supplemented by descriptions of recommended techniques and analytical procedures and test certificates that show the performance of individual columns are sometimes supplied. The help of manufacturers and laboratory suppliers in furnishing copies of their technical literature is gratefully acknowledged.

Table 1
LIQUID PHASES FOR GAS CHROMATOGRAPHY

In toto some 700—1000 liquid phases for gas chromatography have been mentioned in the literature, and firms supplying gas chromatographic materials may list 100—200; this multiplicity of phases is frequently arranged on a scale of chromatographic selectivity using the so-called McReynolds' constants;[1] McReynolds' constants are the differences between Kovats retention indices for selected test compounds determined on any gas chromatographic liquid phase and on a reference liquid phase, squalane; the higher the value of the McReynolds' constants, the greater the chromatographic polarity of the phase; listings of McReynolds' constants are given in the original paper, in later articles by several authors, and in the catalogs of scientific suppliers

The McReynolds' scheme is the best system available for defining the relative selectivity of gas chromatographic liquid phases, but is has certain limitations; briefly stated, these are (1) McReynolds' constants are determined at 120°C, yet columns are rarely used at this temperature; (2) the constants are determined with phase loadings of 20%, although most studies employ loadings of less than 15%; (3) the retention of compounds partly depends on the type of diatomaceous earth supports and the surface treatment; (4) serious objections have been raised to using *n*-alkanes as standards in the determination of Kovats indices; (5) the Kovats index can be affected by the condition of the column

The available evidence suggests that the number of liquid phase for gas chromatography in current use exceeds a desirable level and sets of standard or preferred phases have been prepared; the following five liquid phases, for example, are stated to be among the most popular and should accomplish about three quarters of gas chromatographic analyses[2]

Poly(100%-dimethoxysilane) such as OV-101, SP-2100, and SE-30 GC grade

Liquid phases of this type have been used for the analysis of almost every kind of organic compound; they are very popular for hydrocarbon analysis and are the most thermally stable of the siloxane liquid phases

Poly(50%-methyl/5%-phenylsiloxane) such as OV-17 or SP-2250

OV-17 is a general purpose phase and is the second most popular liquid phase; it is suitable for the analysis of aromatic and heteroaromatic compounds, drugs, steroids, and trimethylsilyl derivatives; siloxanes of this type are available in a range of phenyl content from 5—75%; increasing phenyl content produces increased retention of naphthenic compounds such as *cis*-hydrindane and polarizable organic compounds such as aromatic, halogen, olefinic, and acetylenic compounds

Poly(alkylene oxides), particularly Carbowax® 20 M, poly(ethylene oxide) also called polyethylene glycol

Liquid phases of this type are useful for the analysis of most organic compounds, solvent mixtures, flavor ingredients, etc.; they are particularly useful for compounds carrying hydroxyl groups, such as alcohols, but are not recommended for the analysis of alkanes below about C_4—C_9; the poly(ethylene oxides) are susceptible to degradation in the presence of trace oxygen at higher temperatures; other useful phases are the reaction products of Carbowax® 20 M with terephthalic acid (Carbowax® 20 M-PTA) or with 2-nitroterephthalic acid (FFAP, SP-1000, AT-1000, or OV-351)

Poly(cyanoalkylsiloxanes), particularly those of medium cyanoalkyl content, such as OV-225, SP-2300, Silar® 5CP, and AN 600

Polymers of this general structure are available in a range of cyanoalkyl content from 5—100%; increasing cyano content results in increased retention of primary alcohols, ketones, and esters with respect to ethers; increased retention is also seen for compounds containing pi-electrons such as olefinic and aromatic compounds, while decreased retention is seen for alkanes

Poly(50%-methyl/50%-3,3,3-trifluoropropylsiloxane) such as OV-202 and SP-2401

OV-202 and SP-2401 have structures similar to OV-210 and OV-215, but the former are oils and provide rather more efficient chromatographic separations than the latter, which are viscous oils or gums; compared with OV-17, decreased retention is found for naphthenes such as *cis*-hydrindane and polarizable organic compounds, while increased retention is found for alcohol groups; the trifluoropropyl siloxanes are recommended when a different order of retention of oxygen compounds is desired than that observed with other liquid phases.

REFERENCES

1. **McReynolds, W. O.,** *J. Chromatogr. Sci.,* 8, 685, 1970.
2. **Yancey, J. A.,** *J. Chromatogr. Sci.,* 24, 117, 1986.

Table 2
SUPPORT MATERIALS FOR GAS CHROMATOGRAPHY

Manville
P.O. Box 5108
Denver,
Colo. 80217

Chromosorb® Supports

Chromosorb® supports prepared from diatomite are available in an untreated form, referred to as nonacid washed (NAW); some supports are available with the acid-washed treatment (AW); hydrochloric acid is used to remove soluble iron from the diatomite surface; the support is then washed to nearly neutral with deionized water; the acid washing reduces surface activity of the support; supports may be treated with dimethylchlorosilane (DMCS) in order to convert silanol groups on the surface to silyl esters, thereby greatly reducing the surface activity

Chromosorb® High Performance (HP)

This grade of support is an acid-washed and silanized flux-calcined diatomite; it shows superior inertness and column efficiency and a short column conditioning time; the HP grade was developed for the analysis of steriods, bile acids, alkaloids, pharmaceuticals, and medical and toxicological compounds; Chromosorb® W and Chromosorb® G are available in the HP grade

Chromosorb® A

Chromosorb® A is a support for preparative work whose structure does not easily break down with handling and which has a good loading capacity

Chromosorb® G

Chromosorb® G is a flux-calcined support used for separating polar compounds; its low surface area, hardness, and good handling properties make it a good replacement for the light friable Chromosorb® W; because of its smaller surface area and higher density a 5% liquid phase coating on Chromosorb® G corresponds approximately to a 12% coating on Chromosorb® W

Chromosorb® P

Chromosorb® is a pink diatomite support prepared from firebrick, which is used principally for hydrocarbons and moderately polar compounds on the analytical or preparative scale; the surface of Chromosorb® P is more adsorbent than those of other Chromosorb® supports; Chromosorb® P provides high column efficiencies; its high reactivity toward polar compounds can be decreased by masking the surface with a liquid phase or by deactivating the surface

Chromosorb® R

Chromosorb® R 6470-1 is a special support developed to coat the inside walls of open tubular capillary columns; it has an ultrafine particle size

Chromosorb® W

This support is a flux-calcined diatomite used to separate polar compounds which have relatively low adsorption properties; it is fairly fragile and should be carefully handled

Chromosorb® 750

Chromosorb® 750 is claimed to be the most inert, highly efficient support available; prepared from a pure diatomite crude, acid washed, and silane treated, it is designed for biomedical and pesticide analyses; its mechanical strength is similar to that of Chromosorb® G

Chromosorb® Century Series

Chromosorb® century series porous polymers are cross-linked resins; the resins have a uniform rigid structure of a distinct pore size and are infusible and insoluble in most solvents; porous polymers will separate a variety of gases and compounds with boiling points up to about 250°C, and will analyze aqueous solutions containing organic compounds capable of hydrogen bonding, such as amines, alcohols, glycols, and free acids; when these polymer beads are packed in long columns and operated at subambient temperatures they can provide separations similar to those on molecular sieve columns; columns of Chromosorb® polymer beads should be conditioned with inert gas (helium, nitrogen, etc.) with flow for at least 2 hr at a temperature 25°C below the recommended isothermal temperature; this conditioning process removes light organic materials such as monomers, solvents left in the polymer matrix from the manufacturing process, and any materials adsorbed from the atmosphere

Chromosorb® 101

Chromosorb® 101 is a solid adsorbent of the type used in gas-solid chromatography; it has a relatively low affinity for hydroxyl-containing compounds; as the hydroxyl content increases, or the hydroxyl groups are moved closer together, the retention time decreases; water is eluted before methanol and ethylene glycol (bp 190°C) before *n*-amyl alcohol (bp 137°C); the retention time of C_4 diols decreases as the hydroxyl groups move closer together; Chromosorb® 101 shows no tailing with oxygenated compounds; it will effectively separate hydrocarbons, alcohols, fatty acids, esters, aldehydes, ketones, ethers, glycols, vinylchloride, styrene, acrylonitrile, and chlorinated solvents, and can be used to analyze alcohols in blood, cosmetics, beverages, metabolites in urine, quaternary ammonium compounds in biological fluids, and to determine the carboxylic acid content of nonvolatile esters

Chromosorb® 102

Chromosorb® 102 is a porous styrene-divinylbenzene copolymer, with a porous structure of an open-cell variety so that samples can readily penetrate its pores; columns prepared from it can be operated isothermally up to 250°C and at somewhat higher temperatures during a temperature-programmed run; Chromosorb® 102 performs in a way similar to that of a conventionally packed column with a high liquid phase loading; retention times on the column are relatively high, but they can be reduced by coating the material with a liquid phase; Chromosorb® 102 will separate light and permanent gases, as well as lower molecular weight compounds such as acids, alcohols, glycols, ketones, esters, and hydrocarbons; it will also analyze propellants such as CO_2, nitrous oxide, chloropentafluoroethane, octafluorocyclobutane, propane, dichlorodifluoromethane, isobutane, and fluorotrichloromethane; Chromosorb® 102 can be used as packing material for liquid chromatography, where it has the advantage of being compatible with mobile phases having a pH range from 0—14

Chromosorb® 103

Chromosorb® 103 is a polyaromatic porous polymer material developed for amines and basic compounds; it will not handle acidic materials and glycols, which are totally adsorbed; methylamine is readily separated from light gases such as ammonia; among diamines, retention times decrease as the amino groups move closer together, i.e., 1,2-propanediamine precedes 1,3-propanediamine; t-butylamine elutes before t-butyl alcohol; water cannot be separated easily from ethylamine, but is easily separated from butylamine; Chromosorb® 103 can be used to quantitatively determine ethanol in beer and to detect ppm levels of ammonia, arsine, butylamine, ethylenediamine, ethylamine, ethanolamine, dimethylamine, diisopropylamine, diethylamine, isopropylamine, methylamine, propyleneimine, and phosphine

Chromosorb® 104

Chromosorb® 104 is an acrylonitrile-divinylbenzene-type resin with a highly polar surface, which is efficient for gas analysis at subambient, ambient, and higher temperatures; it will effectively separate sulfur-containing compounds at low levels, aqueous ammonia and hydrogen sulfide at low levels, isomeric xylenols, alcohols, ketones, nitriles, aldehydes, and hydrocarbons; retention times are longer on Chromosorb® 104 than on other Chromosorb® century series polymers, and it has the highest polarity in the series; the selectivity of Chromosorb® 104 also differs from that of other century series polymers

Chromosorb® 105

Chromosorb® 105 is a white, nonfriable, insoluble, cross-linked polyaromatic resin; it separates formaldehyde from water and methanol, acetylene from lower hydrocarbons, and most other classes of organic compounds of different polarity having a boiling point of up to 200°C; this support can be used to analyze quaternary ammonium compounds in biological fluids and for the determination of the carboxylic acid content of esters; Chromosorb® 105 is inert and has a high retentive capacity to collect volatiles from headspace samples and aqueous solutions

Chromosorb® 106

Chromosorb® 106 is a cross-linked sytrene-divinylbenzene-type resin; it is a hard granular solid with a nonpolar surface and, thus, it retains nonpolar organic compounds longer in a gas chromatographic column than polar compounds; Chromosorb® 106 is stable up to 250°C; it separates gases and low boiling compounds and will efficiently resolve C_2-C_5 fatty acids from C_2-C_5 alcohols; it can also be employed as an adsorbent to trap organics from air or water

Chromosorb® 107

Chromosorb® 107 has different retention characteristics from other century series polymers and separates several classes of compounds, such as alcohols, esters, and ethers; it elutes water, even at high concentrations, after sulfur dioxide; Chromosorb® 107 is the preferred adsorbent to trap and analyze vinyl acetate

Chromosorb® 108

Chromosorb® 108 is a moderately polar cross-linked acrylic ester-type resin similar to Chromosorb® 107 in surface characteristics, but with a smaller surface area; with proper handling it is stable up to 250°C; Chromosorb® 108 gives excellent separation of gases and aqueous compounds; water is retained longer on this polymer than on other Chromosorb® polymers apart from Chromosorb® 104; the retention characteristics of Chromosorb® 108 differ from those of other century series polymers

Chromosorb® T

Chromosorb® T is a support for gas chromatography made from Teflon 6 powder; Teflon 6 is a higher molecular weight TFE-fluorocarbon resin; TFE resin is insoluble in common solvents and is resistant to chemical attack; Teflon 6 is used as a support when a highly inert surface is required to avoid tailing problems encountered with compounds such as water, hydrazine, and sulfur dioxide that are adsorbed and exhibit tailing when used with diatomite supports; temperature affects the physical properties of TFE resin; because of its static charge and softness it is necessary to handle Chromosorb®T gently; chilling the resin can improve its handling properties

Alltech Associates, Inc.
Applied Science Labs.
2051 Waukegan Road
Deerfield,
Ill. 60015

Gas Chrom® SUPPORTS

Gas Chrom®Q

Gas Chrom® Q is a white diatomaceous earth that has been screened, acid and base washed, neutralized, and silanized; it must be handled carefully during coating and column packing; Gas Chrom® Q shows maximum inertness and lowest adsorption, and is recommended for the analysis of steriods, pesticides, alkaloids, pharmaceuticals, toxic pollutants, amines, and other labile compounds; the miximum recommended stationary phase load is 15 or 10% for gum phases

Gas Chrom® QII

Gas Chrom® QII is an analog of Gas Chrom® Q made from salt water diatoms; it is closely similar to Chromosorb® W-HP, Supelcoport®, and the Anakroms®; for most applications either Gas Chrom® Q or Gas Chrom® QII can be used, but there are some cases where one or the other is preferred; Gas Chrom® QII generally gives higher efficiency, lower back pressures, and has better coating characteristics for polar liquid phases; Gas Chrom® Q is more inert to thermally labile and sensitive compounds; the higher surface energy of Gas Chrom® QII gives more efficient coating, especially with polar phases

Gas Chrom® S,A,P,Z Series

Fresh water white diatomaceous earth is the starting material for these supports; Gas Chrom® S is treated to remove the bulk of inorganic contaminants and is then water washed to neutrality; Gas Chrom® A is made by acid washing Gas Chrom® S to remove the bulk of the inorganic contaminants and is then water washed to neutrality; Gas Chrom® P is made by base washing Gas Chrom® A to remove organic contaminants; it is then water washed to neutrality; Gas Chrom® Z is made by treating Gas Chrom® P with dimethyldichlorosilane to eliminate active sites on the surface of the support

Gas Chrom® R

The red-colored Gas Chrom® R is prepared from C-22 firebrick material; made from diatomaceous earth, it differs from the white supports in being calcined without a flux, has a larger surface area, and is generally more adsorptive; Gas Chrom® R can achieve the highest efficiency of all diatomaceous supports and is particularly suitable for the analysis of hydrocarbons and other nonpolar compounds

Gas Chrom® Porous Polymers

Gas Chrom® porous polymers are low in residual monomers and contaminants, resulting in faster conditioning times with lower column bleed; they have a low affinity for water, making them ideal for the determination of low levels of water in organic solvents or the determination of organics in water; Gas Chrom® 220 is a low polarity porous polymer; its properties are similar to those of Porapak® Q and Chromosorb® 102; the temperature limit is 250—300°C; Gas Chrom® 254 is slightly more polar than Gas Chrom® 220; it is a good substitute for Porapak® P and Chromosorb® 101; the temperature limit is 275—310°C

Chrompack
P.O. Box 8033
4330 EA Middelburg,
The Netherlands

CP® Wax® for Amines

For the gas chromatographic separation of amines, which are relatively highly polar compounds, a polar stationary phase is preferred in order to obtain sufficient selectivity; a phase of this type was obtained by increasing the basic properties of a polyethylenglycol with a molecular weight of approximately 40,000 with a basic reagent and a temperature treatment; the stationary phase thus obtained, CP™ Wax® for amines, has a high efficiency, a high degree of inertness, is thermostable, and shows low bleeding; for gas chromatography a film of 0.2 μm CP™ Wax® 51 for amines is deposited on the wall of a fused silica capillary column; the minimum and maximum operating temperatures are, respectively, 60 and 240°C isothermally and 260°C in a temperature-programmed mode; below 60°C the stationary phase becomes a solid with low separation efficiency; amines elute from the column as sharp symmetrical peaks

Table 3
PACKINGS FOR HIGH PERFORMANCE LIQUID CHROMATOGRAPHY

Manville
P.O. Box 5108
Denver,
Colo. 80217

Chromosorb® LC Packings

Chromosorb® LC grade packings were developed for specialized applications, but all are characterized by high column efficiency, stability, ease of packing, and fast and accurate separation

Chromosorb® LC-1

Chromosorb® LC-1 is a diatomite-based support with a recommended maximum liquid phase loading of 4%; for optimum efficiency the mobile phase should be saturated with the liquid phase and a precolumn should be used; the support gives fast and efficient separation of many polar compounds when used in a liquid-liquid mode; applications include aromatic amines, benzyl alcohols, phenols, and carbamates

Chromosorb® LC-2

Chromosorb® LC-2 is a pellicular packing with active porous silica layers permanently bonded to an inert spherical solid glass core; the maximum liquid phase loading should be about 2%; although developed for liquid-solid analysis, Chromosorb® LC-2 may also be used as a support for liquid-liqud chromatography; it is versatile in separating various types of compound by selection of a solvent system of the right composition; typical applications are nitroanilines, aromatic amines, pesticides, plasticizers, aromatic compounds, phenols, barbiturates, amides, steriods, vitamins, and ureas

Chromosorb® LC-3

Chromosorb® LC-3 is a pellicular packing with alumina bonded to the surface of diatomite; it can be used for the analysis of polycylic hydrocarbons, alkaloids, amines, aldehydes, ketones, surfactants, plasticizers, and nitroanilines

Chromosorb® LC-4

Chromosorb® LC-4 is a pellicular reversed-phase packing with a monomolecular layer of octadecyltrichlorosilane chemically bonded to Chromosorb® LC-2; the organic phase is permanently bonded with a relatively stable Si–O–Si–type bond; the aliphatic alkyl group on the surface gives good separations of fused ring aromatic hydrocarbons, quinones, herbicides, barbiturates, steroids, and food additives

Chromosorb® LC-5

Chromosorb® LC-5 is a pellicular reversed-phase packing with a monomolecular layer of diphenyldichlorosilane bonded to Chromosorb® LC-2; the phenyl group gives excellent reverse-phase separation of pharmaceutical and biomedical compounds.

Chromosorb® LC-6

Chromosorb® LC-6 is a porous, irregularly shaped microparticulate silica adsorbent; it has high resolving power and will separate similar polarity compounds with an isocratic solvent system or wider polarity range mixtures using gradient elution; applications include pesticides, alcohols, plasticizers, steroids, fat-soluble vitamins, anticonvulsants, amines, organic intermediates, essential oils, and phenols

Chromosorb® LC-7

Chromosorb® LC-7 is a reversed-phase microparticulate HPLC column packing prepared by bonding an octadecyl group to Chromosorb® LC-6 via a Si–0–C link; this C_{18} nonpolar phase functions in either partition or ion-pair chromatography modes; this packing performs well under isocratic or gradient elution conditions with a wide range of solvent systems; it will analyze anthraquinones, fused ring aromatics, water-soluble vitamins, estrogens, herbicides, nucleosides, catecholamines, benzodiazepines, beverage additives, urine, analgesics, antibiotics, fatty acids, fire retardents, dyes, narcotics, anticonvulsants, food preservatives, and trace organics in water

Chromosorb® LC-8

Chromosorb® LC-8 is a microparticulate HPLC column packing obtained by bonding a cyano-type group to Chromosorb® LC-6; Chromosorb® LC-8 complements Chromosorb® LC-7 for reversed-phase HPLC and can also be used in the normal phase mode; applications include pesticides, esters, flavors, essences, nucleic acid bases, dye intermediates, pharmaceuticals, drug metabolites, soft drink additives, and catecholamines

Chromosorb® LC-9

Chromosorb® LC-9 is a microparticulate HPLC column packing obtained by bonding an amino-type functional group on Chromosorb® LC-6, the chemical reaction being controlled so as to provide a reproducible organic loading; this packing performs well under isocratic or gradient elution conditions with a wide range of solvent systems; applications include carbohydrates, strongly polar compounds, nitro compounds, aromatic compounds, high fructose syrup, sugars, amines, nucleotides, carboxylic acids, and polyhydroxy compounds

E. Merck
Frankfurter Strasse 250
D-6100 Darmstadt 1
Federal Republic of Germany

LiChrosorb®

LiChrosorb® adsorbents for HPLC are totally porous, with a fractured particle form; they are finely classified in the 5-, 7-, and 10-μm range

	Average particle size (μm)	Description	Specific surface area (m^2/g)
LiChrosorb® Si60	5,7,10	Neutral silica gel	500
LiChrosorb® Si 100	5,7,10	Neutral silica gel	300
LiChrosorb® Alox T	5,10	Basic aluminum oxide 150 (type T)	70
LiChrosorb® NH_2	5,7,10	Amino function on the end of a hydrocarbon chain; hydrolysis resistant	250
LiChrosorb® CN	5,7,10	Cyano function on the end of a hydrocarbon chain; hydrolysis resistant	250
LiChrosorb® DIOL	5,7,10	Vicinal hydroxyl function on the end of a hydrocarbon chain; hydrolysis resistant	250
LiChrosorb® RP-2	5,7,10	Silica gel chemically modified with dimethyldichlorosilane; hydrolysis resistant	250
LiChrosorb® RP-8	5,7,10	Hydrolysis-resistant hydrocarbon phase (C_8) chemically bound to silica gel	250

	Average particle size (μm)	Description	Specific surface area (m^2/g)
LiChrosorb® RP-18	5,7,10	Hydrolysis-resistant hydrocarbon phase (C_{18}) chemically bound to silica gel	150
LiChrosorb® KAT ion exchanger	10	Silica gel with sulfonic acid groups	

MERCK

Perisorb®

Perisorb® adsorbents for HPLC are surface porous with a low specific surface area; they therefore have only a slight influence on the separation behavior of the main column when used as a precolumn; their particle size is 30—40 μm

	Structure of the chromatographically active layer	Remarks
Perisorb® A	Silica gel 60	Maximum loading with liquid phase 3%
Perisorb® RP-2	Silica gel silanized with dimethyldichlorosilane	
Perisorb® RP-8	Hydrocarbon phase (C_8) chemically bound to silica gel; hydrolysis resistant	
Perisorb® RP-18	Hydrocarbon phase (C_{18}), chemically bound to silica gel; hydrolysis resistant	
Perisorb® CX-S	Silica gel with sulfonic acid groups	Strong acidic ion exchanger, stable between pH 1 and 9
Perisorb® PA-6	Polyamide 6	Polycaprolactam layer approximately 2 μm

Macherey-Nagel
Düren,
Germany

Nucleosil®

Nucleosil® is a spherical, totally porous silica gel; it is available in two different pore diameters, 50 Å (Nucleosil® 50) and 100 Å (Nucleosil® 100); for steric reasons only Nucleosil® 100 is modified to give maximum possible coverage of the silica gel with organic groups; due to its narrow pore structure, Nucleosil® 50 is recommended for adsorption chromatography, while Nucleosil® 100 can be employed for both adsorption and partition chromatography; Nucleosil® packing materials are mechanically stable and can be used in all fields of HPLC; both unmodified and modified Nucleosil® types are obtainable in different particle sizes with a narrow particle distribution; modified Nucleosil® types comprise those which are primarily for reversed-phase chromatography, namely, C_{18}, C_8, and phenyl, and those with chemically bonded polar phases such as CN, NO_2, NH_2, $N(CH_3)_2$, and OH; Nucleosil® types SA and SB are ion exchangers which are stable in the pH range 1—9 and do not swell; they exhibit constant permeability, even when the strength of the ions varies and/or when the pH of the eluent is altered

Polygosil®

Polygosil® is an irregularly shaped silica gel available with a pore diameter of 60 Å; it is suitable for both adsorption and partition chromatography; Polygosil® is available in the unmodified form and with chemically bonded phases; these packing materials are fractionated with a narrow particle size distribution; due to the relatively small pore diameter and, therefore, the relatively low ion exchange loading capacity, Polygosil® ion exchangers are not produced

The Separations Group
P.O. Box 867
Hesperia,
Calif. 92345

Vydac® HPLC Packings

Vydac® TP Silica

TP silica is spheroidal with 300-Å pores, a low surface area of 80 m^2/g, and a pore volume of 0.6 cm^3/g; its wide pores make it suitable for the separation of large biomolecules, while its low surface area means that as a medium capacity reversed phase it can be used for the chromatography of very hydrophobic small molecules

Vydac® HS Silica

HS silica is spherical with 80-Å pores, a surface area of 500 m^2/g, and a pore volume of 0.8 cm^3/g; the high surface area permits a high loading of bonded phase, allowing the separation of relatively hydrophilic compounds; the manufacturing process for these two silicas yields high purity silica, low in metal and sulfate content; the rigid, physically rugged, spheroidal particles do not fracture in the column, thus reducing the back pressure and increasing the lifetime of HPLC columns

Vydac® 201 HS

201 HS is a high capacity reversed-phase material; an 18-carbon chain is chemically bonded to HS silica which is then endcapped with trimethylsilane; a high loading of the bonded phase provides excellent separation of relatively polar compounds such as the B vitamins, organic acids, biological bases, and nucleosides

Vydac® 201 TP

201 TP is a medium capacity reversed-phase material; an 18-carbon chain is chemically bonded to TP silica producing a medium level phase loading; very nonpolar compounds such as polynuclear aromatic hydrocarbons, aflatoxins, and α and β carotenes can be separated

Vydac® 214 TP

214 TP is a short chain reversed-phase material consisting of a four-carbon chain chemically bonded to TP silica which is then endcapped with trimethylsilane to prevent adsorption of polar compounds; the large pore packing material is ideal for large biomolecules and gives excellent separations of peptides and proteins

Vydac® 218 TP

218 TP is a long chain reversed-phase material consisting of an 18-carbon chain chemically bonded to TP silica which is then endcapped with trimethylsilane to prevent adsorption of polar compounds; this large pore packing material is ideal for large biomolecules and separates proteins and peptides

Vydac® 219 TP

219 TP is an aromatic reversed-phase material consisting of a diphenyl group chemically bonded to TP silica which is then endcapped with trimethylsilane to prevent adsorption of polar compounds; it gives alternative selectivity for proteins and peptides and provides excellent separation for them

Vydac® Ion Exchange Packings

Vydac® microparticulate ion exchange materials are available with anion and cation exchange groups; these are chemically bonded to TP silica

301 TP

301 TP is a strong anion exchange material with a quaternary amine functional group for the analysis of nucleotides and organic acids

401 TP

401 TP is a strong cation exchange material with a sulfonic acid functional group for analysis of amines, purines, pyrimidines, and organic bases

Vydac® Solid Core Materials

Vydac® SC (Solid Core)

Vydac® SC is made by chemically coating a solid glass support with silica to which various phases are chemically bonded; it has a surface area of 8 m^2/g and a particle size distribution centered at 30 μm

101 SC	Silica adsorbent
201 SC	Reversed phase (C_{18})
301 SC	Anion exchange
401 SC	Cation exchange
501 SC	Polar (CN) phase
601 SC	Polar (NH_2) phase

Vydac® 3040L Column

The Vydac® 3040L column is a high performance anion exchanger column that allows the rapid purification of oligonucleotides as large as 40 nucleotide residues; it has a DEAE-type ion exchange function with slight hydrophobic characteristics; the 3040L column is stable in buffers from pH 2—9; excellent separation of oligonucleotides differing by a single nucleotide residue is possible; adjustment of pH modifies the relative retention of different nucleotide residues and the presence of organic solvent in the eluent results in good peak shape

Chrompack
P.O. Box 3
4300 AA Middelburg,
The Netherlands

CP Spher® Columns

CP Spher® provides excellent performance due to a very narrow distribution of spherical particles of 10 μm in size; features are extremely good peak symmetry, batch-to-batch reproducibility, and high surface coverage; CP Spher® Si is available for normal phase operation, while CP Spher® C8 and CP Spher® C18 are available for reversed-phase operation; a CP Spher® C8 column, 25 cm × 4.6 mm I.D., provides optimum separation and resolution of amino acid-*o*-phthalaldehyde derivatives

CP MicroSpher® Columns

The use of CP MicroSpher® with spherical 3-μm particles gives columns of very high efficiency, good peak symmetry, and batch-to-batch reproducibility; CP MicroSpher® Si is used for normal phase separations and CP MicroSpher® C18 for reversed-phase separations

Applied Science Labs.
2051 Waukegan Road
Deerfield,
Ill. 60015

Versapack®

Versapack® packings were developed to mimic μBondapak® materials from Waters Associates; the characteristics of the starting silica were chosen to match the characteristics of μBondapak® in shape, size, surface area, pore size, and pH; the shape of Versapack® silica produces columns with exceptional permeability, often generating backpressure 40% less than irregularly shaped materials; Versapack® C18 is made by bonding C18 functions to the silica monomerically up to a carbon lead of 10%; the remaining silanol functions are then deactivated by endcapping with trimethylsilyl groups; the material produced has the same selectivity as μBondapak® C18

HAMILTON COMPANY P.O. BOX 10030 RENO, NEV. 89520

PRP-1 Reversed-Phase Resin

PRP-1 reversed-phase resin is a rigid, macroreticular styrene-divinylbenzene copolymer; it is particularly suited for reversed-phase separations, but functions equally well for gel permeation chromatography (molecular weights 2,000—100,000) and normal phase separations; PRP-1 is a spherical adsorbent with a 10-μm nominal particle size, resulting in very low back pressures; it can be used over the pH range 1—13 and works efficiently with salt concentrations up to 0.5 *N*; PRP-1 reversed-phase columns have been used for the separation of nucleosides and bases, antiepileptic and sulfa drugs, chlorophenols, carboxylic acids, catecholamines, analgesics, water-soluble vitamins, phenoxyacetic acids, and ion pairing of organic and inorganic anions

Table 4
COLUMNS FOR AMINO ACID ANALYSIS

Alltech Associates, Inc.
Applied Science Labs.
2051 Waukegan Road
Deerfield,
Ill. 60015

A number of resins and columns have been developed specifically for amino acid analysis

Ion Exchange Resins

The high speed amino acid resin (HSAAR) column contains a polymer resin that is stable from pH 0—14 and is classified to ±0.5 μm; it exhibits up to 100,000 plates per meter and affords a fast and reliable method of amino acid determination; postcolumn derivatization of amino acids with ninhydrin or *o*-phthalaldehyde may be used; the pressure on the column should not exceed 200 atm at an eluent flow rate of 0.7 mℓ/min at 60°C; the high resolution amino acid resin (HRAAR) column is designed for the analysis of physiological samples such as urine and plasma; analyses can be completed in less than 2 hr in a reliable and accurate manner; most common physiological amino acids can be separated using the column, and derivatization with *p*-phthalaldehyde permits determination in the 10—100-pmol range; pressure on the column should not exceed 200 atm using an eluent flow rate of 0.6 mℓ/min at 38°C

Analysis Using Precolumn Derivatization

The derivative formed by reaction between an amino acid and *o*-phthalaldehyde is a highly fluorescent indole which is amenable to reversed-phase chromatography; two different types of reversed-phase columns, both using a highly endcapped octyl silica based on Adsorbosphere® silica, 5 μm, have been developed, C8 material was used, one reason for which was its fast equilibration time at the end of the gradient; the OPA-HS column was designed for high speed analysis and uses chemically optimized C8 material; this column is suitable for protein acid hydrolyzate applications; mobile phases are simple and various parameters can be altered to control the retention times of many amino acids; the OPA-HR column was developed for applications where a high degree of resolution is required; the packing material used is an optimized C8 much like that used in the OPA-HS, but the extended bonding protocol used enhances the material's selectivity toward amino acids commonly found in physiological fluids

Pierce Chemical Company
P.O. Box 117
Rockford,
Ill. 61105

High Speed AA511 Column

The high speed AA511 column is packed with a special sulfonated polystryene-divinylbenzene copolymer designed to give high speed and high resolution analyses; analysis of amino acid mixtures obtained from protein hydrolyzates requires less than 30 min

High Efficiency Resins

Pierce high efficiency resins are manufactured under closely controlled conditions using new techniques for sizing and processing; these cation exchange resins provide separation of amino acids in protein hydrolyzates and physiological fluids in the shortest possible time; they exhibit superior resolving characteristics with a shorter column; the following resin types are available

PC-1A— a sulfonated divenylbenzene-polystyrene 8% cross-linked resin, bead diameter 14 ± 2 μm; it is recommended for earlier model analyzers with limited eluent pressure capability

PC-4A— a sulfonated divinylbenzene-polystyrene 8% cross-linked high resolution resin, bead diamter 9.0 ± 0.5 μm, recommended for single column microbore column methods

PC-6A— a sulfonated divinylbenzene-polystyrene 8% cross-linked resin, bead diamter 11 ± 1 μm; it is recommended for single column analysis of hydrolyzates or physiologicals using standard bore columns or analyzers capable of operating at eluent pressures up to 1000 psi

Beckman Instruments Inc.
Altex Scientific Operations
1716 Fourth Street
Berkeley,
Calif. 94710

Spherogel® Amino Acid Column

When using ion exchange methods to separate complex amino acid mixtures, the Spherogel® amino acid column gives excellent high resolution performance; the column is packed with a specifically developed cross-linked resin

Ultrasphere® C_{18}-IP Column

The Ultrasphere® C_{18}-IP column was developed for the separation of ionic compounds on a hydrophobic packing; under ion-pair conditions phosphothreonine and phosphoserine show baseline resolution and good peak shape

Ultrasphere® C_{18} 3-μm Column

Columns packed with smaller diameter particles have higher efficiencies per unit length; short columns packed with 3-μm particles therefore give comparable resolution to a 5-μm column; using an Ultrasphere® C_{18} 3-μm column the *o*-phthalaldehyde derivatives of 33 amino acids can be separated with excellent resolution in less than 15 min

Varian Instrument Group
611 Hansen Way
Palo Alto,
Calif. 94303

MicroPak® Amino Acid Analysis Columns

MicroPak® amino acid analysis columns are designed for the resolution of complex mixtures of free amino acids and are used in conjunction with ninhydrin or *o*-phthalaldehyde detection; the stationary phase is a strong cation exchange resin available in the sodium form for analysis of protein hydrolyzates and in the lithium form for biological fluid analysis

Benson Company
P.O. Box 12812
Reno,
Nev. 89510

Benson resins for amino acid analysis give optimum peak resolution, low operating pressure, and long column life, for systems using standard bore (6.9 mm I.D.) or microbore (2.5 mm I.D.) columns; the resins are manufactured to exacting specifications and are very easy to pack into columns; the expected operating pressure for microbore column systems is about 1000 psi

Resin	Methodology
BH-4	Protein hydrolysates; single- or dual-column systems
BP-B5	Protein hydrolysates (basic groups); physiological fluids; peptides
BP-AN6	Protein hydrolysates or acidic-neutral groups; physiological fluids; single- or dual-column systems

Resins for Microbore Columns

B-X8, 8 ± 1 μm	Protein hydrolysates; physiological fluids; single- or dual-column systems
D-X8.25, 8 ± 1 μ	Protein hydrolysates; single-column systems
A-X8.75, 5 ± 1 μm	Protein hydrolysates; single-column systems

PICO-TAG® Work Station

Waters Associates
Milford,
Mass. 01757

The PICO-TAG® work station is designed for the batchwise hydrolysis of proteins and peptides together with precolumn derivatization of the resulting primary and secondary amino acids before their HPLC separation; the compact unit comprises an adjustable thermostatically controlled oven for hydrolysis and a vacuum/nitrogen manifold for rapid, efficient, sample drying, reagent removal, and vacuum sealing; in addition to sample preparation for amino acid analysis, the PICO-TAG® work station can be used to dry solvents and buffers from HPLC fractions; it is faster than rotary evaporation due to the simultaneous processing of multiple samples

Samples are processed in a reaction vial, the work station accommodating four vials; each reaction vial holds up to 12 sample tubes and a work station oven thus has a capacity of 48 samples; fewer contaminants are found due to the addition of HCl into the reaction vial instead of the sample tube, and the HCl removal is much faster; a slider valve on the cap of the reaction vial permits a vacuum to be applied and the vial to be sealed by pushing a button; precolumn derivatization is based on the use of phenylisothiocyanate, which has been used for over 30 years as the primary reagent in protein sequence analysis; the reaction of phenylisothiocyanate with amino acids forms the stable phenylthiocarbamyl amino acids; the reagent reacts equally well with primary and secondary amines; unlike the reactions of ninhydrin or *o*-phthalaldehyde following hypochlorite oxidation, phenylisothiocyanate forms the same chromophore with proline as it does with primary amino acids; very similar response is obtained for all amino acids except lysine, which gives more response due to the formation of a disubstituted product; high sensitivity (low picomole) analysis of both primary and secondary amines is obtained

Table 5
COLUMNS FOR THE ANALYSIS OF OPTICAL ISOMERS

GAS CHROMATOGRAPHY

Chrompack
P.O. Box 8033
4330 EA Middelburg,
The Netherlands

The chiral polymer liquid phase XE-60-S-valine-α-phenylethylamide is used for coating flexible fused silica capillary columns; a 50-m × 0.25-mm chiral column has a high efficiency, while the thermal and optical stability of the liquid phase are very good; the phase is very versatile, having been used for the enantiomeric separation of amino acids, amines, hydroxy acids, amino alcohols, chiral alcohols, and carbohydrates; amino acid enantiomers are chromatographed as their *N*-trifluoroacetyl isopropyl esters, the order of elution being determined by the configuration of the valine moiety; the 1-enantiomers are thus retarded

Chirasil-L-Val®

Chirasil-L-Val® columns are fused silica columns coated with the chiral stationary phase based on L-valine-*tert*-butylamide; Chirasil-L-Val® is synthesized by the reaction of L-valine-*tert*-butylamide with a copolymer of dimethyl- and carboxyalkylmethylpolysiloxane; the phase has a high maximum temperature, 220°C isothermal and 240°C with temperature programming; this allows faster separations and higher boiling compounds can be analyzed; separation of the enantiomers of optically active drugs, α-amino acids, α-amino alcohols, glycols, aromatic and aliphatic α-hydroxycarboxylic acids, sulfoxides, and amines is possible; the column also allows quantitative amino acid determination by enantiomer labeling

Chirasil-D-Val®

In addition to the Chirasil-L-Val® columns, its ''mirror image'', a Chirasil-D-Val® column is also available; Chirasil-D-Val® differs from Chirasil-L-Val® only by a changed elution sequence of L and D compounds, caused by its D-enantiomeric configuration; this is of particular interest when the optical purity of D-enantiomeric compounds has to be determined; the smaller peak of the L-compound elutes before the larger peak of the D-main enantiomer, in contrast to Chirasil-L-Val® where the D-enantiomer elutes before the L-enantiomer

LIQUID CHROMATOGRAPHY

Macherey-Nagel
Neumann-Neander Strasse 6-8
P.O. Box 307
D-5160 Düren,
Germany

Resolvosil®-BSA-7

Resolvosil® is a chiral HPLC column for optical resolution based on bovine serum albumin (BSA) covalently bonded to wide-pore Nucleosil®; the column is compatible with mobile phases consisting of aqueous buffers of pH between 5 and 9; phosphate and borate buffers are suitable; small amounts of alcohols can be used as organic solvent modifiers; retention on the column is readily regulated via the mobile phase parameters, pH, ionic strength, and adjustment of surface tension using 1-propanol; retention is drastically reduced by 1—2% of 1-propanol and more than 5% is rarely required; the low capacity of the column limits the amount that can be injected; the performance of the column is based on the selective interactions of proteins with low molecular weight compounds, and it has proved successful for optical resolution as well as for the determination of enantiomeric purity; compounds which have been studied include aromatic amino acids, amino acid derivatives, aromatic sulfoxides, coumarin derivatives, and benzoin and benzoin derivatives

Astec, Inc.
P.O. Box 197
Whippany, NJ 07981

CYCLOBOND™ I (β-cyclodextrin bonded)

Cyclodextrins, known for their ability to form inclusion complexes in their hydrophobic cavity, have long been used in solution form. CYCLOBOND™ I is the first commercially available bonded form of this technique for HPLC. CYCLOBOND™ I is the beta form (7 glucose units) coupled to a high purity 5 μm silica gel through unique chemistry to form stable, nonhydrolytic bonds. The toroid structure formed by the glucose units has a diameter of 8 Å with the secondary hydroxyl groups on the larger circumference and the primary hydroxyl groups on the smaller side. The interior cavity is therfore relatively hydrophobic. The surface of the media has been fully silanized to remove all background interference and increase stability of the support

J. T. Baker Chemical Co.
222 Red School Lane
Phillipsburg,
N.J. 08865

Bakerbond® Chiral Phases

Bakerbond® chiral-phase HPLC columns can resolve thousands of racemic mixtures, each analysis requiring only a few minutes; unique bonding chemistry provides better enantiomeric discrimination than was ever possible before; the chiral-phase columns are remarkably stable, being used for hundreds of hours with no loss of selectivity and a minimal decrease in resolving ability; the phases available are

DNBPG (ionic) ([R]-*N*-3,5-dinitrobenzoylphenylglycine)
DNBPG (covalent) ([R]-*N*-3,5-dinitrobenzoylphenylglycine)
DNBLeu (covalent) ([S]-*N*-3,5-dinitrobenzoylleucine)

Bakerbond® chiral stationary phases use a combination of π-acidity, hydrogen bonding, and steric interaction to obtain separation; maximum resolution is achieved by manipulation of the mobile phase or derivatization of compounds to optimize enantiomeric difference in these weak interaction; with the DNBPG phase it should be possible to separate racemic mixtures including secondary benzyl alcohols, mandelic acid analogs, aryl phosphonates, α-indanol and α-tetralol analogs, propranolol and analogs, aryl sulfoxides, aryl phhthalides, aryl actams, aryl succinimides, aryl hydantoins, bis-β-naphthol analogs, aryl acetamides, cyclic alcohols, and nitrogen heterocycles; the DNBLeu phase is useful for bulky hindered compounds such as aromatic phthalides, benzodiazepinones, substituted pyrolidone esters, and aromatic amides

LKB-Produkter AB
Box 305
S-16126 Bromma,
Sweden

Enantiopac® Chiral Column

The EnantioPac® HPLC cartridge column provides a fast, simple, and reliable method for the direct racemic separation of optical isomers, including amines, acids, and other compounds without previous derivative preparation; based on a novel approach, the plasma protein α_1-acid glycoprotein (α_1-AGP) is immobilized on silica microparticles and used as the bonded phase; the support can thus be loaded with up to ten times the amount of α_1-AGP than can be achieved using conventional attachment procedures; this provides a greatly increased number of chiral interaction sites with resultant high enantioselectivity for many classes of compounds

THIN LAYER CHROMATOGRAPHY

Macherey Nagel

Chiralplate®

Chiralplate® is a TLC glass plate coated with a reverse-phase silica gel and impregnated with a chiral reagent and Cu (II) ions; the separation of optically active isomers is based on ligand exchange; separations achievable include amino acids, *N*-methylamino acids, *N*-formylamino acids, thiazolidine derivatives, dipeptides, and lactones; the plates can be used to control the optical purity of L-DOPA and D-penicillamine; for optimal results the plates should be activated for 15 min at 100°C prior to spotting; the separation time for amino acids is generally less than 30 min

Table 6
CHROMATOGRAPHY PAPERS

Whatman Ltd.
Springfield Mill
Maidstone,
Kent, ME14 2 LE,
England

Pure Cellulose Paper Grades

1 Chr	A thin paper, 0.18 mm thickness with a flow rate of 130 mm/30 min (water); white, smooth with a normally hard surface, it is recommended for general purpose chromatography
2 Chr	A slightly slower paper than 1 Chr with a thickness of 0.18 mm and a flow rate of 115 mm/30 min (water); it has a white, smooth, normally hard surface and is recommended for optical or radiometric scanning
3 Chr	A paper 0.36 mm thick with a flow rate of 130 mm/30 min (water); it has a white, rougher surface than 3 MM Chr, and has a higher wet strength than other papers, except for 3 MM Chr, which is similar; it is often used for separating inorganics and for electrophoresis
3MM Chr	A relatively thick (0.33-mm) paper with a flow rate of 130 mm/min (water); it has a white surface and is widely used for electrophoresis
4 Chr	This is the fastest thin paper (0.21 mm thick) with a flow rate of 180 mm/30 min (water); it has a white, smooth, normally hard surface used for routine and/or repetative chromatography where relatively low loadings are used
17 Chr	A very thick (0.92-mm) paper with a flow rate of 190 mm/30 min (water); it is extremely absorbent, will accept heavy loadings, and is recommended for preparative paper chromatography and electrophoresis
20 Chr	A paper 0.16 mm thick with the slow flow rate of 85 mm/30 min (water); with a white, smooth, normally hard surface it is recommended for the separation of samples of unknown composition
31 ET Chr	This is a thick (0.50-mm) extremely fast paper with a flow rate of 225 mm/30 min (water); it has a white, soft surface, its principal use being in the electrophoresis of large molecules

Ion Exchange Cellulose Grades

DE 81	DEAE cellulose paper; this is a weakly basic anion exchanger with diethylaminoethyl functional groups; a thin paper (0.20 mm) with a flow rate of 95 mm/30 min and an ion exchange capacity of 1.7 μeq/cm²; DE 81 is used for anions (acids and ampholytes above their isoelectric point); it is employed for separating nucleotides and related compounds including complex nucleotide "fingerprints" by two-dimensional procedures
P 81	Cellulose phosphate paper; a strong cation exchanger with the high ion exchange capacity of 18.0 μeq/cm²; a thin paper (0.23 mm) with a flow rate of 125 mm/30 min, it is useful for the separation of biogenic amines, antibiotics, histamines, and certain metals (Fe, Cu) at very low concentrations

Adsorbent Paper (Silica Gel Loaded)

SG 81	This paper combines cellulose and large pore silica gel, permitting its use for separations in which both partition and adsorption are important mechanisms; a moderately thick (0.27-mm) paper with a flow rate of 110 mm/30 min, SG 81 has definite advantages over TLC for the quantification of phospholipids; it will efficiently separate polar lipids, steroids, phenols, dyes, natural pigments, and keto-acids

Schleicher & Schuell Inc.
10 Optical Avenue
Keene,
N.H. 03431

Analytical Papers

2043a	Specially smooth uniform surface for most chromatography work, 110,[a] 1.30[b]
2043b	For most chromatography work including circular chromatography and evaluation by elution, 110, 1.64
591-C	Thin, smooth-textured paper for electrophoresis and one- and two-dimensional chromatography, 90, 1.04
589 White Ribbon-C	Suitable for most electrophoresis and chromatography, including circular chromatography, very low ash content, 88, 1.12
589 Black Ribbon-C	For chromatography when R_f values are well separated, very low ash content, 80, 1.07
589 Orange Ribbon-C	For general chromatography and electrophoresis, very low ash content, 80, 1.49
589 Red Ribbon-C	For ascending and circular filter chromatograms, a dense paper, 60, 0.87
507-C	A thin, ultradense, hard paper with a smooth surface, suitable for separating small amounts of materials with similar R_f values, 60, 0.88
589 Blue Ribbon-C	A dense paper which will resolve compounds with similar R_f values

Preparative Papers

470-C	A thick, smooth paper for electrophoresis and ascending chromatography of amino acids, 240, 8.20
903-C	A soft, medium thick paper for electrophoresis and chromatography, 238, 4.35
598-C	A moderately thick, open-textured paper, 145, 3.08
589 Green Ribbon-C	A thick, rough-textured paper, 128, 2.65
593-C	A moderately thick paper for chromatography and electrophoresis of amino acids and peptides, 105, 2.75
576-C	A thin paper suitable for photometric analysis and for use with small samples, 50, 0.81

[a] Capillary rise, average rise in millimeters of distilled water in 30 min.
[b] Absorbancy, difference in weight of a 10 × 10-cm sheet weighed dry and reweighed after immersion for 10 sec in distilled water.

Schleicher & Schuell Inc.
Keene,
N.H. 03431

Micropolyamide TLC sheets

Micropolyamide TLC sheets are used to obtain sharp separations of dansyl-amino acids and for NH_2-terminal amino identification; the 25-μm-thick micropolyamide membrane is available supported by fluorescent-free aluminum or plastic; sensitivity is 10^{-12} mol per amino acid for the former and 10^{-3} mol for the latter.

Table 7
ADSORBENTS FOR THIN LAYER CHROMATOGRAPHY

E. Merck,
Frankfurter Strasse 250
D-6100 Darmstadt 1
Federal Republic of Germany

One can prepare TLC layers on which chromatographic analysis is to be performed from the loose, powdered adsorbents that are available, or they can be brought ready to use on backings of glass, aluminum, or plastic;

ready-to-use layers are of better quality and are to be preferred for quantiative *in situ* determinations and for complicated separations; the following adsorbent types are available as loose materials and/or as ready-to-use adsorbents

Aluminum oxide, 60, 90, 150	Kieselguhr
Cellulose	Silica gel 60/kieselguhr
PEI cellulose	NH_2
Florisil	Polyamide 11
Silica gel 40, 60, 100, 200, 500, 1000	RP-2, RP-8, RP-18
Silica gel 60 silanized (RP-2)	Si 50000

Loose-powdered adsorbents are free of organic binder; they either contain no additive, indicated by H in the product name, or contain gypsum as binder, indicated by G; in order to obtain sufficient stability, adhesion, and abrasion resistance, all ready-to-use layers contain an organic binder; an F in the name indicates the addition of fluorescent indicators, the number following indicating the excitation wavelength and a subsequent s showing acid stability of the indicator; a number directly following the adsorbent designation indicates the average pore diameter in angstroms; chemically modified adsorbents can be recognized by letter and number combinations such as RP-2, RP-8, RP-18, or NH_2; RP signifies reversed phase, the number indicating the length of the hydrocarbon chain used for the modification; the exception is RP-2 which is a dimethylsilyl modifications; the modification NH_2 has a terminal amino group, i.e., a weak basic ion exchange function; for analytical purposes layer thicknesses of 0.1—0.25 mm are usual, whereas thicknesses of 0.5—2 mm are used for preparative separations; the vast majority of all TLC separations on Merck plates are today carried out on silica gel 60 layers

Table 8
PREFORMED LAYERS FOR TLC

Whatman Ltd.,
Srpingfield Mill,
Maidstone,
Kent ME14 2LE
England

Reversed Phase TLC Plates

These TLC plates give good resolution and reproductivity as a result of carefully controlled particle size distribution, layer thickness, purity and bonded phase chemistries. Correlation with HPLC columns provides a good starting point for methods development.

C_{18}, C_8, and C_2 Reversed Phase Plates

These plates can be used with a wide range of spotting solvents. Shorter carbon chains are used for increased polarity and affinity for aqueous solvents; longer chains for greater retention.

Diphenyl Reversed Phase Plates

Diphenyl reversed phase plates are easy to use and durable and can be used with highly aqueous solvent systems. They are ideal for proteins, peptides, amino acids, and other biological samples.

Multi-K Dual Phase TLC Plates

The plates allow combined reversed phase/adsorption chromatography. They are ideal for mixed polarity or for samples with a high contamination of lipophilic material.

Microcrystalline Cellulose TLC Plates

Microcrystalline cellulose plates have good surface uniformity and outperform fibrous cellulose plates in selectivity and speed. They display exceptional linearity and uniformity and are free from binder. Their abrasion resistance gives durability.

TLC Plates for Adsorption Chromatography

Two high-purity silica gels are used for normal phase separations (K6 with pore diameter of 60 Å, K5 with pore diamter of 150 Å). The gels have performance superior to silica gel G, showing better resolution and higher sensitivity. The plates are more durable and reproducible, they withstand aggressive reagents, and can be charred. Plates for adsorption chromatography are suitable for moderately strong polar compounds, such as lipids, carboyhdrates, drugs, and proteins.

High Performance Thin Layer Chromatography Plates (HPTLC)

HPTLC plates consist of a 4.5-μm particle size silica gel plus inert binder in a uniform 200-μm layer on glass. The gel has a narrow particle size distribution and homogeneity. The dense uniform layer provides stable baselines in densitometry. HPTLC plates give short development distances and times, together with low band diffusion. Microsamples can be analyzed.

Analtech Inc.
75 Blue Hen Drive
P.O. Box 7558
Newark,
Delaware 19807

High Performance Thin Layer Uniplates (HPTLC)

These plates provide high resolution separations in under 5 min with migration distances as small as 5 to 7 cm. Adsorbent layers 150 μm thick are prepared from high efficiency silica gel of 10 μm average particle diameter and narrow particle size distribution. The smooth homogeneous surface of HPTLC uniplates gives high signal-to-noise ratios for densitometric scanning. High performance uniplates are available in silica gel, cyano (CN) and amino (NH_2) adsorbents.

High Performance Reversed Phase Uniplates

These reverse phase adsorbents share the characteristics of silica gel HPTLC plates: high efficiency, 150 μm layer thickness and smooth surface for densitometry. Three nonpolar bonded supports are available, RP18, RP8, and RP2. These fully silanized silica gels are compatible with all organic solvents. However, due to the hydrophobic nature of the RP18 adsorbent, aqueous wettability is limited to about 60% water in the mobile phase. Wettability is greater with the slightly more polar shorter chain RP series plates. Selectivity can be altered for a given sample by selecting a reversed-phase plate of proper nonpolarity.

Reversed Phase Uniplates

Reversed phase separation (RPS) uniplates are made of a hydrocarbon impregnated silica gel. The hydrocarbon is adsorbed on the silica gel surface and is not chemically anchored to the particle. The behavior of these plates parallels that of most C_{18} HPLC packings. RPS plates will accept 100% water as the mobile phase or spotting solvent. This property is advantageous when using polar compounds such as proteins and peptides. Mixtures of water or buffers plus a polar organic modifier such as methanol, acetonitrile, or tetrahydrofuran can be used with the plates; however, these cannot tolerate nonpolar organic solvents.

Unisil Silica Gel G and GF Uniplates

These plates are prepared using a high purity silica gel with a calcium sulfate binder. They offer exceptional discrimination for aflatoxins compared to other adsorbent layers. The letter F indicates the presence of a shortwave indicator fluorescing at 254 nm. These uniplates are available binder free; the adsorbent contains a small amount of colloidal silica which helps adherence to the glass plate.

MODIFIED SILICA GEL UNIPLATES

The addition of some specific material to the base adsorbent often aids in the analysis of particular compounds. The additive material may be dissolved in the solution used to prepare the spreading slurry, or the finished TLC plate can be impregnated with the agent

Silica Gel H Uniplates with Potassium Oxalate

The addition of 1% potassium oxalate during slurry preparation permits increased discrimination of polyphosphoinositides.

Silica Gel H Self-Charring Uniplates

A common method for visualizing organic compounds on a silica gel TLC plate is charring, accomplished by spraying the plate with dilute sulfuric acid and heating. Silica gel H uniplates containing ammonium sulfate eliminate the use of sulfuric acid. The developing solvent is evaporated and the dry plate is heated at 150° to 200°C for 30 to 60 min to produce charring.

Silica Gel H Uniplates with Magnesium Acetate

These plates are prepared with 5, 7.5, or 10% magnesium acetate as the slurry liquid. The addition of magnesium acetate decreases the acidity of the layer, permitting increased discrimination of phospholipids.

Silica Gel GF Uniplates with 0.1 *N* Sodium Hydroxide

The use of 0.1 *N* sodium hydroxide to prepare the slurry makes these plates more basic and produces improved separation of organometallics and some acidic compounds.

Silica Gel GHL and GHLF Uniplates Impregnated with Silver Nitrate for Argentation Chromatography

Silica gel GHL and GHLF uniplates are impregnated with 5 to 20% silver nitrate which modifies the separation characteristics of the adsorbent layer to allow increased discrimination of compounds containing carbon-carbon double bonds.

Alumina G and GF Uniplates

Alumina G uniplates are produced using a slurry containing calcium sulfate hemihydrate as a binder.

CELLULOSE PLATES

Avicel Microcrystalline Cellulose Uniplates

Avical uniplates are coated with Avical microcrystalline cellulose, which consists of regenerated alpha cellulose particles of nearly spherical shape and 50 μm mean diameter. Compounds separated on this cellulose tend to form more compact spots than on fibrous cellulose layers, although MN-300 cellulose may resolve some compounds better.

MN-300 Cellulose Uniplates

These plates are coated with MN-300 cellulose, which is a native-fibrous cellulose consisting of relatively long "cut" fibers 20 to 30 μm long.

MN-300 PEI Cellulose Uniplates

These plates are coated with polyethyleneimine (PEI) impregenated MN-300 cellulose, capacity 0.5 mEq/g. They are used for the separation of nucleic acid materials.

DEAE Cellulose Uniplates

Diethylaminoethyl (DEAE) cellulose carries positive and negative charges at neutral and acid pH and is employed to separate negatively charged molecules by ion exchange chromatography. Due to the hydrophilic character of the cellulose substrate, DEAE cellulose will achieve ion exchange separations of proteins and nucleic acids. DEAE cellulose uniplates are available with mixed layers containing DEAE cellulose and unmodified cellulose. Mixed layer plates have lower ion exchange capacity and provide greater chromatographic mobility for negatively charged molecules.

Section VI
Indexes

AUTHOR INDEX

A

B

C

D

E

F

G

H

M

N

O

T

U

V

W

Y

Z

COMPOUND INDEX

A

B

E

F

G

H

I

K

L

N

O

P

S

T

U

V

W

X

SUBJECT INDEX

A

B

C

D

E

F

G

H

I

K

L

M

N

O

P

Q

R

S

T

U

V

W

Z